Interviewer Effects from a Total Survey Error Perspective

Statistics in the Social and Behavioral Sciences Series

For more information about this series, please visit: https://www.crcpress.com/go/ssbs

Interviewer Effects from a Total Survey Error Perspective

Edited by
Kristen Olson, Jolene D. Smyth,
Jennifer Dykema, Allyson L. Holbrook,
Frauke Kreuter, and Brady T. West

CRC Press
Taylor & Francis Group
Boca Raton London New York

CRC Press is an imprint of the
Taylor & Francis Group, an **informa** business
A CHAPMAN & HALL BOOK

Chapman & Hall / CRC Press
Taylor & Francis Group
6000 Broken Sound Parkway NW, Suite 300
Boca Raton, FL 33487-2742

First issued in paperback 2021

ISBN 13: 978-1-03-224151-7 (pbk)
ISBN 13: 978-0-367-89631-7 (hbk)

Visit the Taylor & Francis Web site at
http://www.taylorandfrancis.com

and the CRC Press Web site at
http://www.crcpress.com

To those who taught us about, and inspired our work on, interviewers, including

Charlie Cannell
Don Dillman
Jack Fowler
Bob Groves
Paul Lavrakus
Jim Lepkowski
Jon Krosnick
Lois Oksenberg
Nora Cate Schaeffer
Rainer Schnell

Contents

Preface

On three snowy days in February 2019, scholars, students, and practitioners from across the United States and Europe gathered at the University of Nebraska-Lincoln (UNL) in Lincoln, Nebraska, for a workshop featuring intense discussion of what we know – and don't know – about survey interviewers and standardized interviewing. Despite a blizzard that dropped eight inches of snow on the city just days before, 60 hardy workshop participants arrived with coats, scarves, gloves, snow boots, and an unflappable resolve to share their knowledge about interviewers. We spent two days discussing a series of invited papers and contributed posters, and a third day synthesizing and documenting what we learned and setting an agenda for future research and collaboration.

The participants included academics and practitioners, seasoned scholars and students, federal statistical agency methodologists and statisticians, survey managers at large government contractors and small academic survey shops, self-employed professionals, and recently retired researchers. This heterogeneity was critically important for meeting two of our primary goals. First, we wanted to reflect the remarkable growth in research on interviewers and interviewing over the last decade or so. Given this goal, we sought a set of invited papers from an international group of established and early career scholars and practitioners who had been actively contributing to that body of work about interviewers. Second, we sought to have representation from the next generation of scholars as well as a broad range of survey practitioners. Here, we had an open call for contributions from early career scholars, students, and survey practitioners who actively participated in the workshop by presenting posters and serving as discussants. Contributions from both sets of participants are also reflected in this volume.

Each participant in the workshop had a formal role, either as an author of a paper or poster, a discussant on a paper, or a rapporteur. As such, all participants were actively engaged in the workshop. The participants were Daniela Ackermann-Piek, Paul Beatty, Grant Benson, David Biagas, Alexandra Breslin, Theresa Camelo, Alexandru Cernat, Fred Conrad, Steve Coombs, Jim Dahlhamer, Jessica Daikeler, Julie de Jong, Vidal Díaz de Rada, Jennifer Dykema, Brad Edwards, John Eltinge, Kim Ethridge, Holly Fee, Sabine Friedel, Dana Garbarski, Nikki Gohring, Patrick Habecker, Jessica Herzing, Allyson L. Holbrook, Margaret Hudson, Jerreed Ivanich, Birgit Jesske, Tim Johnson, Robin Kaplan, Evgenia Kapousouz, Alian Kasabian, Jennifer Kelley, Frauke Kreuter, Ron Langley, Lisa Lin-Freeman, Geert Loosveldt, Peter Miller, Zeina Mneimneh, Rod Muilenburg, Tiffany Neman, Linh Nguyen, Kristen Olson, Yfke Ongena, Fiona Pashazadeh, Angelica Phillips, Heather Ridolfo, Joseph Rodhouse, Antje Rosebrock, Megan Ruxton, Michael Schober, Nora Cate Schaeffer, Nick Schultz, Silvia Schwanhäuser, Sharan Sharma, Jolene Smyth, Rachel Stenger, Rodney Terry, Jerry Timbrook, Brady West, Jamie Wescott, Anna Wiencrot, Celine Wuyts, and Erica Yu.

We also appreciate exceptional contributions and support from the following offices and individuals at UNL: the College of Arts and Sciences, and especially Associate Dean for Research and Partnerships Kirk Dombrowski; the Office of Research and Economic Development, and especially Director of Research Strategy and Infrastructure Jennifer Nelson; the Department of Sociology, and especially Department Chair Julia McQuillan; and the Social and Behavioral Sciences Research Consortium, and especially Director Dan Hoyt. All travel and management of the funds was administered at the highest level of quality by Breana Garretson and Sara Mattson in the College of Arts and Sciences Oldfather Business Cooperative.

We want to draw special attention to the contributions of the team of graduate research assistants (GRAs) at the University of Nebraska-Lincoln. Angelica Phillips, Rachel Stenger, and Jerry Timbrook went above and beyond their normal GRA duties to prepare materials for the workshop, guide visitors around Lincoln, and help out the organizing team in all possible ways. Angelica Phillips and Rachel Stenger also facilitated the timely completion of this volume by helping with detailed formatting of each chapter. To these three students and their invaluable contributions, we say thank you.

The workshop and this edited volume were made possible through generous funding from the National Science Foundation (SES-1758834), the Charles Cannell Fund in Survey Methodology of the Survey Research Center at the University of Michigan, and the Rensis Likert Fund for Research in Survey Methodology at the University of Michigan. These grants made it possible to cover all expenses for the workshop, including travel for each participant, and provided support for our research assistants to help produce this volume. UNL's Office of Research and Economic Development and the UNL College of Arts and Sciences also generously provided supplies and contributed funds for the workshop. This support was invaluable.

Finally, Rob Calver at CRC Press was an exceptionally supportive publisher. We appreciate the frequent check-ins and feedback to make sure we were on track. We also thank each of our employing institutions for the support and academic freedom to work on this project.

The editors of the volume strongly believe in supporting and training the next generation of scholars in survey methodology. As such, all royalties from this volume will be donated to the Charles Cannell Fund for Survey Methodology at the University of Michigan.

Kristen Olson
Jolene D. Smyth
Jennifer Dykema
Allyson L. Holbrook
Frauke Kreuter
Brady T. West

Contributors

Daniela Ackermann-Piek
GESIS – Leibniz Institute for the Social
  Sciences
Mannheim, Germany
and
University of Mannheim
Mannheim, Germany

Yasmin A. Altwaijri
King Faisal Specialist Hospital and
  Research Centre
Riyadh, Kingdom of Saudi Arabia

Christopher Antoun
University of Maryland
College Park, Maryland

Michael Bosnjak
Leibniz Psychology
Trier, Germany

Alison W. Bowers
Virginia Polytechnic Institute and State
  University
Blacksburg, Virginia

Alexandru Cernat
University of Manchester
Manchester, United Kingdom

Young Ik Cho
University of Wisconsin-Milwaukee
Milwaukee, Wisconsin

Frederick G. Conrad
University of Michigan
Ann Arbor, Michigan

Steve M. Coombs
University of Wisconsin-Madison
Madison, Wisconsin

James Dahlhamer
National Center for Health Statistics
Hyattsville, Maryland

Jessica Daikeler
University of Mannheim
Mannheim, Germany
and
GESIS – Leibniz Institute for the Social
  Sciences
Mannheim, Germany

Julie A. de Jong
University of Michigan
Ann Arbor, Michigan

Jennifer Dykema
University of Wisconsin-Madison
Madison, Wisconsin

Brad Edwards
Westat
Rockville, Maryland

Dorothy Farrar Edwards
University of Wisconsin-Madison
Madison, Wisconsin

Dana Garbarski
Loyola University Chicago
Chicago, Illinois

Marieke Haan
University of Groningen
Groningen, the Netherlands

Patrick Habecker
University of Nebraska-Lincoln
Lincoln, Nebraska

Allyson L. Holbrook
University of Illinois at Chicago
Chicago, Illinois

Lisa Holland
University of Michigan
Ann Arbor, Michigan

Ryan Hubbard
Westat
Rockville, Maryland

Margaret L. Hudson
University of Michigan
Ann Arbor, Michigan

Andrew L. Hupp
University of Michigan
Ann Arbor, Michigan

Jerreed Ivanich
Johns Hopkins Bloomberg School of
    Public Health
Baltimore, Maryland

Timothy P. Johnson
University of Illinois at Chicago
Chicago, Illinois

Michael Josten
Conflict Management Consulting
    (CMC)
Munich, Germany

Robin Kaplan
Bureau of Labor Statistics
Washington, DC

Evgenia Kapousouz
University of Illinois at Chicago
Chicago, Illinois

Jennifer Kelley
University of Essex
Colchester, United Kingdom
and
University of Michigan
Ann Arbor, Michigan

Julie M. Korbmacher
Max Planck Institute for Social Law and
    Social Policy
and
BIDAQ – Bayerisches Institut für Daten,
    Analysen und Qualitätssicherung
Munich, Germany

Yuliya Kosyakova
Institute for Employment Research
Nuremberg, Germany
and
University of Mannheim
Mannheim, Germany
and
University of Bamberg
Bamberg, Germany

Frauke Kreuter
University of Maryland
College Park, Maryland
and
University of Mannheim
Mannheim, Germany
and
Institute for Employment Research
Nuremberg, Germany

Ulrich Krieger
University of Mannheim
Mannheim, Germany

Geert Loosveldt
KU Leuven
Leuven, Belgium

Aaron Maitland
National Center for Health Statistics
Hyattsville, Maryland

Nancy A. Mathiowetz
University of Wisconsin-Milwaukee
Milwaukee, Wisconsin

Peter V. Miller
Northwestern University
Evanston, Illinois

Zeina N. Mneimneh
University of Michigan
Ann Arbor, Michigan

Daniel Nielsen
University of Michigan
Ann Arbor, Michigan

Kristen Olson
University of Nebraska-Lincoln
Lincoln, Nebraska

Yfke Ongena
University of Groningen
Groningen, the Netherlands

Fiona Pashazadeh
University of Manchester
Manchester, United Kingdom

Heidi Reichert
University of Michigan
Ann Arbor, Michigan

Joseph W. Sakshaug
Institute for Employment Research
Nuremberg, Germany
and
Ludwig Maximilian University of Munich
Munich, Germany
and
University of Mannheim
Mannheim, Germany

Nora Cate Schaeffer
University of Wisconsin-Madison
Madison, Wisconsin

Michael F. Schober
New School for Social Research
New York City, New York

Heather Schroeder
University of Michigan
Ann Arbor, Michigan

Rob K. Schultz
University of Wisconsin-Madison
Madison, Wisconsin

Silvia Schwanhäuser
Institute for Employment Research
Nuremberg, Germany
and
University of Mannheim
Mannheim, Germany

Jolene D. Smyth
University of Nebraska-Lincoln
Lincoln, Nebraska

Hanyu Sun
Westat
Rockville, Maryland

Jamie Wescott
RTI International
Research Triangle Park, North Carolina

Brady T. West
University of Michigan
Ann Arbor, Michigan

Celine Wuyts
KU Leuven
Leuven, Belgium

Ting Yan
Westat
Rockville, Maryland

H. Yanna Yan
University of Michigan
Ann Arbor, Michigan

Erica Yu
Bureau of Labor Statistics
Washington, DC

Benjamin Zablotsky
National Center for Health Statistics
Hyattsville, Maryland

Carla Zelaya
National Center for Health Statistics
Hyattsville, Maryland

About the Editors

Kristen Olson, Ph.D., is Leland J. and Dorothy H. Olson Professor and Vice Chair of the Department of Sociology at the University of Nebraska-Lincoln, USA.

Jolene D. Smyth, Ph.D., is Associate Professor in the Department of Sociology and the Director of the Bureau of Sociological Research at the University of Nebraska-Lincoln, USA.

Jennifer Dykema, Ph.D., is Distinguished Scientist and Senior Survey Methodologist at the University of Wisconsin Survey Center, USA.

Allyson L. Holbrook, Ph.D., is Professor of Public Administration and Psychology at the University of Illinois at Chicago, USA.

Frauke Kreuter, Ph.D., is Director of the Joint Program in Survey Methodology at the University of Maryland, College Park, Professor of Statistics and Methodology at the University of Mannheim, Germany, and Head of the Statistical Methods Research Department (on leave) at the Institute for Employment Research in Nuremberg, Germany.

Brady T. West, Ph.D., is Research Associate Professor in the Survey Research Center at the Institute for Social Research on the University of Michigan-Ann Arbor campus, USA.

Section I

History and Overview

1

The Past, Present, and Future of Research on Interviewer Effects

Kristen Olson, Jolene D. Smyth, Jennifer Dykema,
Allyson L. Holbrook, Frauke Kreuter, and Brady T. West

CONTENTS

1.1 Introduction

Interviewer-administered surveys are a primary method of collecting information from populations across the United States and the world. Various types of interviewer-administered surveys exist, including large-scale government surveys that monitor populations (e.g., the Current Population Survey), surveys used by the academic community to understand what people think and do (e.g., the General Social Survey), and surveys designed to gauge public opinion at a particular time point (e.g., the Gallup Daily Tracking Poll). Interviewers participate in these data collection efforts in a multitude of ways, including creating lists of housing units for sampling, persuading sampled units to participate, and administering survey questions (Morton-Williams 1993). In an increasing number of surveys, interviewers are also tasked with collecting blood, saliva, and other biomeasures, and asking survey respondents for consent to link survey data to administrative records (Sakshaug 2013). Interviewers are also used in mixed mode surveys to recruit and interview nonrespondents after less expensive modes like mail and web have failed (e.g., the American Community Survey and the Agricultural Resource Management Survey; de Leeuw 2005; Dillman, Smyth and Christian 2014; Olson et al. 2019). In completing these varied tasks, interviewers affect survey costs and coverage, nonresponse, measurement, and processing errors (Schaeffer, Dykema and Maynard 2010; West and Blom 2017).

Errors introduced by interviewers can take the form of bias or variance. Early research found that interviewers vary in how they administer survey questions and that their effects were similar to sample clusters in both face-to-face (Hansen, Hurwitz and Bershad 1961; Kish 1962) and telephone surveys (Groves and Magilavy 1986; Mathiowetz and Cannell 1980). In particular, similar to a design effect for cluster samples, interviewers increase the variance of an estimated mean as a function of their average workload ($\bar{b}$) and the intra-interviewer correlation (IIC) (the degree of within-interviewer correlation in measurements): $1 + (\bar{b} - 1)IIC$. IIC values range from very small (0.001) to large (0.20) and larger (Elliott and West 2015; Groves and Magilavy 1986; O'Muircheartaigh and Campanelli 1998), with median values typically below 0.02. Given an IIC of 0.02 and a workload of 50 interviews per interviewer, the variance of an estimated mean is almost doubled and standard errors are increased by 40%. Thus, a fundamental goal of research on interviewers is understanding what contributes to (and how to minimize) the IIC.

Even if the IIC is small, interviewer characteristics and behaviors can still bias responses. Interviewer sociodemographic characteristics (e.g., sex, race, age, education), personality traits, experience, attitudes and expectations, or even the paralinguistic qualities of their voices may be associated with responses to survey questions or indicators of survey error (e.g., an indicator of a sampled unit responding to a survey request; Schaeffer et al. 2018). Inasmuch as these are fixed characteristics of interviewers, they typically bias estimates. In addition, research based on coding information about the interviewer–respondent interaction shows that interviewers' behaviors such as misreading questions, probing directively, and acting non-neutrally may affect respondents' behaviors and survey error (e.g., Fowler 2011; Ongena and Dijkstra 2006).

The goal of the *Interviewers and Their Effects from a Total Survey Error Perspective Workshop* in February 2019 was to convene an international group of leading academic, government, and industry researchers and practitioners to discuss methods, practical insights, and research findings on interviewers. Specifically, the workshop aimed to (1) synthesize and expand knowledge about the impacts of interviewers on multiple error sources, (2) evaluate study design and estimation approaches for studying interviewer effects, and (3) produce an agenda for future studies of interviewers. After two days of presentations and posters, workshop participants spent the third day discussing and identifying areas where more work is needed. This chapter introduces an edited volume consisting of chapters written by workshop participants. In order to do so, we first provide an overview of research and practice related to interviewers at different stages of the survey process. Second, we situate the chapters from this volume within the existing literature. Finally, we identify areas for future research that arose from the third day of focused discussion.

1.2 Training, Managing, and Monitoring Interviewers

Standardized interviewing aims to limit the effect of the interviewer on the resultant survey data. Although standardized interviews are the gold standard, a strict implementation of standardization may not exist in practice because "standardized" interviewer training, monitoring, and feedback systems vary widely across survey organizations (Viterna and Maynard 2002). Standardized interviewing commonly includes asking questions exactly as written, recording answers as provided, following up inadequate answers nondirectively, and acknowledging adequate answers (Fowler and

Mangione 1990). Yet some of the central tenants of standardized interviewing – including reading questions verbatim – are inadequately operationalized in actual training materials. Additionally, survey practitioners often make decisions about interviewer training with little to no empirical evidence regarding effectiveness. Furthermore, there has been almost no update to standardized training philosophy or materials since Fowler and Mangione's (1990) canonical volume. In Chapter 3, Schaeffer et al. tackle this important issue, updating basic training of General Interviewing Techniques for question administration based on decades of research on interviewer–respondent interaction and survey practice. Among other things, their update tackles thorny issues of how to recognize a codable answer for different question types (thus acknowledging the critical role of characteristics of survey questions in determining what makes an answer codable), how to maintain respondent engagement, and what common conversational practices can be allowed in the interview.

While studies of interviewers and their effects often have implications for survey operations, it was clear during the workshop that this research has not been fully adopted by survey organizations. We have few recent descriptions of how survey organizations select, train, and monitor survey interviewers, especially for smaller survey organizations. Yet many operational concerns are researchable, and such research could yield both theoretical and practical insights. Miller and Mathiowetz (Chapter 2) describe how Charlie Cannell's seminal work expanded theoretical insights into interviewer–respondent interaction, how this interaction can positively or negatively affect data quality, and what kind of practical training methods may address these concerns. Unfortunately, few studies have systematically examined properties of interviewer training protocols. While Daikeler and Bosnjak's (Chapter 4) meta-analysis of interviewer training includes many studies on training efficacy for avoiding refusals, few studies evaluate training efficacy for administering survey questions. The workshop identified many areas for research on recruiting, training, and monitoring interviewers, including how to identify successful interviewers from a pool of applicants, how the pool of interviewers has changed over time and across modes of data collection, and the influence of the supervisor on the interviewer. Future research is also needed to address the mix of training that should be devoted to skill development, to shaping interviewer perceptions about their tasks, what combination of in-person and online methods are best for delivering content, the optimal length of training for different aspects of the survey process, and interviewer training when standardization is difficult to maintain.

Many have questioned the utility of standardization, particularly for sensitive, complicated, or difficult questions or questionnaires (e.g., Conrad and Schober 2000; Maynard et al. 2002; Schaeffer 1991; Schober and Conrad 1997; Suchman and Jordan 1990). Interviewers often break from standardization on these types of items and questionnaires. Based on qualitative in-depth interviews with survey interviewers and their reactions to vignettes, Kaplan and Yu (Chapter 5) describe interviewers' approaches to administering sensitive and difficult questions, providing unique insights into why deviations from standardization occur. Innovations in technology, increased computing resources in the field, and hands-on monitoring of interviewers using paradata have dramatically changed how we monitor and train interviewers, particularly for in-person interviewing (e.g., Edwards, Maitland and Connor 2017; Olson and Wagner 2015). Edwards, Sun, and Hubbard (Chapter 6) report on an attempt to intervene in real time when interviewers break from standardization for difficult questions as detected via monitoring using computer-assisted recorded interviewing (CARI). These two chapters highlight the need to train interviewers on follow-up methods such as probing and clarification, and the value of reinforcing this training through real-time monitoring. Throughout the workshop, participants echoed the need for more information on how survey organizations monitor and provide feedback to interviewers and what methods are most efficacious for in-person and telephone interviews.

Although in-depth interviews and CARI recordings provide unique insights into interviewers' behavior during the survey data collection process, many organizations lack resources to conduct such studies and instead use paradata and interviewer observations to evaluate the survey process after data collection is finished. Schwanhäuser et al. (Chapter 7) use paradata on interview duration and indicators from the survey data to detect potential falsification by interviewers; this study is unique in that the interviewer-level falsification indicators are compared to data from the field that identified three *actual* falsifiers, revealing where the statistical analysis-based falsification indicators help to identify problems and where they fail. West et al. (Chapter 8) use data from two large surveys to assess whether post-survey interviewer observations about the survey process are associated with indicators of measurement error. The sets of observations that are most important to collect for monitoring and evaluation purposes were difficult to discern given the tremendous heterogeneity in the types of survey questions asked and observations collected across studies. More work is needed to align these observations across survey organizations, to facilitate replication of effects and potentially benefit secondary data users.

1.3 Interviewer Effects Across Contexts and Modes

The survey context also matters. Although interviewers are instructed to conduct interviews in private, many interviews are conducted in the presence of known others, violating respondents' need for privacy when reporting about sensitive topics (Mneimneh et al. 2015). In community-based studies where in-community interviewers are recruited who may know the respondent personally, these effects may be amplified. Alternatively, third-party presence may facilitate recall when questions are particularly difficult. Mneimneh, de Jong, and Altwaijri (Chapter 9) examine whether conducting an interview with various family or non-family members present affects reports of health behaviors and attitudes in the in-person Saudi National Mental Health Survey. Habecker and Ivanich (Chapter 10) examine how youth's reports of internalizing and externalizing behaviors and other sensitive topics are affected when the youth are interviewed by a member of the community who is known to them in a community-based participatory research study in an American Indian community. Combined, these studies reveal the need for more research into interview privacy and the importance of interviewer training on how to maximize privacy.

Additionally, the mode and/or device for the interview – in-person, landline phone, cellular phone, or audio computer-assisted self-interviewing; and interviewer input into a desktop or laptop, tablet, or smartphone – provides important context and potential for variation in interviewer-related errors (e.g., Childs and Landreth 2006; Timbrook, Smyth and Olson 2018). Notably, the mode or device for the interaction changes the nature of the interaction between interviewers and respondents. In Chapter 12, Ongena and Haan examine differences in the interviewer–respondent interaction across telephone and in-person interviews; in Chapter 13, Schober et al. examine differences in interviewer–respondent interaction across voice and text message–based interviews. Conrad et al. (Chapter 11) replace human interviewers with avatars, examining race-of-interviewer effects when the interviewer is virtual, but their voice is that of an audio-recorded human. These chapters raise important questions about interviewing: Are there contexts in which the presence of the interviewer (live or otherwise) is an important feature of the survey or where tasks are critical for the interviewer to consider? Can virtual interviewers provide some of the benefits of live interviewers while minimizing interviewer error? Are text message–based interviews or avatars particularly beneficial for specific populations?

1.4 Interviewers and Nonresponse

While most of the research identified above focuses on measurement error, interviewers affect other error sources. For example, interviewers' nonresponse rates vary extensively (e.g., Campanelli, Sturgis and Purdon 1997; Groves and Couper 1998) – due to both heuristic cues from their voices (e.g, Groves et al. 2008; Schaeffer et al. 2018) and their behaviors during the recruitment interaction (e.g., Couper and Groves 2002; Schaeffer et al. 2013) – and substantially contribute to nonresponse error variance (e.g., West, Kreuter and Jaenichen 2013; West and Olson 2010). Interviewer flexibility and tailoring have been linked to successful recruitment, although there are only limited examples of how tailoring is operationalized (e.g., Groves and Couper 1998). Research about tailoring generally relies on interviewer reports with measures that vary across studies. To address this, Ackermann-Piek, Korbmacher, and Krieger (Chapter 14) predict survey contact and cooperation with the same set of covariates across four different studies conducted by the same survey organization. Because they find little replication in associations between the covariates and the nonresponse outcomes across studies, they emphasize the importance of real-time monitoring of interviewers.

Interviewer flexibility may not always be a good thing. In Chapter 15, Wescott discusses a case management model for a telephone survey in which interviewers "own" cases and make their own decisions about when to call cases. While interviewers report being more satisfied with the autonomy afforded by the case management model, the model yields lower productivity than using a call scheduler. More work is needed to understand how interviewer autonomy and insights from a field data collection approach may be integrated into a telephone survey organization to increase interviewer engagement and ultimately retain interviewers.

Survey interviews increasingly ask respondents to provide blood, saliva, urine, or other biomeasures or for permission to link their survey data to administrative data (e.g., Jaszczak, Lundeen and Smith 2009; Sakshaug 2013; Sakshaug et al. 2012). In Chapter 16, Pashazadeh, Cernat, and Sakshaug use nurse characteristics and paradata to predict different stages of nonresponse for nurses' attempts to collect biomeasures for a general population survey. These stages – participating in the nurse visit, consenting to a blood sample, and obtaining a blood sample – reveal substantial variation in the nonresponse outcomes related to the nurses, and that the predictors of nonresponse vary across the stages. This work and additional workshop discussion suggest that we need more research on the antecedents and consequences of interviewer variation in the ability to successfully collect auxiliary measures.

1.5 Interviewer Pace and Behaviors

Four chapters examine interview pace and behaviors. Holbrook et al. (Chapter 17) test what interviewer and question characteristics predict interviewer reading speed (IRS) and what effect IRS has on response latencies and indicators of respondent comprehension and mapping difficulties. Garbarski et al. (Chapter 18) examine how the time taken to administer and answer questions is associated with interviewer, respondent, and question characteristics. Kelley (Chapter 19) examines how well question administration time thresholds generated from timing paradata can be used to identify question misreadings, testing three methods of setting thresholds. Olson and Smyth (Chapter 20) address the longstanding finding that interviews get faster later in the field period by examining changes in interviewers' behaviors over time.

Although the studies ostensibly examine the same phenomenon – pace – their conceptualizations and operationalizations vary (e.g., interview duration, question duration, interviewer speaking time or speed, response latencies, words per second, and questions per minute; see also Chapter 21, in which Dahlhamer et al. use average number of seconds per question across a questionnaire). Other studies examining pace use behavior or interaction coding to examine interviewer and respondent behaviors (e.g., Fowler 2011; Ongena and Dijkstra 2006), producing evidence of similar variation in how behavior coding is implemented (from live interviews, recordings, or transcripts, and focused on respondents, interviewers, or both) and the variety of operational and analytical decisions often made in such research. Such decisions include assigning codes at the question level or the conversational turn level (Olson and Parkhurst 2013); coding the entire questionnaire or a subset of items (e.g., see review in Ongena and Dijkstra 2006); using codes individually (e.g., Fowler 2011; Mathiowetz and Cannell 1980) or in combination (e.g., to measure rapport as in Garbarski, Schaeffer, and Dykema 2016); examining behaviors when they initially occur or at any point during the question–answer interaction; examining behaviors individually or sequentially; and dealing with overlapping speech, interruptions, and other normal conversational events. Even simple issues of how many interviews to code and (acceptable) reliability of the codes vary across studies.

Despite this heterogeneity, there are common patterns observed across these studies. First, interviewers speed up with more experience (Holbrook et al., Chapter 17; Garbarski et al, Chapter 18; Olson and Smyth, Chapter 20; Olson and Bilgen 2011; Olson and Peytchev 2007). Second, question characteristics drive this phenomenon. Figure 1.1 shows the percent of variance in question duration attributable to interviewers, respondents, and questions across multiple studies with question-level duration as the outcome. Some models are estimated as three-level multilevel models, so the residual variance accounts for question-level variation. Other models are estimated as cross-classified models, where question-level variation is explicitly estimated as part of the model. Across these four studies, interviewer- and respondent-level variation is small and the question-level (or residual that incorporates questions) variation is large.

There is a consistent tendency for longer questions and questions written at higher grade levels to have longer durations (e.g., Garbarski et al. Chapter 18; Couper and Kreuter 2013; Olson and Smyth 2015). Yet other question characteristics, including question placement, question sensitivity, type of question (attitude, factual, knowledge), response option format, presence of definitions, emphasis, instructions, parentheticals, battery items, and measures from survey evaluation tools such as QUAID (Graesser et al. 2006) or SQP (Saris and Gallhofer 2007), are inconsistently parameterized or have inconsistent associations

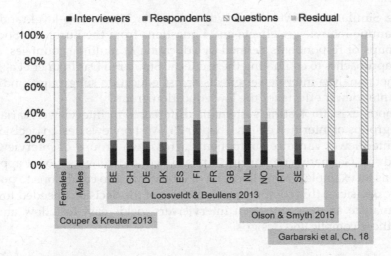

FIGURE 1.1
Variance in question duration due to interviewers, respondents, and questions.

with duration. Pace reflects interviewer behaviors and influences respondent behaviors, although understanding the mechanisms for these connections still requires more work. One clear direction from the workshop for future research was to identify a common set of dependent variables related to pace and behaviors, a common set of question characteristics, and a common set of respondent and interviewer characteristics, parameterize these identically, and evaluate whether this standardization "solves" some of the inconsistencies across these different observational studies. In addition, with advances in text analysis and searches, the relative cost and error trade-offs of human coders versus computer coding for this type of research is also of interest.

1.6 Estimating Interviewer Effects

Given the nesting of respondents within interviewers, following Kish's ANOVA-based model (Kish 1962), hierarchical or random effects models have long been used for the study of interviewer effects (e.g., Dijkstra 1983; Hox 1994; O'Muircheartaigh and Campanelli 1998). Recent applications of hierarchical models expand the complexity of these models, including hierarchical cross-classified random effects models to study how properties of survey questions themselves may affect and be affected by interviewers (e.g., Couper and Kreuter 2013; Holbrook et al. 2016; Olson and Smyth 2015) or to separately estimate area effects from interviewer effects (e.g., O'Muircheartaigh and Campanelli 1998, 1999).

Multilevel models are flexible and can be used to infer whether interviewer effects differ across subgroups of items, respondents, and interviewers. In Chapter 21, Dahlhamer et al. use cross-classified multilevel models to disentangle area effects from interviewer effects on over 100 outcomes in the National Health Interview Survey. They then meta-analyze these interviewer effects across question and interviewer characteristics, finding that longer questions and those with higher reading levels have larger interviewer effects, as do questions on more complex and difficult topics. Interviewers who administer questions at a faster pace also have larger interviewer effects than those who administer questions at a slower pace. Similarly, Loosveldt and Wuyts (Chapter 22) use two-level random effects models to examine interviewer effects on 14 questions from the European Social Survey across subgroups of respondents defined by education in multiple countries, comparing two analytic approaches to estimating these effects. Similar to Dahlhamer et al., Loosveldt and Wuyts meta-analyze interviewer effects across education subgroups, finding consistently larger interviewer effects for the low education group.

Other methods exist for testing systematic differences in interviewer variance across independent groups of interviewers. In Chapter 23, West reviews design decisions needed to compare interviewer variance components across two groups of interviewers, using a unique study in Germany comparing standardized and conversational approaches to interviewing as an example. From allocation of interviewers to conditions to power analyses to analytic decisions, this comprehensive review of the decisions needed to effectively design and analyze an experiment on interviewers yields easy-to-follow and practical insights into these complicated designs.

1.7 Closing Thoughts

Survey research is adaptive, as reflected by the content and range of chapters in this volume. Even though interviewers have been central to data gathered to understand society since the beginning of survey research, we know surprisingly little about them. Namely, what interviewer characteristics exist across studies and across organizations, how interviewers perceive their job, and how we can recruit and retain high-quality interviewing staff. Yet these are important issues – interviewers can do a lot of harm to survey data if they try to do so (Chapter 7) but also inadvertently introduce error into data even when they are attempting to follow their training. Understanding the challenges and constraints interviewers face will facilitate understanding the mechanisms underlying interviewer-related survey errors.

Most of the chapters in this volume use observational data. Observational research is constrained by the data that a research team has available. Through the workshop and studies featured in this volume, we have the opportunity to refine our conceptualizations and operationalizations and create more consistency across operational implementations and research studies. Examples of key concepts/topics that would benefit from a shared set of definitions include "conversational interviewing," "standardized interviewing," "interviewer effects," "interviewer experience," "interview(er) pace," and "question characteristics" (and the types therein).

More work is needed to examine interviewers cross-nationally. In some countries, a more flexible form of interviewing is standard practice. How this relates to "conversational interviewing" in United States nomenclature (e.g., Conrad and Schober 2000), as well as the concomitant effects on survey data, are unknown. The heterogeneity in interviewer-related variance across countries (Loosveldt and Wuyts, Chapter 22) reveals the need for understanding interviewing practice, monitoring, and supervision in a multinational context. This volume contains many studies conducted outside of the United States (e.g., Chapters 4, 7, 8, 9, 12, 14, 16, 19, 22, and 23). Yet a deep understanding of how interviewers operate in different contexts is an area for future research.

As researchers and practitioners, we call on survey organizations to make information on interviewers available in public-use analytic data files. Table 1.1 contains a list

TABLE 1.1

A List of Recommended Information About Interviewers to Include in Public-use Data Files and Methodology Reports

---

Interviewer ID
Interviewer characteristics
 Demographics (gender, race, age, educational status)
 Personality assessments
 Experience (within survey, within organization, across organizations)
 Certification test scores
 Number of other jobs they are currently working
 Performance metrics and problems on other studies
 Ever been fired for other performance issues on other studies
 Number of hours work on other studies
Interviewer expectation and attitudinal measures
 Ratings of the importance of a completed interview
 Ratings of the importance of obtaining high-quality data
 Description of how sensitive and difficult questions are approached
 Ratings of other variables related to job satisfaction and engagement
Interviewing training variables
 Content of training and training methods (e.g., round-robin, type and content of at-home study, hours/days of training by topic)
 Participation in any specialized training (e.g., for refusal avoidance)
 Number of trainings attended
Interview process variables
 Sanitized post-survey observations
 Number of other projects interviewers involved in current project worked on during current project
 Whether and how interviewers were matched to respondents
 Measures of field performance, including ICC information for variables
 Adaptive design features and implementation
 Interviewers' notes about survey questions
Organization and study-specific characteristics
 Information on recruiting, hiring, training, and attrition
 Monitoring, feedback, and falsification detection activities
 Description of the supervisory structure of the interviewer corps, e.g., number of interviewers per supervisor

---

of concepts that workshop participants considered important to examine and include in future research studies. At the bare minimum, an anonymized interviewer ID variable on data files would allow analysts to estimate interviewer variance components. Additional data on interviewers, extending beyond simply demographics and experience, would facilitate understanding the mechanisms by which interviewers affect survey data. At the interviewer level, these include measures of work productivity and quality, measures of how thinly spread the interviewer is across multiple projects and/or organizations, attitudinal and expectation measures from the interviewers themselves about their job, as well as measures from field performance. At the study or organizational level, information about the amount, type, and content of interviewer training as well as details about the supervision and monitoring practices and feedback provided to interviewers from supervisors or monitors would greatly enhance studies of interviewer effects. Detailed information about the content of general interviewing techniques training versus study-specific training also would provide insights into how these individual decisions cumulate to affect survey errors related to interviewers. Some of this information could be included in methodology reports. Although many organizations may consider information about training, supervision, and monitoring to be proprietary, more complete disclosure is certainly needed to understand these rarely studied, yet critically important, survey practices.

Finally, dissemination and integration of research on interviewers into survey practice is hard. Many new practices may face cultural opposition at organizations simply because it is not the way that work has been done in the past. Clients, survey project managers, and supervisors are often risk averse to trying something new, even if integrating recommendations based on research can improve survey practice. Furthermore, most survey field professionals do not have time to read the latest research, and survey methodologists are often disconnected from field operations. These factors conspire to make translation of research into practice difficult.

We suggest a few ways forward. First, many professional association meetings contain both survey field professionals and methodological researchers. Carving out space for these two disparate groups to discuss mutually interesting problems can facilitate research translating into practice and practice informing research questions. For several years the Midwest Association for Public Opinion Research has sponsored a workgroup in which researchers and practitioners talk about important topics on interviewers. In 2019, the discussion focused on interviewer training, for instance. Second, methodologists who work at organizations with survey shops or who contract for research with survey shops are well positioned for translation of research. Seminars or brown bag discussions with senior field managers at the organizations that collected the survey data about findings related to practice could inspire some changes in survey practice that would improve data quality. Furthermore, recognizing contributions other than simply research articles – for instance, data being available in the public domain, availability of code, availability of interviewer debriefing reports – or short articles that provide case studies of survey practices that worked or did not work would ease translation of work from one research team to another. Although these are hard, we think they are worthwhile future pursuits.

Acknowledgments

The workshop and this edited volume were supported by the National Science Foundation (SES-1758834), the Charles Cannell Fund in Survey Methodology of the Survey Research Center at the University of Michigan, and the Rensis Likert Fund for Research in Survey Methodology at the University of Michigan. Any opinions, findings, and conclusions or recommendations expressed in this text are those of the authors and do not necessarily reflect the views of the National Science Foundation.

References

Campanelli, P., P. Sturgis, and S. Purdon. 1997. *Can you hear me knocking: An investigation into the impact of interviewers on survey response rates*. London: Survey Methods Centre at SCPR.

Childs, J. H. and A. Landreth. 2006. Analyzing interviewer/respondent interactions while using a mobile computer-assisted personal interview device. *Field methods* 18:335–351.

Conrad, F. G. and M. F. Schober. 2000. Clarifying question meaning in a household telephone survey. *Public opinion quarterly* 64:1–28.

Couper, M. P. and R. M. Groves. 2002. Introductory interactions in telephone surveys and nonresponse. In *Standardization and tacit knowledge: Interaction and practice in the survey interview*, ed. D. W. Maynard, H. Houtkoop-Steenstra, N. C. Schaeffer, and J. van der Zouwen, 161–177. New York: John Wiley & Sons.

Couper, M. P. and F. Kreuter. 2013. Using item-level paradata to explore response time and data quality. *Journal of the royal statistical society, A* 176:271–286.

de Leeuw, E. 2005. To mix or not to mix data collection modes in surveys. *Journal of official statistics* 21:233–255.

Dijkstra, W. 1983. How interviewer variance can bias the results of research on interviewer effects. *Quality and quantity* 17:179–187.

Dillman, D. A., J. D. Smyth, and L. M. Christian. 2014. *Internet, phone, mail, and mixed mode surveys: The tailored design method*. Hoboken, NJ: John Wiley & Sons.

Edwards, B., A. Maitland, and S. Connor. 2017. Measurement error in survey operations management. In *Total survey error in practice*, ed. P. P. Biemer, E. De Leeuw, S. Eckman, B. Edwards, F. Kreuter, L. E. Lyberg, N. C. Tucker, and B. T. West, 253–277. Hoboken, NJ: John Wiley & Sons.

Elliott, M. R. and B. T. West. 2015. "Clustering by interviewer": A source of variance that is unaccounted for in single-stage health surveys. *American journal of epidemiology* 182:118–126.

Fowler, F. J. 2011. Coding the behavior of interviewers and respondents to evaluate survey questions. In *Question evaluation methods: Contributing to the science of data quality*, ed. J. Madans, K. Miller, A. Maitland, and G. Willis, 7–21. Hoboken, NJ: John Wiley & Sons.

Fowler, F. J. and T. W. Mangione. 1990. *Standardized survey interviewing: Minimizing interviewer-related error*. Newbury Park, CA: Sage Publications.

Garbarski, D., N. C. Schaeffer, and J. Dykema. 2016. Interviewing practices, conversational practices, and rapport: Responsiveness and engagement in the standardized survey interview. *Sociological methodology* 46:1–38.

Graesser, A. C., Z. Cai, M. M. Louwerse, and F. Daniel. 2006. Question understanding aid (QUAID): A web facility that helps survey methodologists improve the comprehensibility of questions. *Public opinion quarterly* 70:3–22.

Groves, R. M. and M. Couper. 1998. *Nonresponse in household interview surveys*. New York: John Wiley & Sons, Inc.

Groves, R. M. and L. J. Magilavy. 1986. Measuring and explaining interviewer effects in centralized telephone facilities. *Public opinion quarterly* 50:251–266.

Groves, R. M., B. C. O'Hare, D. Gould-Smith, J. Benki, and P. Maher. 2008. Telephone interviewer voice characteristics and the survey participation decision. In *Advances in telephone survey methodology*, ed. J. M. Lepkowski, C. Tucker, J. M. Brick, E. D. de Leeuw, L. Japec, P. J. Lavrakas, M. W. Link, and R. L. Sangster, 385–400. Hoboken, NJ: John Wiley & Sons.

Hansen, M. H., W. N. Hurwitz, and M. A. Bershad. 1961. Measurement errors in censuses and surveys. *Bulletin of the international statistical institute* 38:351–374.

Holbrook, A. L., T. P. Johnson, Y. I. Cho, S. Shavitt, N. Chavez, and S. Weiner. 2016. Do interviewer errors help explain the impact of question characteristics on respondent difficulties? *Survey practice* 9:no pp.

Hox, J. J. 1994. Hierarchical regression models for interviewer and respondent effects. *Sociological methods and research* 22:300–318.

Jaszczak, A., K. Lundeen, and S. Smith. 2009. Using nonmedically trained interviewers to collect biomeasures in a national in-home survey. *Field methods* 21:26–48.

Kish, L. 1962. Studies of interviewer variance for attitudinal variables. *Journal of the American statistical association* 57:91–115.

Mathiowetz, N. A. and C. F. Cannell. 1980. Coding interviewer behavior as a method of evaluating performance. *ASA Proceedings of the section on survey research methods*.

Maynard, D. W., H. Houtkoop-Steenstra, N. C. Schaeffer, and J. van der Zouwen. 2002. *Standardization and tacit knowledge: Interaction and practice in the survey interview*. New York: John Wiley & Sons, Inc.

Mneimneh, Z. M., R. Tourangeau, B.-E. Pennell, S. G. Heeringa, and M. R. Elliott. 2015. Cultural variations in the effect of interview privacy and the need for social conformity on reporting sensitive information. *Journal of official statistics* 31:673–697.

Morton-Williams, J. 1993. *Interviewer approaches.* Cambridge: University Press.

O'Muircheartaigh, C. and P. Campanelli. 1998. The relative impact of interviewer effects and sample design effects on survey precision. *Journal of the royal statistical society, A* 161:63–77.

O'Muircheartaigh, C. and P. Campanelli. 1999. A multilevel exploration of the role of interviewers in survey non-response. *Journal of the royal statistical society, A* 162:437–446.

Olson, K. and I. Bilgen. 2011. The role of interviewer experience on acquiescence. *Public opinion quarterly* 75:99–114.

Olson, K. and B. Parkhurst. 2013. Collecting paradata for measurement error evaluations. In *Improving surveys with paradata: Analytic uses of process information,* ed. F. Kreuter, 43–72. Hoboken, NJ: John Wiley & Sons.

Olson, K. and A. Peytchev. 2007. Effect of interviewer experience on interview pace and interviewer attitudes. *Public opinion quarterly* 71:273–286.

Olson, K. and J. D. Smyth. 2015. The effect of CATI questions, respondents, and interviewers on response time. *Journal of survey statistics and methodology* 3:361–396.

Olson, K., J. D. Smyth, R. Horwitz, S. Keeter, V. Lesser, S. Marken, N. Mathiowetz, J. McCarthy, E. O'Brien, J. Opsomer, D. Steiger, D. Sterrett, J. Su, Z. T. Suzer-Gurtekin, C. Turakhia, and J. Wagner. 2019. *Transitions from telephone surveys to self-administered and mixed-mode surveys.* Oakbrook Terrace, IL: American Association for Public Opinion Research.

Olson, K. and J. Wagner. 2015. A field experiment using GPS devices to monitor interviewer travel behavior. *Survey research methods* 9:1–13.

Ongena, Y. P. and W. Dijkstra. 2006. Methods of behavior coding of survey interviews. *Journal of official statistics* 22:419–451.

Sakshaug, J. W. 2013. Using paradata to study response to within-survey requests. In *Improving surveys with paradata: Analytic uses of process information,* ed. F. Kreuter, 171–190. Hoboken, NJ: John Wiley and Sons.

Sakshaug, J. W., M. P. Couper, M. B. Ofstedal, and D. R. Weir. 2012. Linking survey and administrative records: Mechanisms of consent. *Sociological methods & research* 41:535–569.

Saris, W. E. and I. N. Gallhofer. 2007. *Design, evaluation, and analysis of questionnaires for survey research.* Hoboken, NJ: John Wiley and Sons.

Schaeffer, N. C. 1991. Conversation with a purpose – or conversation? Interaction in the standardized interview. In *Measurement errors in surveys,* ed. P. Biemer, R. M. Groves, L. Lyberg, N. A. Mathiowetz, and S. Sudman, 367–391. New York: John Wiley & Sons, Inc.

Schaeffer, N. C., J. Dykema, and D. W. Maynard. 2010. Interviewers and interviewing. In *Handbook of survey research,* ed. P. V. Marsden and J. D. Wright, 437–470. Bingley, UK: Emerald Group Publishing.

Schaeffer, N. C., D. Garbarski, J. Freese, and D. W. Maynard. 2013. An interactional model of the call for survey participation: Actions and reactions in the survey recruitment call. *Public opinion quarterly* 77:323–351.

Schaeffer, N. C., B. H. Min, T. Purnell, D. Garbarski, and J. Dykema. 2018. Greeting and response: Predicting participation from the call opening. *Journal of survey statistics and methodology* 6:122–148.

Schober, M. F. and F. G. Conrad. 1997. Does conversational interviewing reduce survey measurement error? *Public opinion quarterly* 61:576–602.

Suchman, L. and B. Jordan. 1990. Interactional troubles in face-to-face survey interviews. *Journal of the American statistical association* 85:232–253.

Timbrook, J., J. Smyth, and K. Olson. 2018. Why do mobile interviews take longer? A behavior coding perspective. *Public opinion quarterly* 82:553–582.

Viterna, J. and D. W. Maynard. 2002. How uniform is standardization? Variation within and across survey research centers regarding protocols for interviewing. In *Standardization and tacit knowledge: Interaction and practice in the survey interview*, ed. D. W. Maynard, H. Houtkoop-Steenstra, N. C. Schaeffer, and J. van der Zouwen, 365–397. New York: John Wiley & Sons.

West, B. T. and A. G. Blom. 2017. Explaining interviewer effects: A research synthesis. *Journal of survey statistics and methodology* 5:175–211.

West, B. T., F. Kreuter, and U. Jaenichen. 2013. "Interviewer" effects in face-to-face surveys: A function of sampling, measurement error, or nonresponse? *Journal of official statistics* 29:277–297.

West, B. T. and K. Olson. 2010. How much of interviewer variance is really nonresponse error variance? *Public opinion quarterly* 74:1004–1026.

2

The Legacy of Charles Cannell

Peter V. Miller and Nancy A. Mathiowetz

CONTENTS

2.1 Summary of Cannell's Key Contributions

Between 1960 and 1990, Charles Cannell and his colleagues documented significant survey response errors, created a model of the survey response process, developed a theory of the interview and a set of interviewing techniques to encourage accurate reporting, and designed methods for observing interviewer and respondent behavior. These integrated accomplishments have had a noteworthy impact on thinking about interviewer–respondent interaction, but they deserve more attention from students of the survey interview now. In this chapter, we summarize Cannell's work, examine its evolution, and suggest ways in which it can inform current research and practice.

2.1.1 Documenting Errors in Self-Reports

In order to judge if interviewing procedures lead to better self-reports of behavior, we need to understand when self-reports are more trustworthy. Cannell's productive collaboration on "validity studies" with the U.S. National Center for Health Statistics (NCHS) early in the 1960s built the foundation for subsequent research on interviewing. The studies were record check efforts: interviews with people who were known – by virtue of information in medical records – to have been hospitalized or to have visited a physician. Respondents in these investigations failed to report these documented experiences in predictable ways. For example, as the amount of time between the date of the interview and the medical encounters increased, reporting of hospitalizations and doctor visits declined (cf. the much earlier work on memory by Ebbinghaus 1913). Shorter hospital stays were less likely to be reported than longer ones. Hospitalizations associated with embarrassing health conditions were less likely to be reported (Cannell, Fisher, and Bakker 1965; Cannell and Fowler 1963). Other investigations funded by NCHS around this time focused on reporting of chronic health conditions, comparing survey reports to medical records (Balamuth and Shapiro 1965; Madow 1967). These studies also found that health experiences were under-reported.

Such findings led to the following general inferences: (1) self-reports of experiences will decline in accuracy as the experiences themselves recede in time; (2) self-reports of experiences will decline in accuracy as their subjective salience to the respondent declines; (3) self-reports of experiences will be less accurate if they are embarrassing ("socially undesirable"). Interviewing procedures that led to more reporting of experiences that are difficult to remember or uncomfortable to report can be said to have produced more accurate data. These generalizations were in line with contemporaneous work by other investigators outside the health research sphere (Ferber 1966; Neter and Waksberg 1964).

The under-reporting concept was further extended to phenomena lacking external validation information. As Marquis, Cannell, and Laurent (1972:1) wrote, "Since higher reporting does seem to suggest more accurate reporting, the number of reported events has been used as a dependent variable in later studies where better validity data have not been available." This benchmark formed the basis for judging the quality of data produced by different interviewing approaches.

2.1.2 Modeling the Survey Response Process

Cannell developed a theory of the response process. Survey methodologists will recognize components of the cognitive approach to survey response (Tourangeau, Rips, and Rasinski 2000) in a model originally sketched by Cannell, Marquis, and Laurent (1977:53) and revised in Cannell, Miller, and Oksenberg (1981). As can be seen in Figure 2.1, the response process begins with the respondent's comprehension of the question. Beyond the consideration of question wording and sentence structure complexity, Cannell and colleagues' concept of comprehension also involves broader issues of interpretation, factoring in consideration of the intended meaning of questions.

Step 2 of the model, labeled here as cognitive processing, involves retrieval, beginning with the identification of the relevant information needed for searching one's memory. Step 3 involves the process of evaluation of the retrieved information with respect to its accuracy. Step 4 involves the assessment of that same information with respect to other goals (which may lead to inaccurate reporting), and Step 5 identifies the actual response.

An interesting feature of the model is that it suggests where the response process can go astray, resulting in less than complete or accurate information. Cannell and his colleagues believed that deviation from the linear phases 1 through 5 could be due to insufficient retrieval effort or motivation (possibly embarrassment) on the part of the respondent, or due to extraneous situational cues (e.g., unhelpful interviewer behavior). Cannell came to rely on the idea of "interference" in memory (McGeoch 1932; Tulving 1972), which suggests, in the survey context, that events are not simply forgotten, but are hard to retrieve because succeeding events have "buried" them. Encouraging more retrieval effort in the

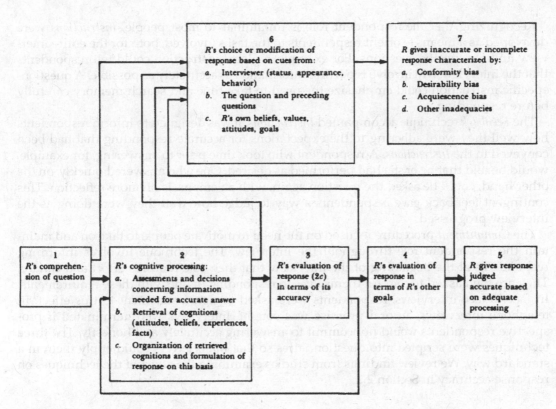

FIGURE 2.1
Diagram of respondent's (R) question-answering process. *Source:* Cannell, Miller, and Oksenberg (1981).

interview, therefore, is a worthwhile approach. Both cognitive and motivational difficulties can hamper respondents from progressing linearly through the steps to accurate response. The model's depiction of a "detour" to inaccurate answering is the intellectual forerunner of the concept of "satisficing" in survey response (Krosnick 1991). How questions are designed and how interviewers administer them are keys to respondents' staying on the "straight and narrow" path.

2.1.3 Inventing New Interviewing Techniques

Cannell's approach to the interview was rooted in his extensive observation of interviewers while leading field operations in the U.S. Department of Agriculture and at the University of Michigan Survey Research Center. He developed interviewing procedures designed to encourage respondents to stay on the path to accurate reporting. These procedures are based on a theory of the interview. In essence, Cannell argued that the interview involves not only the communication in questions and probes, but also various sorts of "meta-communication" – intentional and unintentional – that inform respondents about how they should undertake their tasks. Shaping the meta-communication to encourage accurate responding and to avoid inadvertent encouragement of error are the aims of interviewing procedures that Cannell and his colleagues developed. These techniques included ones designed to teach respondents what is needed to meet question demands (*Instructions* and *Feedback*) and one designed to motivate respondents to expend needed effort (*Commitment*).

Recognizing that the respondent role is unfamiliar to most people, *Instructions* were developed as a means to orient respondents to the tasks involved, both for the entire interview and for specific questions. For example, a global instruction could tell respondents that the interview will involve reporting information as accurately as possible. A question-specific instruction could emphasize the need to take time and search memory carefully before responding.

The *Feedback* technique accompanied *Instructions*. It was designed to inform respondents how well they were adhering to the expectations for accurate responding that had been conveyed in the *Instructions*. A respondent who took time prior to answering, for example, would be told that he or she had performed as desired. One who answered quickly, on the other hand, could be asked the question again, with an emphasis on more reflection. This contingent feedback gave respondents a way to judge how well they were doing as the interview progressed.

The *Commitment* procedure focused on the need to motivate people to take on and maintain the respondent role throughout the interview. The technique involves informing respondents at the beginning of the interview that their role requires effort and candor. They are then asked to commit themselves to responding in line with these requirements. In face-to-face interviews, respondents were asked to sign a statement to this effect. In telephone interviews, an oral promise was sought. Interviews were terminated if prospective respondents would not commit to answering accurately and honestly. The three techniques were scripted into questionnaires so that interviewers could apply them in a standard way. We review findings from studies examining the effect of the techniques on response accuracy in Section 2.2.

2.1.4 Creating Methods of Interview Observation

The theory of interviewing described above rested on novel empirical analyses of interview interactions. Cannell and his colleagues (most notably Floyd J. (Jack) Fowler and Kent Marquis, later Lois Oksenberg) set out to gather accurate descriptions of the interviewing process so as to inform both theory and practice with respect to the sources of error in the interview. Although behavior coding is often cited as initially developed for use in the evaluation of interviewers, the early studies suggest that its focus was primarily on understanding the interaction in the survey process.

Cannell and Fowler (1964) first uncovered evidence of interviewer effects on reporting of health experiences. They observed that reporting was affected by the size of the interviewers' workload and when during the field period they conducted interviews – later interviews elicited less information. In addition, they saw that the interviewers' pace of speech negatively affected reporting (see also Holbrook, et al., Chapter 17, and Olson and Smyth, Chapter 20, this volume). Such observations laid the foundation for later controlled experiments. Two subsequent studies (Cannell, Fowler, and Marquis 1968; Marquis and Cannell 1969) are foundational investigations into the survey interaction process in face-to-face data collection efforts. The first study used third-person observers and reinterviews with respondents a day later to gather their impressions of the interviewer and the interview; the second study used tape recordings of the interviews. In the first effort, despite hypothesizing that interviewers' and respondents' knowledge and attitudes would be highly correlated with the quality of data, the authors found that behavior in the interview was the main variable that correlated with the index of reporting quality. In addition, Cannell, Fowler, and Marquis (1968) found that the behaviors of the interviewer and the respondent were best described as "balanced." More verbal output by one actor was matched by output from the other. This relationship suggested that systematically changing the interviewer's speech output would change the respondent's activity level, thereby increasing both the amount and the quality of reported health information (a hypothesis studied later by Marquis, Cannell, and Laurent 1972).

Tape recording was employed in the second study (Marquis and Cannell 1969). Further, the coding scheme was expanded, including codes to capture information about probes (both the use and an evaluation of the quality of the probes) and the use of feedback by either the interviewer or the respondent. Three relevant findings from this study are: (1) a large proportion of the interviewer and respondent behavior is extraneous interaction, not helpful to the interview process; (2) interviewer feedback was one of the most frequent behaviors by the interviewers and was indiscriminate with respect to the quality of the response provided; (3) the coding scheme identified problems with questions (poor syntax, complex instructions, parenthetical phrases) and respondent difficulty in understanding or retrieving the information of interest. The findings from these studies led Cannell and colleagues to focus on taming communication in the interview in service of measurement goals. Over time, methods for recording and analyzing interview interaction (behavior coding) were applied to other measurement problems, including question pretesting and interviewer evaluation.

2.2 Evidence for the Effectiveness of Interviewer *Feedback, Instructions,* and *Commitment*

2.2.1 Early Studies

Marquis (1970) conducted the first experiment on the interviewing techniques for his dissertation. He found that interviewer *Feedback* statements provided *only* after respondents reported instances of morbidity (e.g., chronic health conditions) led to significant increases in such reports, compared to a control condition in which interviewers provided no feedback. The experimental evidence supported the idea that social reinforcement could improve response accuracy, on the assumption – supported by record check studies (e.g., Madow 1967) – that more reporting is more accurate.

In this study, as in many subsequent experiments conducted by the Cannell team, the interviewers were trained to administer both experimental and control interview treatments, thus serving as their own controls. The alternative design would be to assign different interviewers to different experimental treatments. Each design embodies trade-offs. The first design risks interviewers inferring that a particular method is preferred and behaving so as to support the suspected hypothesis, or misapplying the techniques, thereby attenuating the experimental effects. The second design risks confounding experimental effects with interviewer differences. Cannell and colleagues' choice of the within-interviewer design rests on their practice of not revealing the hypothesis and carefully training a single set of interviewers to follow each questionnaire script strictly. Cases were randomly assigned to treatments. The interviewers and the respondents were all white. The sample was households selected from neighborhoods with similar socioeconomic characteristics and age ranges, a procedure designed to control extraneous demographic variance that could muddy the comparison of interviewing treatments.

The next experiment (Marquis, Cannell, and Laurent 1972) was the first to examine a questioning technique that evolved in later studies into *Instructions*. The experimental treatments involved the length of questions asked, reinforcement (*Feedback*), and reinterviews. Question length was manipulated by adding what was intended to be redundant verbiage – not more memory cues or socio-emotional material. The idea, following research by Matarazzo, Wren, and Saslow (1965), was that – as noted in interview observation studies described above – more lengthy speech by the interviewer would lead to more lengthy responses, which, in turn, would yield more health information. This notion, from research on behavior modification, as was contingent reinforcement (feedback), is an example of how Cannell and colleagues melded theories from social psychology with empirical induction.

Respondents in this study were members of a prepaid health insurance plan. Medical record information from respondents' clinic visits for 6 months in 1968 was to serve as the validating criterion for interview reports and allow for detection of both under- and over-reporting. The study's findings were mixed and not easy to explain. One reason for the mixed findings may have been the quality of the record information; there was some concern that the record information was not accurate. The study budget limited the sample size and aspects of the experimental execution. Further, the experimental findings were moderated by education.

Respondents with higher and lower education reacted differently to the question length and reinforcement treatments. Reinforcement seemed to reduce error for respondents with less than high school education but not for those with more years of education. The reverse was true for longer questions. All in all, the study seemed to produce more questions than answers. What carried forward was the continued belief that interviewer reinforcement could be a useful tool 'and that some form of question lengthening was worth further exploration. The study's authors reasoned that question lengthening might have cueing effects, might suggest that the inquiry is important or that taking time to think about the question is a good idea. The length of the question itself might be a way to provide time for the respondent to think (Blair, et al. 1977; Cannell, Marquis, and Laurent 1977:70).

The next several years saw further refinement of the question length and feedback techniques and the addition of the *Commitment* procedure to the suite of interviewing approaches examined in experimental studies. The *Commitment* procedure is a global motivational tool for the interview. It seeks an explicit agreement from the respondent that he or she will undertake actions to reach the goal of rendering complete and accurate information. The procedure was tested in an experiment that examined effects on health experience questions. The experiment included an expanded definition of accurate reporting. The dependent measures were intended to capture the amount of information provided by respondents, as well as the specificity of the response (e.g., number of mentions to open questions, number of doctor visits reported, number of symptoms reported for the pelvic region of the body, specific dates for medical appointments or for which the respondent was limited in his/her activities, and evidence of checking external sources of information before responding) (Oksenberg, Vinokur, and Cannell 1979a).

Instructions grew out of the earlier work on long questions. Cannell and colleagues decided that, rather than speculating about the possible reasons why long questions seemed to have a useful effect, it made sense to insert messages at the beginning of the questionnaire (e.g., a global instruction about the need for accuracy, completeness, and precision) and also just prior to selected questions to inform respondents how to achieve those results. The procedural *Instructions* focused on, for example, telling respondents to take time and search memory before answering a query.

Instructions were studied along with *Commitment* and *Feedback* in a separate experiment (Oksenberg, Vinokur, and Cannell 1979b). The questionnaires in both studies became a script, beginning with requests for *Commitment*, then global *Instructions*, and then questions preceded by *Instructions* and followed by contingent *Feedback*. The findings of the studies are summarized in Cannell, Miller, and Oksenberg (1981). In brief, the combined forces of the three techniques showed promising results when compared with a control condition containing only questions to be administered. The dependent variables were, again, measures of the amount of information respondents provided and the precision with which it was rendered. Another, smaller experiment (Miller and Cannell 1977) comparing one treatment with commitment, instructions, and feedback with a control that did not include these techniques found higher reporting of behaviors likely to be under-reported (e.g., time watching television, "X-rated" movie attendance) and lower reporting of behaviors likely to be over-reported (e.g., book reading) in the experimental condition. Miller and Cannell (1982), in a national telephone study, found that *Commitment* and *Instructions* produced more reporting of socially *undesirable* attitudes toward television watching, as well as more responses to open questions and reports of health conditions, while the *Feedback* treatment, when combined with the other techniques, did not add to these effects.

In the late 1970s, the University of Michigan's Survey Research Center (SRC) undertook a methodological study to evaluate the use of random-digit dialing (RDD) sampling and telephone data collection for the collection of health information, specifically to test the feasibility of conducting the National Health Interview Survey (NHIS) by telephone (Thornberry 1987). As part of the mode comparison study (comparing the telephone survey conducted by SRC to the in-person NHIS collected by the U.S. Bureau of the Census for the National Center for Health Statistics), the telephone sample cases were randomly assigned to one of two treatments. The control group was intended to replicate the procedures used by the Census Bureau in which interviewers were limited to a predefined set of behaviors – asking the question as worded and using standard interviewing procedures such as neutral probes. In the experimental treatment, *Commitment*, *Instructions*, and *Feedback* were incorporated into the questionnaire. The researchers examined rates of reports of various health behaviors, with the hypothesis that higher rates of reporting of health incidences were indicative of higher quality data. The study found that for several health behaviors (bed days, work loss days, days with reduced activity, acute conditions, chronic conditions), the experimental form of the questionnaire led to increased levels of reporting.

The authors acknowledged that there was not, at the time, "unequivocal acceptance of the underreporting hypothesis" (Thornberry 1987:23). Marquis, who worked with Cannell on a number of the studies reviewed here, attacked the record check evidence that was used to support the idea that many events are under-reported (Marquis 1978; Marquis, Marquis, and Polich 1986). He and colleagues dismissed findings of "reverse" record check studies, in which the sample of respondents is chosen from records of events – such as the hospitalization and doctor visit studies conducted by Cannell and Fowler discussed earlier – arguing that this design cannot discover over-reports of these experiences. They analyzed the results of "full" record check studies – where the sample and the record evidence are independent – of sensitive behavioral events. They argued that these studies produced more evidence of over-reporting of sensitive experiences than under-reporting. They further argued that the main issue with reports of sensitive information is reliability, not response bias. In response to such criticisms, Cannell and colleagues examined alternative explanations for the increased reporting of health events resulting from the experimental treatments, for example, over-reporting due to telescoping. After considering alternative explanations, the researchers concluded the "most tenable hypothesis is that the experimental techniques facilitated accurate reporting on health variables in this study" (Thornberry 1987:25). We will return to this debate below.

2.2.2 Subsequent Record Check Investigations

A validation study of the economic questions included in the Panel Study of Income Dynamics (PSID) provided the opportunity to test the interviewing techniques within a "full" record check study as well as examine several hypotheses concerning the nature of response error that Cannell and colleagues had identified in the earlier health studies (Duncan and Mathiowetz 1985). The nature of the record check study, where responses for all respondents were examined by linking to company records for a single manufacturing entity, allowed for the assessment of both the extent and direction of response error. The study found weak support for a reduction in response error associated with the experimental interviewing techniques, most notably for the reporting of less well-known fringe benefits.

In another record check study, Belli and Lepkowski (1996) looked at the effect of feedback alone. The study design had experimental and control conditions, with interviewer feedback programmed into the experimental questionnaire. But Belli and Lepkowski did not replicate earlier interviewing experiments in this paper. Instead, they relied on a behavior coding analysis of a sample of approximately 500 interviews, apparently selected from both the experimental and control conditions. The dependent variable – constructed by subtracting the number of health care events recorded in records from those reported in interviews – also differed from earlier studies. The outcome variable also was skewed due to undercoverage of health care events in the records employed. The authors suggested that feedback given after responses to immediately preceding questions had a negative impact on reporting accuracy (leading to over-reports). It is difficult to compare this finding to those of earlier studies, which relied on comparisons of experimental treatments as a whole, focusing on the potential effect of a number of feedback statements across questions. While we do not view the findings of this study as persuasive, it seems that behavior coding could be employed as a supplement to between-treatment analysis in future studies, to look at effects of instructions and feedback at a micro, question-by-question level.

2.2.3 Summary and Critique

From 1960 until 1990, the Cannell perspective on response error, and interviewing techniques designed to attack this error, evolved and was tested in a variety of settings. The perspective is a coherent, integrated approach to interviewing. The evidence overall suggests that the techniques may lead to better self-report data. But some objections to the response error assumptions that underlie the research persisted during this time, and the record of effects of interviewing techniques is inconsistent across studies.

The Marquis, Marquis, and Polich (1986) paper raised the question of what can be inferred from a record check study. They sought to make the case that "socially undesirable" phenomena are not under-reported in surveys. Their work dismissed "reverse" record check evidence of reporting bias out of hand. It also attacked "full design" record check evidence that found under-reporting bias (e.g., Madow 1967), speculating that the questionnaires in those studies were flawed, leading to mismatches between the survey and record evidence.

Marquis, Marquis, and Polich's (1986) findings are not definitive, because the execution of their record check study is not fully transparent and their suppositions about evidence contrary to their position are not convincing. Further, they may have found "over-reporting" of events – support for their case – due to record errors or match errors, as they acknowledge. In addition, their dismissal of "under-reporting" findings in the studies they examined as artifacts of poor questionnaire design ignores possible questionnaire flaws in studies that support their hypothesis. Still, this paper casts sharp light on the design and execution of record check studies. The malleability of survey response matches to record data has been noted elsewhere (e.g., Miller and Groves 1985).

Since Cannell and colleagues' work utilized evidence from record check studies to frame tests of methods to improve survey reporting, the limitations of record check studies raised by Marquis, Marquis, and Polich (1986) may be seen to cast doubt on the inferences from their interviewing research. But outcome measures in the interviewing studies include ones that are believed to be both under- and over-reported, and attitudinal measures not subject to record "validation." It is more important to recognize the limitations in the ability of record check studies to establish "truth" and carefully design record data-survey report comparisons to reflect these limitations.

Beyond the issues raised by assumptions about response validity, the record of effects of the *Feedback* technique is inconsistent. The original experiment (Marquis 1970) showed noteworthy effects of what Marquis called "social reinforcement." A subsequent investigation (Cannell, Marquis, and Laurent 1977) found a more complicated picture, in which effects of *Feedback* were present for lower education respondents, but not for their counterparts with more education. Oksenberg, Vinokur, and Cannell (1979b) examined *Feedback* effects when combined with *Instructions* and with both *Instructions* and *Commitment*, for both low and high education respondents. In this study, *Feedback* effects were present for both education groups. Miller and Cannell (1982) found that there were no incremental *Feedback* effects when it was combined with *Instructions* and *Commitment* in a national telephone study – the first attempt to employ the techniques in this mode. In subsequent telephone research, feedback was always combined with the other techniques, so its independent effects could not be discerned. This inconsistent record of effects of *Feedback* is a call for further research into the value of what seems intuitively to be a useful approach to orienting respondents to interview tasks.

2.3 More Recent Cannell-Inspired Research

2.3.1 Interview Observation

The use of interview observation focused almost solely on interviewer evaluation during the 1970s (Cannell, Lawson, and Hauser 1975; Mathiowetz and Cannell 1980) but then emerged as a pretesting technique with publications by Morton-Williams (1979), Morton-Williams and Sykes (1984), and Oksenberg, Cannell, and Kalton (1991). The expanded use of behavior coding as an evaluation tool for pretesting provided researchers with a systematic, objective, and quantitative means by which to assess survey questions.

The coding scheme used by Oksenberg, Cannell, and Kalton (1991) is relatively simple, consisting of three interviewer behaviors, all related to how the initial question was asked (exactly, slight change, or major change) and seven respondent behaviors (interruption with answer, request for clarification, adequate answer, qualified answer, inadequate answer, don't know, and refusal to answer). All of the respondent's behaviors apart from the provision of an adequate answer were seen as indications of problems with the question. In contrast, the coding scheme used by Marquis and Cannell (1969) included 12 interviewer behaviors, 7 respondent behaviors, and 9 behaviors (e.g., laughing) that could be exhibited by either the interviewer or the respondent.

What began as a simple coding scheme for evaluating interviewers, and then later, the quality of questions, has grown into a far more nuanced and detailed examination of the interviewer–respondent interaction, for example, in conversational analysis (e.g., Maynard, et al. 2002). Today we see behavior coding used to understand issues ranging from the construction of questions as part of pretesting activities (e.g., Presser, et al. 2004) to use as a metric to understand data quality in time use studies (e.g., Freedman, et al. 2013).

2.3.2 Interviewing Methods

More recent data collection research demonstrates the potential effectiveness of *Commitment* for online, self-administered surveys. Cibelli (2017) experimented with the use of *Commitment*, and *Commitment* and *Feedback,* in two studies as part of her dissertation work. With the use of a simple binary *Commitment* request, where the respondent either agreed or did not agree to commit to working hard to provide accurate information in an online labor force survey, she found a significant decline in item nonresponse and straightlining, as well as the provision of more accurate data as compared to a control group that was not offered a *Commitment* statement. However, a second experiment, in which the nature of *Commitment* was operationalized in a request to agree to five distinct behaviors (careful reading of the question; being precise; using records; providing as much information as possible; and answering honestly), coupled with *feedback* and contextual recall cues, did not result in markedly more accurate or higher quality data as compared to a control group.

2.4 Conclusions

Where do we find ourselves now with respect to Cannell's influence on research and practice? Both the social interaction and the cognitive models of the survey interview process continue to inform the construction of questionnaires and our understanding of the sources of response error (e.g., Schaeffer and Dykema 2011). Behavior coding has become a tool widely used by questionnaire designers (e.g., Presser, et al. 2004). Research on the interviewing techniques that Cannell pioneered, however, has not diffused throughout current survey practice.

Cannell's legacy to survey researchers is a perspective on response error and a set of tools for reducing it. The perspective has currency today – it should guide design of questionnaires and instructions for interviewer behavior. As we have seen, Cannell advocated theorizing about likely reporting error in the measurements planned for each survey. Such hypotheses should direct how we ask questions and how we train interviewers to ask them. Scripting questionnaires with orientation for respondents about the need for accuracy and reinforcing behavior that contributes to this goal is an approach that should be tested and widely employed. Implementing techniques, such as *Commitment*, to motivate respondents to do the necessary work to provide accurate information is particularly important in an era in which "surveys" are myriad and easy to dismiss. Rather than an invitation to refuse participation, a *Commitment* request at the start of an interview, followed by *Instructions* and *Feedback* that teach the respondent role and reinforce appropriate behavior, can distinguish our surveys from trivial requests. Cannell's approach can help to legitimate survey requests in the public eye.

Although the applicability of the social interaction model of the survey process may not be obvious for self-administered online data collection, any data collection activity is a process with actors and interaction among those actors. Some actors may be distant (the unseen researcher), some may be inanimate (smartphone; see Schober, et al., Chapter 13, this volume), but the interaction among these actors contributes to determining the quality of survey data. We believe that the world of online data collection is ripe for further research with respect to how *Instructions*, *Feedback*, and *Commitment* could improve the overall quality of such data. The perspective can be applied in Web, interactive voice response (IVR), and short message service (SMS or "texting") surveys. In fact, implementing the techniques in such contexts may be more effective, since automated communication will faithfully execute interviewing instructions that human interviewers do not always follow. If surveys employ avatar interviewers (Cassell and Miller 2007; see Conrad, et al., Chapter 11, this volume), nonverbal reinforcement procedures, attempted but not validated in earlier research (Marquis 1970), can be reliably implemented. In short, the addition of more communication media for conducting surveys opens up new possibilities for testing and adopting methods that Cannell pioneered.

References

Balamuth, E., and S. Shapiro. 1965. *Health interview responses compared with medical records*. Washington, DC: PHS Vital and health statistics, Series D, No. 5.

Belli, R., and J. Lepkowski. 1996. Behavior of survey actors and the accuracy of response. In Conference proceedings: *Health survey research methods*, ed. R. Warnecke, 69–74. Hyattsville, MD: National Center for Health Statistics, DHHS Pub. No. (PHS) 96-1013.

Blair, E., S. Sudman, N. Bradburn, and C. Stocking. 1977. How to ask questions about drinking and sex: Response effects in measuring consumer behavior. *Journal of marketing research*, 14(3):316–21.

Cannell, C., G. Fisher, and T. Bakker. 1965. *Reporting of hospitalization in the health interview survey*. Washington, DC: PHS vital and health statistics, Series 2, No. 6.

Cannell, C., and F. Fowler. 1963. *A study of the reporting of visits to doctors in the national health survey*. Research Report. Ann Arbor, MI: Survey Research Center.

Cannell, C., and F. Fowler. 1964. A note on interviewer effects on self-enumerative procedures. *American sociological review*, 29(2):269.

Cannell, C., F. Fowler, and K. Marquis. 1968. *The influence of interviewer and respondent psychological and behavioral variables on the reporting in household interviews*. Washington, DC: PHS Vital and Health Statistics, Series 2, No. 26.

Cannell, C., S. Lawson, and D. Hauser. 1975. *A technique for evaluating interviewer performance*. Ann Arbor, MI: Survey Research Center.

Cannell, C., K. Marquis, and A. Laurent. 1977. *A summary of studies of interviewing methodology*. Washington, DC: PHS Vital and Health Statistics, Series 2, No. 69.

Cannell, C., P. Miller, and L. Oksenberg. 1981. Research on interviewing techniques. In *Sociological methodology 1981*, ed. S. Leinhardt, 389–437. San Francisco, CA: Jossey-Bass.

Cassell, J., and P. Miller. 2007. Is it self-administration if the computer gives you encouraging looks? In *Envisioning the survey interview of the future*, ed. F.G. Conrad, and M.F. Schober, 161–78. New York: John Wiley & Sons.

Cibelli, K. 2017. *The effects of respondent commitment and feedback on response quality in online surveys.* Ph.D. Dissertation, The University of Michigan.

Duncan, G., and N. Mathiowetz. 1985. *A validation study of economic survey data.* Ann Arbor, MI: The Institute for Social Research.

Ebbinghaus, H. 1913. *Memory: A contribution to experimental psychology.* New York: Teachers College, Columbia University.

Ferber, R. 1966. *The reliability of consumer reports of financial assets and debts.* Urbana, IL: Bureau of Economic and Business Research, University of Illinois.

Freedman, V., J. Broome, F. Conrad, and J. Cornman. 2013. Interviewer and respondent interactions and quality assessments in a time diary study. *Electronic international journal of time use research*, 10(1):55–75.

Krosnick, J. 1991. Response strategies for coping with the cognitive demands of attitude measures in surveys. *Applied cognitive psychology*, 5(3):213–36.

Madow, W. 1967. *Interview data on chronic conditions compared with information derived from medical records.* Washington, DC: PHS vital and health statistics, Series 2, No. 23.

Marquis, K. 1970. Effects of social reinforcement on health reporting in the household interview. *Sociometry*, 33:203–15.

Marquis, K. 1978. *Record check validity of survey responses: A reassessment of bias in reports of hospitalizations.* Santa Monica, CA: Rand Corporation.

Marquis, K., and C. Cannell. 1969. *A study of interviewer-respondent interaction in the urban employment survey.* Research Report. Ann Arbor, MI: Survey Research Center, University of Michigan.

Marquis, K., C. Cannell, and A. Laurent. 1972. *Reporting of health events in household interviews: Effects of reinforcement, question length, and reinterviews.* Washington, DC: PHS vital and health statistics, Series 2, No. 45.

Marquis, K., S. Marquis, and J. Polich. 1986. Response bias and reliability in sensitive topic surveys. *Journal of the American statistical association*, 81(394):381–9.

Matarazzo, J., A. Wren, and G. Saslow. 1965. Studies of interview speech behavior. In *Research in behavior modification: New developments and clinical implications*, ed. L. Kresner, and U.P. Ullman, 179–210. New York: Holt, Rinehart, and Winston.

Mathiowetz, N., and C. Cannell. 1980. Coding interviewer behavior as a method of evaluating performance. *Proceedings of the section on survey research methods*, Alexandria, VA: American Statistical Association.

Maynard, D., H. Houtkoop-Steenstra, N. Schaeffer, and J. van der Zouwen. 2002. *Standardization and tacit knowledge: Interaction and practice in the survey interview.* New York: John Wiley and Sons.

McGeoch, J.A. 1932. Forgetting and the law of disuse. *Psychology review*, 39(4):352–70.

Miller, P., and C. Cannell. 1977. Communicating measurement objectives in the interview. In *Strategies for communication research*, ed. P.M. Hirsch, P. Miller, and F.G. Kline, 127–51. Beverly Hills, CA: Sage Publications.

Miller, P., and C. Cannell. 1982. A study of experimental techniques for telephone interviewing. *Public opinion quarterly*, 46(2):250–69.

Miller, P., and R. Groves. 1985. Matching survey responses to official records: An exploration of validity in victimization reporting. *Public opinion quarterly*, 49(3):366–80.

Morton-Williams, J. 1979. The use of 'verbal interaction coding' for evaluating a questionnaire. *Quality and quantity*, 13(1):59–75.

Morton-Williams, J., and W. Sykes. 1984. The use of interaction coding and follow-up interviews to investigate comprehension of survey questions. *Journal of market research society*, 26:109–27.

Neter, J., and J. Waksberg. 1964. A study of response errors in expenditures data from household interviews. *Journal of the American statistical association*, 59(305):18–55.

Oksenberg, L., C. Cannell, and G. Kalton. 1991. New strategies for pretesting survey questions. *Journal of official statistics*, 7(3):349–65.

Oksenberg, L., A. Vinokur, and C. Cannell. 1979a. The effects of commitment to being a good respondent on interview performance. In *Experiments in interviewing techniques: Field experiments in health reporting, 1971–1977*, ed. C. Cannell, L. Oksenberg, and J. Converse. Washington, DC: Department of Health, Education, and Welfare, DHEW Publication No. (HRA) 78-3204.

Oksenberg, L., A. Vinokur, and C. Cannell. 1979b. The effects of instructions, commitment and feedback on reporting in personal interviews. In *Experiments in interviewing techniques: Field experiments in health reporting, 1971–1977*, ed. C. Cannell, L. Oksenberg, and J. Converse. Washington, DC: Department of Health, Education, and Welfare, DHEW Publication No. (HRA) 78-3204.

Presser, S., M. Couper, J. Lessler, E. Martin, J. Martin, J. Rothgeb, and E. Singer. 2004. Methods for testing and evaluating survey questions. *Public opinion quarterly*, 68(1):109–30.

Schaeffer, N., and J. Dykema. 2011. Questions for surveys: Current trends and future directions. *Public opinion quarterly*, 75(5):909–61.

Thornberry, O. 1987. *An experimental comparison of telephone and personal health interview surveys*. Washington, DC: PHS vital and health statistics, Series 2, No. 106.

Tourangeau, R., L. Rips, and K. Raskinski. 2000. *The psychology of survey response*. New York: Cambridge University Press.

Tulving, E. 1972. Episodic and semantic memory. In *Organization of memory*, ed. E. Tulving, and W. Donaldson, 381–403. New York: Academic Press.

Section II

Training Interviewers

3

General Interviewing Techniques: Developing Evidence-based Practices for Standardized Interviewing

Nora Cate Schaeffer, Jennifer Dykema, Steve M. Coombs,
Rob K. Schultz, Lisa Holland, and Margaret L. Hudson

CONTENTS

3.1 Introduction

Interviewers continue to have an important role in collecting data, particularly in studies that require locating or sampling respondents or administering complex survey instruments. General interviewing techniques (GIT) refer to the practices used by face-to-face and telephone survey interviewers for asking questions and obtaining codable answers from respondents. Since Fowler and Mangione (1990) codified a set of practices for standardization, supplementary or complementary interviewing techniques have been proposed to support motivation (Dijkstra 1987), recall (Belli et al. 2001), and comprehension (Schober and Conrad 1997). However, the accumulating evidence on interaction during interviews has not yet led to a comprehensive updating of interviewing techniques. This chapter describes steps we are taking to update GIT based on studies of interaction. After situating current practice and reviewing our reasons for revisiting GIT at this time, we describe our process and goals as well as gaps in current practice. We then outline key concepts and techniques for the first lessons in a revised GIT training. Our proposals are rooted in the structure imposed by the conversational practices associated with the question–answer sequence and the influence of question form on what constitutes a codable answer.

3.1.1 Brief Historical Context

Although most contemporary research interviewing uses standardized interviewing, early interviewing practices were relatively informal (Converse 1987, 95–97, 335; Williams 1942). By the 1940s and 1950s, there was a movement toward standardization, probably motivated by studies that showed the impact of the interviewer on reliability and validity (Converse 1987, 335; Hyman et al. [1954] 1975; see discussion in Schaeffer 1991). Fowler and Mangione (1990) provided an influential codification of standardization, summarized in Table A3A.1 in the online supplementary materials (Online Appendix 3), the first principle of which is that questions are read as worded. Although there is little documentation about how different survey centers implement standardization, all 12 academic survey centers studied by Viterna and Maynard required that questions be read verbatim (2002, 394), as do "conversational" interviewing methods (Schober and Conrad 1997) and the U.S. Bureau of the Census (undated). Other features of standardization varied at the centers Viterna and Maynard examined. Both the University of Wisconsin Survey Center (UWSC) and the Survey Research Operations (SRO) at the University of Michigan's Survey Research Center (SRC), the centers involved in the project reported here, implement fairly rigorous versions of standardization. (For other versions of standardization see Brenner 1982, 139; Dijkstra 1987; Gwartney 2007, 203.)

3.1.2 Reasons to Revisit GIT

There are several reasons to revisit GIT. First, we now have many studies that describe the actual behavior of interviewers and respondents. These studies have benefited from the vocabulary and techniques of conversation analysis and related disciplines. Second, as studies that code interaction have accumulated, the coding systems embody increasingly refined understandings of what should be labeled an "action" and what those labels should be (e.g., compare Cannell et al. 1989; Dykema, Lepkowski, and Blixt 1997; Olson and Smyth 2015; Ongena and Dijkstra 2007; Schaeffer and Maynard 2008). Third, analytic advances have helped clarify the contribution of interviewers, respondents, and questions to error (e.g., West and Blom 2017). Fourth, it is also possible that a comprehensive framework for interviewing techniques could contribute to developing common vocabularies and practices across research centers, thus reducing "house effects." Fifth, recent research both elaborated our understanding of the characteristics of questions that might be involved (e.g., Alwin 2007; Saris and Gallhofer 2007; Schaeffer and Dykema 2011a) and has shown that those characteristics predict both the participants' behavior and measures of data quality or proxies for it (e.g., Dykema et al. 2020; Garbarski et al., Chapter 18, this volume; Holbrook et al., Chapter 17, this volume; Olson and Smyth 2015; Olson, Smyth, and Cochran 2018; Schaeffer and Dykema 2011b). For our purposes, studies that focus on actions (e.g., question reading) and requirements of actions (e.g., exactly as worded), rather than psychological concepts (e.g., the subjective experience of rapport), provide stronger guidance for interviewer training.

3.1.3 A Process for Revising GIT

This chapter describes the process UWSC and SRO used to revise and update GIT. We also describe the first six modules of our training; work on the remaining modules is continuing. Our organizations began with compatible training protocols but somewhat different vocabularies. As we developed each topic in the training, we reviewed published documentation of interviewer training (e.g., Fowler and Mangione 1990; Gwartney 2007) and interviewer training at our organizations. We also considered relevant experiments and observational studies about survey interviews and questions; research in conversation analysis, linguistics, cognitive psychology, and measurement theory; and our own close observations of interaction in the interview over many years. We drew on extensive experience in training and monitoring interviewers to consider what interviewers could be trained to do in the fast-paced environment of an interview. We tried to reach a consensus on proposed techniques during extensive discussion and review of examples.

3.2 Revisiting Interviewing Techniques

Models of the question–answer process suggest a range of criteria to consider in choosing among candidate interviewing techniques (e.g., Cannell Miller and Oksenberg 1981; Dykema et al. 2020; Ongena and Dijkstra 2007). These models and accompanying research led us to identify a (somewhat aspirational) set of criteria (see Table 3.1). In many cases we had to decide whether to select among interviewing techniques we identified in our current GIT (e.g., always reread the question) or develop new techniques for situations unaccounted for in our current GIT (e.g., rules for responsive follow-up).

3.2.1 General Goals

In selecting interviewing techniques, we aimed to balance the traditional objective of standardization – to reduce interviewer variance – against other goals and to make our reasons explicit. These goals included the following: (1) explicitly acknowledge that a question's response format drives interaction in the interview and tailor training modules (e.g., on how to recognize a codable answer and follow-up) around different question forms (e.g., Dykema et al. 2020; Holbrook, Cho, and Johnson 2006; Olson, Smyth, and Cochran 2018); (2) reduce burden on the respondent and motivate the respondent's engagement by authorizing more responsiveness by the interviewer (e.g., Garbarski, Schaeffer, and Dykema 2016); and (3) fill gaps in current training that left interviewers or quality control supervisors unsure how to proceed. The literature and our transcripts offered the most evidence for the first of these goals, which is emphasized in the discussion below.

To illustrate how some of the criteria and goals were considered in a specific decision, we describe our decisions about how interviewers should repair errors in the initial reading of a question. The current version of training at UWSC and the 1992 edition of the interviewer training at SRO did not train interviewers on how to repair the misreading of a question. The training script (adapted for individual studies) in use at SRO by 2013 read: "If you make a mistake in reading the question, no matter how small, it is your responsibility to start over and re-read the entire question." This guidance may be unclear because what constitutes a "mistake" is ambiguous and it can be difficult for an interviewer to determine quickly what the "entire" question is (e.g., a question may have a preamble).

TABLE 3.1

Criteria Considered in Reviewing Interviewing Techniques

- Interviewer-focused
  - Need: Is the procedure that the interviewer should follow ambiguous or undefined under current practice (e.g., how to correctly repair an inadequate reading of the question)?
  - Ease of training: Can the technique and the situation when it is to be used be explained clearly? Can interviewers recognize the situation that requires the technique easily and reliably? Can interviewers use the technique to respond consistently to the situation? How complex is the technique? Can a simpler technique accomplish the goal? Is the technique consistent with other techniques?
  - Frequency: How common are the situations that require the technique likely to be? Can one technique be used in many situations or are multiple techniques needed?
  - Instrument support: How clearly does the instrument convey the technique (e.g., optional parenthetical statements) to interviewers? Does the technique require support in the instrument itself, for example, in instructions to the interviewer or programming of specific follow-up techniques (e.g., repeating response categories in a battery "if needed" or after every third item)?
- Respondent-focused
  - Rapport and motivation: Will the technique increase or decrease how responsive the interviewer seems? Will the technique increase or decrease the respondent's engagement?
  - Efficient progress and reducing burden: Does the technique lengthen the interview or increase burden for the respondent? Does the technique reduce the occasions on which the interviewer must intervene or follow-up? Is the technique simple for respondent and interviewer to implement? How complex is the technique? Can a simpler technique accomplish the goal? Is the technique consistent with other techniques?
  - Training the respondent by communicating the practices of standardization consistently: Does the technique model adequate role behavior for the respondent? Does the technique model inadequate role behavior for the respondent? Is the technique consistent with other techniques?
- Measurement-focused
  - Reliability: Is the technique likely to increase or decrease interviewer variance? Is the technique likely to increase or decrease the reliability of the respondent's answer?
  - Validity: Is the technique likely to increase or decrease the accuracy of the respondent's answer? Is the technique likely to communicate a point of view or expectation that might influence the respondent's answer?

After reviewing multiple questions and interactions, we formulated three more specific rules for question reading: (1) if you omit, add, or change a word, reread the sentence from the beginning; (2) if you have trouble reading a word the first time, back up to reread from that word (e.g., "Would it be mainly manufacturing, retail, wholeta- wholesale trade, or something else?"); (3) if you make a mistake reading even one response category, reread all of the categories from the beginning. Our recommendation gave priority to clarity in training (which affects reliability) and support for the respondent's cognitive processing (which affects both reliability and validity) rather than efficiency (which may affect the respondent's engagement).

3.2.2 Specific Gaps in Existing Training

Advances in understanding come with new distinctions and terminology, and we hoped to adopt a vocabulary that could be shared across question writers, interviewers, and survey organizations. We borrowed vocabulary from other research traditions (e.g., "acknowledgement" rather than "feedback"), and we tried to make distinctions (e.g., among response formats) that are needed by interviewers, trainers, and supervisors. Because both the interaction in the interview and the techniques interviewers need depend on the response format of the question, we developed labels for the most common question forms (see Table A in Online Appendix 3 for examples). We outline below some of the gaps in practices and techniques that we identified; others remain. For example, ideally the conventions used by those who write and program survey instruments (to indicate which parts of questions can be inserted or repeated "as needed," and so forth) would be fully integrated with the techniques of interviewing, so that instructions to interviewers were standardized across questions, instruments, and organizations.

3.3 Structure of Training and Key Concepts

Table 3.2 proposes a list of topics for training in GIT. Table 3.2 does not include such important topics as recruiting respondents, study-specific training in a particular instrument, or administrative procedures that the interviewer must learn. The topics in Part 2 on the question–answer sequence and question form are the structural core of GIT. In this chapter, we highlight some of our recommendations for Part 2 of the training.

3.3.1 Question–Answer Sequence

Studies of interaction in the interview have long used the question–answer sequence as a basic unit of analysis (e.g., Cannell et al. 1989), and the "question–answer–acknowledgment" sequence has been labeled "paradigmatic" because it is central to the practices of standardization (Schaeffer and Maynard 1996). As indicated in Figure 3.1, the question–answer sequence identifies three specific locations for actions by the interviewer: reading

TABLE 3.2

Suggested Topics for Interviewing Techniques Section of General Interviewing Techniques (GIT)

Part 1. Introduction	
Lesson 1	Introduction
Lesson 2	The Science of Survey Research
Part 2. The Question–Answer Sequence	
Lesson 3	Question Reading
Lesson 4	Question Form
Lesson 5	How to Recognize a Codable Answer
Lesson 6	Acknowledgments
Part 3. Advanced Techniques	
Lesson 7	Follow-up: Basic Techniques
Lesson 8	Follow-up: Advanced Techniques
Lesson 9	The Respondent Says "Don't Know" or Refuses
Lesson 10	The Respondent Asks a Question: Requests for Repetition and Clarification
Lesson 11	Training the Respondent
Part 4. Special Topics	
Lesson 12	Choice Lists, Open Questions, and Field Coding
Lesson 13	Interviewer Instructions
Lesson 14	Groups of Questions: Filter-follow-up, Yes–No Checklists, Batteries, and Rosters

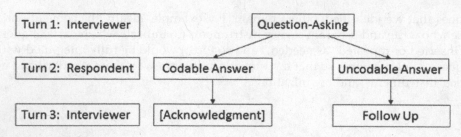

FIGURE 3.1
Basic question–answer sequence.

the question (turn 1); classifying an answer as codable (turn 2) and providing an optional acknowledgment (turn 3); or classifying an answer as an uncodable answer or other talk, such as a request for clarification, (turn 2) and following up (turn 3). This terminology and structural emphasis are fundamental to the training. The concepts may be new to GIT, but they are familiar to both practitioners and researchers.

3.3.2 Response Format (Question Form)

Based on studies of interaction, we incorporated the response format of the question as a second structural core of training (see Table A3A.2 in Online Appendix 3 for examples and descriptions). To reinforce key distinctions consistently, we bypass the traditional classification of questions as "open and closed," in favor of more specific labels. Because each question form projects a different type of answer, codable answers and follow-up actions also differ. The labels for the question forms were chosen to communicate key features to interviewers. "Yes–no" is also a native term for that question form. There are two common forms that ask respondents to select a category. We adopted the labels "selection with ordered categories" and "selection with choice list" to emphasize both the task posed to the respondent (to select) and a key feature of the set of categories offered (an ordered set of categories or list of choices). [The form that we label "selection with choice list" has also been labeled "categorical" or "nominal" (e.g., Holbrook et al. 2007; Krosnick and Presser 2010; Olson and Smyth 2015).] The label "discrete-value question" has long been used at SRO for questions that request (most often) a single number or (less frequently) a label (such as a month) without explicitly offering these as categories for the respondent to select. [Discrete-value questions have also been labeled "identification" (e.g., Dohrenwend 1965) and "open-ended numeric" (Olson and Smyth 2015).] We use the label "open" questions for questions that request an object (e.g., a most important problem) or a narrative (e.g., reasons for leaving a job). Some open questions require verbatim recording, and others are "field coded" (i.e., coded by the interviewer). The discussion of interviewing techniques in this chapter focuses on the three most common question forms: selection questions with ordered categories, discrete-value questions, and yes–no questions.

3.3.3 Codable and Uncodable Answers

Although how to recognize the difference between a codable and uncodable answer would appear to be fundamental to standardized interviewing, we were not able to identify specific instruction on this topic in the trainings we reviewed. Transcripts of interviews reveal a variety of ways in which respondents answer questions (e.g., Brenner 1981; Schaeffer Forthcoming), and how to identify a codable answer is treated more thoroughly in systems of interaction coding than in interviewer training. We adopted three criteria in defining a codable answer. A codable answer must: occur after the respondent has heard the question, answer the question, and match the response format of the question (see Table 3.3).

The task of recognizing a codable answer differs in important ways based on a question's response format. For selection questions, the interviewer must recognize whether the respondent has given a portion or "kernel" of a category that uniquely identifies it (e.g., "very" may uniquely identify "very well," depending on other categories in the set). Answers to yes–no questions may present the greatest challenge, because respondents commonly express these answers in a variety of ways. Following up when the respondent says "yeah," "uh-huh," or "sure," rather than "yes" risks making the interviewer look interactionally incompetent and burdening the respondent. In addition, interviewers probably process some of these answers automatically, possibly without awareness that they have accepted a synonym for "yes" (see, for example, Hak 2002). Although most codable answers for discrete-value questions are single numbers, respondents may also offer synonyms, such as "never" for "zero days last week," or "all of them" or "every day" for "seven days last week." Our training accepts the words and phrases reviewed here as synonyms for codable answers. Although listing synonyms or providing techniques for these common issues might seem to complicate training, interviewers must make such decisions frequently and quickly, and they need direction for how to respond in order to be standardized.

In addition to providing a full codable answer or a "kernel" that uniquely identifies a category, respondents routinely include conversational elements and elaborations that do not contradict the codable answer in their answer turns. These conversational elements include tokens (e.g., "um, yes;" "uh, yes"), repeating an answer (e.g., "yes, yes"), repeating or paraphrasing part of the question (e.g., "yes, I did"), task-related phrases that indicate they are making a selection (e.g., "I would say yes," "put fifty"), uncertainty markers (e.g., "probably yes," "I don't know… disagree," "I guess," "about fifty"), and elaborations

TABLE 3.3

Defining a Codable Answer

A codable answer…
- Occurs after respondent has heard the question
- Answers the survey question
- Matches the response format of the question
  - One of the response categories
  - The format on the screen
- Differs depending on the question's response format
  - Yes–no questions
    - "Yes," "no," specific synonyms for "yes" or "no"
  - Selection questions
    - One of the response categories read to the respondent
    - Unique portion of a response category
    - Reference to a specific category by its position
  - Discrete-value questions
    - A specific, single number (e.g., "zero" for "0 times"), category (e.g., "January" for "month"), or value (e.g., "1948" for "year born")
- Ignores conversational elements
  - Tokens
  - Repetitions of the answer
  - Uncertainty expressions (e.g., "about fifty")
  - Elaborations that do not contradict the codable answer

(which sometimes lead to contradictions) (e.g., "I agree, but, right, on the other hand…" in Table 3.4 line 2 and line 20).

Our rules for identifying a codable answer instruct the interviewer to focus on the kernel of the codable answer in these cases. A codable answer accompanied by tokens, expressions of uncertainty, or elaborations that do not contradict the answer is treated as the respondent's best offer, a rule that accepts these conversational elements as part of "talk," and does not add the burden of additional follow-up. In contrast, the long report that the respondent gives in line 18 in Table 3.4 does not include a unique portion of either response category, and the interviewer uses the basic follow-up technique of rereading the question in line 19 to obtain the codable answer "agree" in line 20.

Identifying what constitutes a kernel of a codable answer for yes–no questions raises additional issues, in part because there are many conversational ways of expressing "yes"

TABLE 3.4

Uses of "Okay," "All Right," and Such

Line Number	Q Number	Turn	Actor	Talk
1	1	1	IV	Okay. All right. And please tell me whether you agree or disagree with each of these statements. The first statement is: People like me don't have any say about what the government does. Would you agree or disagree with that?
2		2	FR	I agree, but, right, on the other hand I'm not intelligent enough to make all those decisions.
3		3	IV	Oh.
4		4	FR	And I've got enough about me to know that I don't want to do it. I don't that responsibility.
5		5	IV	Okay.
6	2	1	IV	All right. And, how about this one: Government agencies usually try to do what is best for the people.
7		2	FR	Well, they should I'll say it that way. I'm not always sure that they do.
8		3	IV	Okay, so would you agree or disagree?
9		4	FR	I agree.
10		5	IV	Okay.
11	3	1	IV	And, I don't think public officials care much about what people like me think. Would you agree or disagree?
12		2	FR	Uh, I agree.
13	4	1	IV	And the government already knows more about me than it needs to.
14		2	FR	I, no, I, I have nothing to hide. Whatever they find out about me they can.
15		3	IV	All right. So, would you agree or disagree?
16		4	FR	I agree.
17	5	1	IV	And, most people who go into public office want to help others.
18		2	FR	That's their theory to get in there, and after they get in there, they forget it. There's too many post office deals and all those other things.
19		3	IV	Right, right. So, I'll read it one more time. Most people who go into public office want to help others. Would you agree or disagree?
20		4	FR	Well, I agree that's why they go in there, but then they forget it.
21	6	1	IV	Okay, and generally speaking, do you usually think of yourself as a republican, democrat, independent, or something else?

Source: Letters and Sciences Survey Center, 1995, AW17 (Female Respondent=FR and Interviewer=IV).

(e.g., "yeah," "yup," "yep," "uh-huh," "mm-hm") and "no" (e.g., "nope," "nah," "mm-mm," and "uh-uh"). There are also variants that repeat part of the verb in the question (e.g., "I did," "I have") or that are marked with degrees of uncertainty (e.g., "probably so," "I suppose so," "I guess so," "I think so," "I believe so"). Answers such as these may be coded "automatically" as "yes" or "no" by some interviewers, but interviewers' practices are likely to vary and may increase interviewer variance (Hak 2002). The variety of ways respondents answer yes–no questions poses core challenges for any sort of research on interviewing: How much detail and how many rules can we expect interviewers to absorb and apply? How consistently are interviewers able to recognize the distinction between conversational practices (e.g., "probably" is as close to "yes" as the respondent is able to come) and research practice (e.g., only "yes" is "yes")? How much tolerance does a respondent have for an interviewer who repeatedly ignores what might seem to the respondent to be the plain conversational meaning of an answer? Lacking guidance from research to answer these questions, our recommendation is that interviewers be trained to accept the synonyms described above, matching conversational and interviewing practice in these cases.

3.3.4 Acknowledgments

A common characterization is that interviewer's action in the third turn (see Figure 3.1) gives the respondent reinforcement or "feedback" on their answer (e.g. Fowler and Mangione 1990, 49). For this reason, interviewers are barred from using phrases in the third turn that might be heard as expressing a point of view on the respondent's answer, and "okay" is sometimes included in that ban (e.g., Gwartney 2007, 198–200). In contrast, studies of interaction recognize that speakers "need acknowledgment from listeners at certain fairly predictable transitions in their own talk" (Boren and Ramey 2000, 269). Studies of how "okay" is used in conversation find that in addition to being used as a "free-standing receipt marker," it frequently appears at transitions and changes of speakership (Beach 1993). "Okay" can simultaneously look backward or forward, that is, "okay" can close some prior action and move toward a next, as when it is used to assess the adequacy of answers in medical interviews (Beach 1995, 274).

The transcript in Table 3.4 illustrates some of these uses of "okay" and related phrases that would usually be discussed in monitoring, such as "all right." "Okay" appears at the beginning of lines 1 and 5 and can be heard as the interviewer acknowledging an answer with "okay." In both cases, the interviewer then projects a move to the next question with "all right." At line 21, this interviewer uses "okay" to both end an elaboration by the respondent and move to the next question. Using a more traditional psychological interpretation, this interviewer's "right, right" (line 19) can plausibly be heard as acquiescing to the respondent's statement and so might be described as "inappropriate feedback." However, we give greater weight to the interactional analysis of the uses of "okay" and similar tokens to manage the flow of the interaction. These tokens convey variously that the interviewer "heard" a response, is moving to a follow-up action, or is moving to a next question. Such interactional uses of "okay," and even "all right," in this transcript seem unexceptional, nonjudgmental, and almost indispensable.

Drawing on interactional analyses like that presented above, we use the label "acknowledgment" (rather than "feedback") for the interviewer's optional response to a codable answer in the third turn of a question–answer sequence. Our recommendations recognize the interactional need for acknowledgments as well as the usefulness of "okay" for managing a return to the current question or move to the next. We focus on the third turn as "acknowledgments" and propose a set of "simple acknowledgments" that are available (but optional) after a codable answer: "Thank you," "okay," repeating the respondent's answer, "I see," "mm-hm," and "uh-huh."

"Response-specific acknowledgments," as we define them, take into account what the respondent has said and often function to "train the respondent" (Fowler and Mangione 1990, 50–53). For example, after the respondent has worked to produce a codable answer to a discrete-value question, the interviewer can acknowledge with "thank you for giving me a single number." In line 10 of Table 3.4, the interviewer could have used a response-specific acknowledgment to help train the respondent, such as "Okay. Thank you for selecting one of our categories." When a respondent thinks aloud as they try to formulate an answer, the interviewer might respond, "Thank you for thinking that through."

Respondents sometimes express thoughts both relevant and irrelevant to the topic of a question, and interviewers routinely respond to these reports and digressions with a follow-up such as "I see. So, would you say you agree or disagree?" (e.g., line 15 in Table 3.4). However, a particularly challenging situation arises when the respondent accompanies a codable answer with an elaboration that conveys the respondent's positive or negative perspective on the answer. For example, when asked, "Are you currently pregnant?" the respondent might answer, "Yes, I'm thrilled." In situations like this when the respondent provides such "emotional information," Dijkstra's "person-oriented" style of interviewing authorizes interviewers to respond empathically without providing information about their own lives (Dijkstra 1987). Our proposal is more modest, to allow interviewers to respond with "I'm glad to hear that" or "I'm sorry to hear that" depending on whether the respondent's elaboration provides a positive or negative evaluation of their circumstance. Thus, in the situation just described, the interviewer could give a simple acknowledgment, move on, or acknowledge the respondent's perspective with "I'm glad to hear that."

Situations where such responses might be appropriate are relatively rare in the transcripts that we have available, so we are cautious in our recommendations. Some interview topics, however, may evoke frequent expressions of emotion, and more study is needed to decide when and how to provide interviewers with responses that recognize the respondent's situation or perspective without adding the interviewer's point of view or personal details.

3.4 Discussion

Standardized interviewing requires interviewers to make frequent and split-second choices among possible alternative actions (e.g., "So would you say 'yes' or 'no'?" versus "So you would say 'yes;' is that correct?" versus "I'll just reread the question…"), many of which have not been evaluated in studies that would permit us to assess their relative impact on reliability or validity. Until we have those studies, since we must train interviewers to do *something*, we apply information from varied sources.

In this chapter we have shown how acknowledging the interactional impact of the question–answer sequence and the response format of the question can lead to proposals for updating current interviewing techniques and adopting new ones. Our articulation of criteria to consider in making decisions about interviewing techniques reveals the uncertainties caused by the lack of evidence (see Daikeler and Bosnjak, Chapter 4, this volume). Experiments about interviewing are extremely challenging to design and execute (see Miller and Mathiowetz, Chapter 2, this volume). Studies can compare packages of techniques (e.g., West et al. 2018) or specific techniques (e.g., Smit, Dijkstra, and van der Zouwen 1997); the number of interviewers and respondents required is substantial (see West et al., Chapter 8, this volume); details of training and quality control are needed, as are manipulation checks to determine whether interviewers actually implemented the techniques; and criteria for determining which technique or set of techniques yields higher quality data are also needed. Even so, our experience in analyzing interaction suggests that characteristics of the specific questions will also be an important influence on the results. The number of each of the various types of yes–no questions, ordered selection questions, and discrete-value questions and the types of those questions (e.g., events and behaviors, evaluations, or judgments) all determine which interviewing techniques will be influential in any given study.

Although we have transcripts from a variety of sources, our collection is dominated by several types of yes–no questions about health and by ordered selection questions about political efficacy and participation in medical research; the questions were administered to very different populations by phone. What we can observe is limited by the questions, interviewers, and samples in our collections. Even with such specialized collections, however, we can begin to systematically examine such topics as how interviewers – including phone interviewers who are regularly monitored – modify interviewing techniques to manage situations presented by respondents. It would be useful to know, for example, whether respondents who give a "formally" inadequate answer such as "I would be for that" are less likely to select "support" if the follow-up is balanced but unresponsive (e.g., "So would you say you support or oppose it?") versus unbalanced but responsive (e.g., "So you would say you support it, is that correct?") (Moore and Maynard 2002, 300–304).

Our recommendations cover a few of the basic topics that help form the infrastructure of measurement in interviewer-administered surveys, but they are far from complete. For example, we have not discussed ways in which interviewers can provide continuity and support the respondent's processing of the task. As an illustration, Table 3.4 shows the interviewer using "and" as a preface to announce the next statement for the respondent to consider at lines 6, 11, 13, and 17. This practice communicates information about the structure of the task to the respondent (see Garbarski, Schaeffer, and Dykema 2016; Haan, Ongena, and Huiskes 2013) in a way that can be efficient and support the respondent's cognitive processing. Other topics we have not discussed in this chapter include follow-up for uncodable answers, and larger topics such as advanced follow-up techniques for specific types of uncodable responses (e.g., reports, requests for repetition or clarification, don't know and refusal responses), batteries (Olson, Smyth, and Cochran 2018), and interviewer instructions (Dykema et al. 2016). We also have not addressed the question of which issues should be focused on during monitoring (quality control).

Finally, although we have focused here on standardized interviewing, the same issues and quandaries face all approaches to measurement that involve interaction. Different approaches to interviewing may, for example, emphasize different criteria, adopt different rules for determining when a question has been adequately answered, or give more interpretative authority to the interviewer. By identifying the choices to be made, the approach we take here potentially offers a more general framework for thinking about research interviewing that can be useful even outside standardized interviewing.

References

Alwin, D. F. 2007. *Margins of error*. Hoboken, NJ: John Wiley & Sons, Inc.

Beach, W. A. 1993. Transitional regularities for 'casual' okay usages. *Journal of pragmatics* 19(4):325–52.

Beach, W. A. 1995. Preserving and constraining options: "Okays" and 'official' priorities in medical interviews. In *The talk of the clinic*, ed. B. Morris, and R. Chenail. 259–89. Hilldale, NJ: Lawrence Erlbaum.

Belli, R. F., W. L. Shay, and F. P. Stafford. 2001. Event history calendars and question list surveys: A direct comparison of interviewing methods. *Public opinion quarterly* 65(1):45–74.

Boren, T., and J. Ramey. 2000. Thinking aloud: Reconciling theory and practice. *IEEE transactions on professional communication* 43(3):261–78.

Brenner, M. 1981. Aspects of conversational structure in the research interview. In *Conversation and discourse*, ed. P. Werth, 19–40. London: Croom Helm.

Brenner, M. 1982. Response-effects of 'role-restricted' characteristics of the interviewer. In *Response behaviour in the survey-interview*, ed. W. Dijkstra, and J. van der Zouwen, 131–65. London: Academic Press.

Cannell, C. F., P. V. Miller, and L. Oksenberg. 1981. Research on interviewing techniques. In *Sociological methodology Vol. 12*, ed. S. Leinhardt, 389–437. San Francisco, CA: Jossey-Bass.

Cannell, C. F., L. Oksenberg, G. Kalton, K. Bischoping, and F. J. Fowler. 1989. New techniques for pre-testing survey questions: Final report. Grant No. Hs 05616. Vol. Final Report, Grant Number HS 05616, National Center for Health Services Research and Health Care Technology Assessment.

Converse, J. N. 1987. *Survey research in the United States: Roots and emergence, 1890–1960*. Berkeley, CA: University of California Press.

Dijkstra, W. 1987. Interviewing style and respondent behavior: An experimental study of the survey interview. *Sociological methods and research* 16(2):309–34.

Dohrenwend, B. S. 1965. Some effects of open and closed questions on respondents' answers. *Human organization* 24(2):175–84.

Dykema, J., J. M. Lepkowski, and S. Blixt. 1997. The effect of interviewer and respondent behavior on data quality: Analysis of interaction coding in a validation study. In *Survey measurement and process quality*, ed. L. Lyberg, P. Biemer, M. Collins, E. De Leeuw, C. Dippo, N. Schwarz, and D. Trewin, 287–310. New York: John Wiley & Sons, Inc.

Dykema, J., N. C. Schaeffer, D. Garbarski, and M. Hout. 2020. The role of question characteristics in designing and evaluating survey questions. In *Advances in questionnaire design, development, evaluation, and testing*, ed. P. Beatty, D. Collins, L. Kaye, J. Padilla, G. Willis, and A. Wilmot, 119–52. Hoboken, CA: Wiley.

Dykema, J., N. C. Schaeffer, D. Garbarski, E. V. Nordheim, M. Banghart, and K. Cyffka. 2016. The impact of parenthetical phrases on interviewers' and respondents' processing of survey questions. *Survey practice* 9:1–10.

Fowler, Jr., F. J., and T. W. Mangione. 1990. *Standardized survey interviewing*. Newbury Park, CA: Sage.

Garbarski, D., N. C. Schaeffer, and J. Dykema. 2016. Interviewing practices, conversational practices, and rapport: Responsiveness and engagement in the standardized survey interview. *Sociological methodology* 46(1):1–38.

Gwartney, P. A. 2007. *The telephone interviewer's handbook: How to conduct standardized conversations*. San Francisco, CA: Jossey-Bass.

Haan, M., Y. Ongena, and M. Huiskes. 2013. Interviewers' questions: Rewording not always a bad thing. In *Interviewers' deviations in surveys*, ed. P. Winker, N. Menold, and R. Porst, 173–94. Frankfurt am Main: Peter Lang.

Hak, T. 2002. How interviewers make coding decisions. In *Standardization and tacit knowledge*, ed. D. W. Maynard, H. Houtkoop-Steenstra, N. C. Schaeffer, and J. van der Zouwen, 449–70. New York: Wiley.

Holbrook, A., Y. I. Cho, and T. Johnson. 2006. The impact of question and respondent characteristics on comprehension and mapping difficulties. *Public opinion quarterly* 70(4):565–95.

Holbrook, A. L., J. A. Krosnick, D. Moore, and R. Tourangeau. 2007. Response order effects in dichotomous categorical questions presented orally: The impact of question and respondent attributes. *Public opinion quarterly* 71(3):325–48.

Hyman, H. H. 1975 [1954]. *Interviewing in social research*. Chicago, IL: The University of Chicago.

Krosnick, J. A., and S. Presser. 2010. Question and questionnaire design. In *Handbook of survey research, second edition*, ed. P. V. Marsden, and J. D. Wright, 263–313. Bingley: Emerald Group Publishing Limited.

Moore, R., and D. W. Maynard. 2002. Achieving understanding in the standardized survey interview: Repair sequences. In *Standardization and tacit knowledge*, ed. D. W. Maynard, H. Houtkoop-Steenstra, N. C. Schaeffer, and J. van der Zouwen, 281–311. New York: Wiley

Olson, K., and J. D. Smyth. 2015. The effect of CATI questions, respondents, and interviewers on response time. *Journal of survey statistics and methodology* 3(3):361–96.

Olson, K., J. D. Smyth, and B. Cochran. 2018. Item location, the interviewer–respondent interaction, and responses to battery questions in telephone surveys. *Sociological methodology* 48(1):225–68.

Ongena, Y. P., and W. Dijkstra. 2007. A model of cognitive processes and conversational principles in survey interview interaction. *Applied cognitive psychology* 21(2):145–63.

Saris, W. E., and I. N. Gallhofer. 2007. *Design, evaluation, and analysis of questionnaires for survey research*. New York: Wiley.

Schaeffer, N. C. 1991. Conversation with a purpose – or conversation? Interaction in the standardized interview. In *Measurement errors in surveys*, ed. P. P. Biemer, R. M. Groves, L. E. Lyberg, N. A. Mathiowetz, and S. Sudman, 367–92. New York: John Wiley & Sons, Inc.

Schaeffer, N. C. Forthcoming. Interaction before and during the survey interview: Insights from conversation analysis. *International journal of social research methodology*.

Schaeffer, N. C., and J. Dykema. 2011a. Questions for surveys: Current trends and future directions. *Public opinion quarterly* 75(5):909–61.

Schaeffer, N. C., and J. Dykema. 2011b. Response 1 to Fowler's chapter: Coding the behavior of interviewers and respondents to evaluate survey questions. In *Question evaluation methods*, ed. J. Madans, K. Miller, A. Maitland, and G. Willis, 23–39. Hoboken, NJ: John Wiley & Sons, Inc.

Schaeffer, N. C., and D. W. Maynard. 1996. From paradigm to prototype and back again: Interactive aspects of cognitive processing in survey interviews. In *Answering questions*, ed. N. Schwarz, and S. Sudman, 65–88. San Francisco, CA: Jossey-Bass.

Schaeffer, N. C., and D. W. Maynard. 2008. The contemporary standardized survey interview for social research. In *Envisioning the survey interview of the future*, ed. F. G. Conrad, and M. F. Schöber, 31–57. Hoboken, NJ: Wiley.

Schober, M. F., and F. G. Conrad. 1997. Does conversational interviewing reduce survey measurement error? *Public opinion quarterly* 61(4):576–602.

Smit, J. H., W. Dijkstra, and J. Van der Zouwen. 1997. Suggestive interviewer behaviour in surveys: An experimental study. *Journal of official statistics* 13(1):19–28.

US Bureau of the Census. Undated. Conducting the CPS interview. Current population survey interviewer's manual. https://www2.census.gov/programs-surveys/cps/methodology/intman/Part_A_Chapter2.pdf?# (accessed September 29, 2019).

Viterna, J. S., and D. W. Maynard. 2002. How uniform is standardization? Variation within and across survey research centers regarding protocols for interviewing. In *Standardization and tacit knowledge*, ed. D. W. Maynard, H. Houtkoop-Steenstra, N. C. Schaeffer, and J. van der Zouwen, 365–401. New York: Wiley.

West, B. T., and A. G. Blom. 2017. Explaining interviewer effects: A research synthesis. *Journal of survey statistics and methodology* 5(2):175–211.

West, B. T., F. G. Conrad, F. Kreuter, and F. Mittereder. 2018. Can conversational interviewing improve survey response quality without increasing interviewer effects? *Journal of the royal statistical society: Series A (Statistics in society)* 181(1):181–283.

Williams, D. 1942. Basic instructions for interviewers. *Public opinion quarterly* 6(4):634–41.

4

How to Conduct Effective Interviewer Training: A Meta-Analysis and Systematic Review

Jessica Daikeler and Michael Bosnjak

CONTENTS

4.1 Introduction

The content and methods of interviewer training are often overlooked factors in minimizing interviewer effects in interviewer-administered surveys (West and Blom 2017). In particular, experimental variation in the content of interviewer training approaches can provide information about the effectiveness of interviewer training and training methods. This chapter evaluates the effectiveness of interviewer training methods using meta-analytic methods to summarize the results of interviewer training experiments.

Interviewers are one of the key actors in the data collection process of interviewer-administered surveys (e.g., Groves et al. 2009; Singer, Frankel, and Glassman 1983). From a total survey error (TSE) perspective (e.g., O'Muircheartaigh and Campanelli 1998; West and Blom 2017), interviewers can influence four sources of survey error – namely, coverage, nonresponse, measurement, and processing error – and interviewer-related error can bias survey estimates, including regression coefficients (e.g., Fischer et al. 2018). Interviewer training is primarily designed to reduce the effects of interviewers on nonresponse and measurement error (e.g., Billiet and Loosveldt 1988; Fowler and Mangione 1990; Lessler, Eyerman, and Wang 2008). Interviewers can be trained to avoid nonresponse, to systematically administer survey questions, to probe only when it is allowed, and to avoid influence by emphasizing certain response options. Kreuter (2008, 371) describes the interviewers' role as "read questions exactly as worded; probe nondirectively; and record answers without interpretation, paraphrasing, or additional inference about the respondent's opinion or behavior."

Ideal interviewer training should focus on two main areas of interviewer activity, namely, gaining respondents' cooperation (reducing nonresponse rates) and adhering to the practices of standardized interviewing (reducing measurement error) (Alcser et al. 2016; Daikeler et al. 2017). Despite the importance of training, experimental examinations of training approaches are rare due to the costs and complexity of designing and implementing such experiments and the separation of methodologists conducting research from field staff who manage interviewers and implement training programs. Experimental variation in fieldwork is expensive, requiring fielding two studies simultaneously, managing them separately but equivalently, and then comparing the outcomes. Alternatively, lab studies can be used but lack generalizability. Furthermore, most people studying interviewers and methods of improving interviewer quality do not actually do the training at the survey organizations. Finally, inexperienced interviewers require extensive training above and beyond simple interviewing techniques, especially when part of an ongoing study with research goals that are not solely methodological. In Germany, one other reason why the effectiveness of general interviewer training is not questioned has to do with the organization of fieldwork. Both large multinational survey programs, such as the Programme for the International Assessment of Adult Competencies (PIAAC; OECD 2014) and the European Social Survey (ESS; Loosveldt et al. 2014), and small survey projects rely on fieldwork agencies to train and manage interviewers. However, the effectiveness and type of this training is still in some cases a "black box."

Interviewer training has always been an integral part of the survey process, yet the available literature on this subject is quite sparse. While there is some research investigating the effect of interview training on specific data quality outcomes (such as unit nonresponse and correct probing, e.g., Durand et al. 2006; Fowler and Mangione 1986) and guidelines for interviewer training (e.g., Alcser et al. 2016; Daikeler et al. 2017), only Lessler, Eyerman, and Wang (2008) provide a comprehensive overview of the literature on interviewer training. However, as their overview is purely narrative, it does not quantitatively evaluate the training concepts and results. This chapter uses meta-analytic methods to estimate the impact of interviewer training approaches on data quality. The aim is to quantify the benefits of interviewer training and, more importantly, to determine what aspects of training (e.g., training length, practice, and feedback sessions) moderate the reduction of interviewer effects.

4.2 Conceptual Development of Research Questions

Classical interviewer training consists of training on recruiting sample members, general interviewing techniques, and study-specific training (Daikeler et al. 2017; Loosveldt et al. 2014). The focus of this chapter is on general interviewer training, that is, the basic, cross-project part of interviewer training that aims to impart the knowledge and skills that a successful interviewer needs to achieve high data quality both when recruiting participants and maintaining standardization in the interview (see West and Blom 2017). Literature on experimental interviewer training has reported to date on the influence of interviewer training on measurement error and nonresponse error; for a literature overview see Appendix Table A1 in the online supplemental materials.

The present study examines the impact of interviewer training on data quality. Specifically, it addresses seven breakdowns of the interviewing process that compromise data quality: (1) unit nonresponse (nonresponse error); (2) item nonresponse (measurement error); (3) items that are incorrectly administered* (measurement error); (4) items that are incorrectly read out (measurement error); (5) responses that are incorrectly probed (measurement error); (6) responses that are incorrectly recorded (measurement error); and (7) inaccurate responses. The aim is to determine whether these seven breakdowns are influenced by interviewer training and what training aspects contribute to the reduction of these errors, and thus to data quality. In the following, we first discuss nonresponse error and then address measurement error.

4.2.1 Effect of Refusal Avoidance Training on Unit Nonresponse Rates

In their examination of survey participation, Groves and McGonagle (2001, 250–251) assert that two interviewer strategies – tailoring behavior to the perceived features of the sample person and maintaining interaction with the sample person – play a crucial role in gaining the cooperation of potential respondents. They posit that "maintaining interaction is the essential condition of tailoring, for the longer the conversation is in progress, the more cues the interviewer will be able to obtain from the householder" (p. 251). Moreover, they argue that the longer the interaction lasts, the harder it is for the sample unit to refuse to participate. Thus, the first research question addressed concerns whether refusal avoidance training improves response rates (see Q1 in Table 4.1).

4.2.2 Effect of Interviewer Training on Data Quality

Especially in the case of measurement error, interviewers can be trained to avoid certain verbal or non-verbal behaviors that may influence respondents. Interviewer effects reflect the tendency that responses collected by one interviewer are more similar to each other than responses from different interviewers (Groves and Magilavy 1986). Reasons for interviewer effects include the activation of social norms by the interviewer's presence (Bosnjak 2017; Miller and Cannell 1982) and systematic errors in administering the survey (e.g., failure to read questions as worded). A typical example is the tendency of white respondents to report more liberal responses to a black interviewer for racial topics (e.g., Schaeffer 1980). Because interviewer training alerts interviewers to the importance of standardized interviewing with the aim of preventing, or minimizing, interviewer effects on survey responses, our second question examines whether measurement error is reduced if interviewers undergo specific questionnaire administration training (see Q2 in Table 4.1).

4.2.3 Effect Size Heterogeneity

Unfortunately, interviewer training is not standardized or homogeneous in terms of duration, content, and training procedures, although initial efforts have been made (AAPOR

---

* Administration means adherence to the interview protocol (adherence to breaks in the interview, correct administration of filter questions, and the order of questions); it includes reading questions but is much broader.

TABLE 4.1

Research Questions Addressed by the Meta-analysis and Systematic Analysis

Q1	Does refusal avoidance training improve survey response rates compared with training that does not include refusal avoidance training?
Q2	Are effects of interviewers on measurement error reduced if the interviewers undergo specific questionnaire administration training (in terms of correctly administered, read, probed, and recorded items; item nonresponse; accurate responses) beforehand compared to interviewers who did not have special training on questionnaire administration?
Q3	Are the effect size distributions heterogeneous?
Q4	What duration of interviewer training reduces (a) unit nonresponse and (b) other examined error sources that affect data quality?
Q5	Are cooperation rates improved by (a) practice and feedback sessions vs. no practice and feedback sessions; (b) interviewer monitoring vs. no interviewer monitoring; (c) supplementary written training material vs. no supplementary training material; (d) listening to audio refusals vs. not listening to audio refusals?

2016; Daikeler et al. 2017; ISO 2012; Viterna and Maynard 2002). Our third research question explores whether the effect sizes are heterogeneous (see Q3 in Table 4.1). Because of the lack of standardization, heterogeneous training outcomes, and thus effect size heterogeneity, can be expected. Heterogeneous distributions of effect size would imply that the success of interviewer training depends more on the content and methods of training or other factors such as the population and study content than on the training itself; alternatively, there may be other variables that confound these effects that cannot be disentangled in these analyses. Accordingly, a number of other training features must be examined more closely in order to be able to make statements on what constitutes successful training.

4.2.4 Training Features That May Improve Data Quality

Interviewer training duration. Learning theory suggests that learning progress typically follows an S-shaped curve, starting slowly, accelerating, and then leveling off (Thorndike 1913). If the learning curve flattens out or becomes horizontal, learning progress stagnates. This phenomenon, which is referred to as a learning plateau (Thorndike 1913, 99), occurs during the acquisition of complex skills such as learning the behaviors and techniques for standardized interviewing. Our research attempts to evaluate training durations that enable interviewers to learn the skills they need to avoid refusals and reduce interviewer effects (see Q4 in Table 4.1).

Interviewer training methods and determinants of effectiveness. According to Knowles, Holton, and Swanson's (2005) adult learning theory (see review in Tusting and Barton 2003), one reason why adults learn differently than children is that they can already draw on a variety of experiences which affect their actions. Hence, learning techniques that incorporate and extend the experiences of the learners are the most effective, e.g., an interviewer might remember that rephrasing a question in the past led to better understanding of the respondent, and in the training rephrasing techniques could then be taught to extend this skill. Furthermore, individuals differ in their preferred learning style; some react more to visual information, some to auditory, and others to kinesthetic information (Kelly 2010). Knowles, Holton, and Swanson (2005) posited that adults primarily would like to learn new skills because they are confronted with a concrete problem in their everyday life; for example, the concrete problem for the interviewer is to handle the question–answer process as correctly and standardized as possible, so they are willing to learn the necessary skills. According to Knowles, Holton, and Swanson (2005) and Tusting and Barton (2003), most adults prefer self-directed, problem-centered learning.

Against this background, a flexible blended-learning approach to adult learning, which combines traditional face-to-face instruction with online learning, seems especially promising (Means et al. 2013). Blended learning combines the advantages of online learning, such as flexibility in terms of time and place, with those of face-to-face instruction, such as direct interaction with trainers and other trainees and live feedback. Table 4.1 (Q5) summarizes this as our fifth research question.

4.3 Data and Methods

A meta-analysis consists of five steps: (1) a comprehensive literature search; (2) checking the eligibility of studies found; (3) coding of relevant data; (4) calculation of training effect sizes; and (5) analysis of variables that moderate effect size (Borenstein et al. 2009; Lipsey and Wilson 2001). Additional information on each of the steps can be found in the online supplementary materials in the Appendix. To be included in our meta-analysis, studies must have employed an experimental design. We accepted both designs that randomly assigned interviewers to treatments versus control conditions and those that used measurements on experimental pretest–posttest designs. For both types of training, it was essential that the control group received either no or only an introductory briefing. The search was limited to literature in English; over 2,000 results had to be excluded because the broad search terms generated literature related to job interviews, linguistic interviews, cognitive and clinical interviews of victims and witnesses, and studies without an experimental design. Only 14 eligible publications were retrieved, however most of the publications reported more than one experiment or effect size. For a search overview, see Figure 4.1. The most common indicator of data quality was the effect of interviewer training on the response rate (24 effect sizes nested in 11 manuscripts), followed by response accuracy (21 effect sizes nested in 2 manuscripts); correct recording of the response (14 effect sizes all from 1 manuscript); item nonresponse (12 effect sizes nested in 3 manuscripts); reading of questions exactly as worded (12 effect sizes all from 1 manuscript); correct probing (6 effect sizes all from 1 manuscript); and correct item administration (4 effect sizes nested in 2 manuscripts). To account for the fact that several effect sizes are nested within studies, we estimated multilevel models for robustness checks. For correct reading, probing, and recording, where we could identify only one study (Fowler and Mangione 1990), we discuss the outcomes qualitatively as well as response accuracy where the true values were unknown.

4.4 Results

4.4.1 What Is the Effect of Interviewer Training on Data Quality?

Refusal avoidance training and survey response rates (Q1 in Table 4.1). Figure 4.2 shows a forest plot summarizing the study-level differences in response rates between trained and

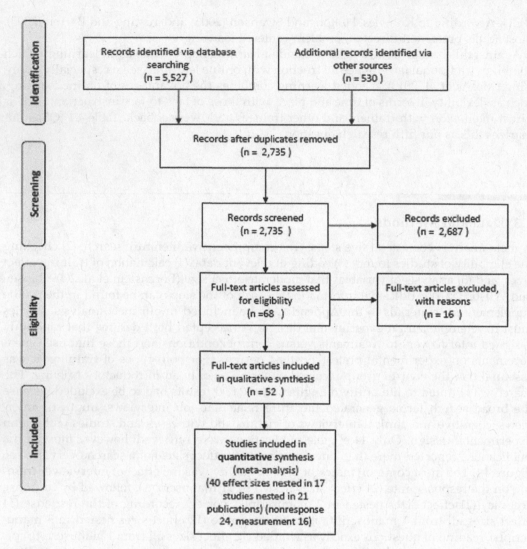

FIGURE 4.1
The literature search process.

untrained interviewers. The x-axis presents the estimated differences in data quality between trained and untrained interviewers. Positive values mean better data quality – for this outcome, higher response rates – for trained interviewers and all confidence intervals (CIs) that do not cross the zero line are significantly different from zero. The y-axis shows all included studies, and each point represents their effect sizes and confidence intervals. The last line of each quality measure with the title "RE model" shows the sampling error weighted mean effect size under the meta-analytic random effects assumption. The effect size distribution in the forest plot indicates that most response rate comparisons show that trained interviewers achieved higher response rates than interviewers without specific refusal aversion training. Surprisingly, there were quite a few zero findings. The sampling error weighted mean effect size estimate, calculated across all 24 effect sizes assuming random effects, was 0.05 (95% CI = 0.0/0.1). This result shows that the response rates achieved by trained interviewers were, on average, five percentage points higher (with a confidence

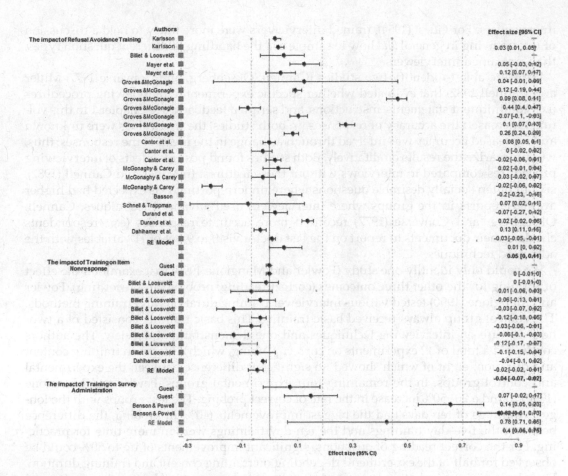

FIGURE 4.2
Forest plots for data quality indicators: trained vs. untrained interviewers.

interval from no effect to 11 percentage points) than those achieved by untrained interviewers. Our first research question (Q1) can therefore be answered in the affirmative.* However, the improvement of five percentage points (CI 0, 11) for nonresponse rates is surprising and indicates that interviewer training has a moderate impact on response rates.

Questionnaire administration training and interviewer effects (Q2 in Table 4.1). Looking next at item nonresponse, our results suggest that trained interviewers achieved significantly higher data quality than untrained interviewers (see Figure 4.2). In particular, we found that, across the three studies, trained interviewers had lower item nonresponse rates than untrained interviewers (4%; 95% CI = −0.07/−0.02).[†]

Across two studies (Benson and Powell 2015; Guest 1954), trained interviewers were more likely to do a set of tasks required of interviewers. In particular, for Benson and Powell (2015), trained interviewers were more likely to administer different question types *and* probe nondirectively *and* not emphasize certain response options than untrained

---

* Using a multilevel model, which accounts for nesting of several effect sizes in one manuscript, confirms this finding (4.5 percentage points difference (CI: −0.07/0.01), with almost 15% of the variance explained on the study level).

† The multilevel model showed for item nonresponse a difference of 2.7% with 63% variance on the study level.

interviewers. For Guest (1954), trained interviewers were more likely to hold a discussion of interviewing in general *and* how to sample *and* the handling of different question types, than untrained interviewers.

We were able to identify two studies (Cannell, Oksenberg, and Converse 1977; Miller and Cannell 1982) that evaluated whether specific experimental interviewing procedures (i.e., commitment statements, instructions, and scripted feedback; see Chapter 1 in this volume) increased the accuracy of responses. In both studies, the true values were unknown and increased accuracy was inferred through a change in the mean of the responses; thus, we summarize the results qualitatively. Both studies found positive effects of interviewing procedures compared to interviews without these features. In Miller and Cannell (1982), six out of ten socially desirable questions where underreporting was expected had higher average reports in the groups where interviewers used advanced techniques. Cannell, Oksenberg, and Converse (1977) recorded more accurate responses (e.g., respondents checked their documents to report on the last doctor visit) to 9 out of 11 variables with the advanced techniques.

We could only identify one study (Fowler and Mangione 1990) that examined the effect of training for the other three outcomes (correct reading, probing, and recording). Fowler and Mangione (1990) tested various interviewer training durations and training methods. The control group always received basic training. This basic training consisted of a two-hour lecture on interviewing techniques and one demonstration interview. The authors conducted a total of 12 experiments on correct probing, which differed in training content and duration, eight of which showed no significant difference between the experimental and control groups. In the remaining four experimental groups, Fowler and Mangione (1990) found a 20–50% increase in the rate of correct probing. The test groups with the longest trainings of ten days had the biggest improvements (40%) in probing, the difference between the five-day trainings and the ten-day trainings were in more time for practicing. For the correct reading of questions, significant improvements of up to 50% could be observed for half of the experiments depending on training content and training duration. Five-day trainings including lectures, demonstrations, a movie, and practicing performed best. For the correct recording of the questions, no improvement was found in nine experiments; the remaining four experiments showed 20–30% more correct recording. The best results were again revealed after the five-day trainings.

Due to the small number of studies, conclusions from these analyses are necessarily suggestive rather than definitive. This overall picture suggests that interviewer refusal avoidance training reduces unit nonresponse rates. Thus, Q1 can be answered in the affirmative. For the other data quality indicators related to measurement error, there is much more limited evidence for improvements. Thus, Q2 requires further investigation.

Effect size heterogeneity (Q3 in Table 4.1). The heterogeneity of training approaches we observed across these studies prompted us to ask whether effect size distributions were heterogeneous (Q3), which would result in further moderator analysis. This question can be answered in the affirmative for the studies on refusal avoidance training ($p \leq .05$), assuming random effects (see Appendix Table A6). To examine whether – and, if so, which – interviewer training features influenced the effect of interviewer training on response rates, we conducted a moderator analysis.

4.4.2 Moderator Analysis: Which Features Render Interviewer Training Successful?

In this section, we present the results for the moderator variables. We report these results for one of the seven data quality indicators (response rates), as eligible studies with a variation on the moderator variables could be identified only for this quality indicator. Specifically, we were interested in whether the duration of interviewer training (Q4), practice and feedback sessions, additional supplementary training material, interviewer monitoring, and blended-learning-based training (Q5) had an impact on the training outcomes. In what follows, we discuss the results for each of the aforementioned data quality indicators.

4.4.2.1 Interviewer Training Duration (Q4 in Table 4.1)

Reduction of unit nonresponse. The duration of interviewer training was found to have only a small but significant overall impact on response rates. We included 22 effect sizes from nine authors. On average, the response rates achieved by interviewers with an average training duration of five to ten hours were 7% higher than those achieved by interviewers that did not receive specialized training (see Figure 4.3). On average, the response rate gap between a three-hour refusal avoidance training and a 7.5-hour refusal avoidance training was only 3%.

Item nonresponse. For preventing item nonresponse, we consider three studies: Dahlhamer et al. (2010), Billiet and Loosveldt (1988), and Guest (1954). The first two studies held an interviewer training of 15 and 16 hours; this training significantly improved item nonresponse rates compared to their control groups. The training in Guest (1954) lasted nine hours and revealed no improvement for item nonresponse, but this difference should not be interpreted causally. We could not evaluate the effect of the duration of training on the administration of items as our data lacked studies that focused on evaluating this outcome. The lack of primary studies furnishing experimental evidence on training duration makes it difficult to derive recommendations for practitioners. Nevertheless, for our fourth research question (Q4), we recommend that between five and ten hours should be devoted to refusal avoidance training.

4.4.2.2 Cooperation Rates and Training Methods (Q5 in Table 4.1)

Training methods also are determining factors for the success of training. Our analysis revealed that, when interviewer training included practice and feedback sessions (n = 17 studies), the response rates achieved by trained interviewers were on average 13% higher than those achieved by the control group (n = 7) (see forest plot in Figure 4.3). This difference was statistically significant. None of the other variables showed any significant differences at the 5% level, but this can also be due to the small number of cases (n = 23), so we report large directional differences. The use of supplementary training manuals (n = 14 studies) improved response rates by 10% compared with training that did not provide supplementary material (n = 10). The usage of audio refusals resulted in a 10% response rate increase (n = 5). But again, these methods were not experimentally varied within a study. Due to the small number of studies, we were not able to look at moderating effects of training for the measurement error indicators.

4.5 Conclusion and Discussion

The aim of this study was to examine the impact of interviewer training on data quality. The results of our meta-analysis and systematic review of experimental effect sizes

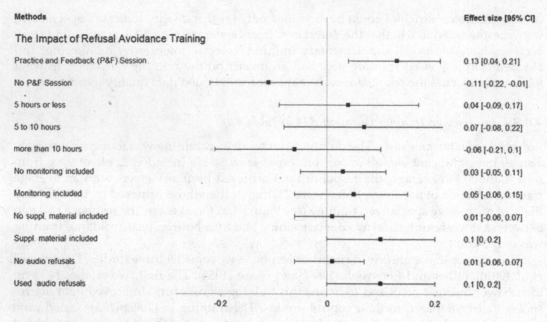

FIGURE 4.3
Forest plots for data quality indicators: moderator analysis.

from 14 different studies is that interviewer training improves data quality targeted to both nonresponse and measurement errors. As the systematic review suggests, we could not identify one specific training feature that affected all data quality indicators because the studies did not examine the same set of features. There is some evidence that different training features such as practice and feedback sessions improved response rates. This shows that application-oriented learning content is effective. Our review also suggests that a refusal avoidance training duration of five to ten hours is recommended.

At least four implications for fieldwork can be concluded from our results. First, training should not focus primarily on persuading reluctant respondents but also on avoiding refusals in the first place. Especially surprising is the albeit significant but moderate improvement in response rates as a result of interviewer training on avoiding refusals. This finding suggests that there are either only a few trainable skills that influence the recruitment of respondents or the studies examined here have not yet used the optimal training methods. It would appear that it is not so much a particular skill on the part of the interviewer that influences the respondent's decision to participate but rather the interaction between the characteristics of the interviewer and those of the potential interviewee (Groves, Cialdini, and Couper 1992; Jäckle et al. 2013; Olson, Kirchner, and Smyth 2016).

Second, interviewer training should continue to focus on adhering to the tenants of standardization during the administration of questions, as the literature review suggests that data quality improvements could be achieved through training in this task. This finding is also in line with studies that have found that interviewers have a substantial impact on measurement bias (Fischer et al. 2018). Third, our results show that training content can best be conveyed by using a wide variety of methods. In particular, practice and feedback sessions should be included in the training program, at least for unit nonresponse avoidance training, as adults learn primarily from experience.

This study has a number of limitations. First, there may be differences in the effectiveness of training across studies due to factors unrelated to training. We did not account for the target population, study content, use of incentives, and so on. A related issue is that we did not identify whether the experiments included a manipulation check to verify that the treatment was being administered as intended. Second, researchers may be interested in the effects of interviewer training on data quality indicators other than those reviewed here. The impact of specific training methods and content on interviewer intraclass correlation coefficients, the bias of estimators, the collection of sensitive information, the collection of biomarkers, and the achievement of high consent rates are questions that remain largely unanswered. However, they must first be addressed by primary research before evidence-based meta-analytic work is possible. Third, we could not include measurement-related outcomes in the quantitative moderator analysis, including item nonresponse, a general adherence to the interview protocol, response accuracy, correctly reading questions, probing responses, and recording responses. However, this is the first meta-analysis to synthesize the effect sizes of refusal aversion training on unit nonresponse. Fourth, gaps in primary research yielded a limited number of studies for each outcome, giving rise to a number of statistical performance problems. The number of studies is very small (n = 14), making general conclusions difficult to impossible, and the effect sizes of three of the seven data quality measures come from the same study and use the same control group. This leads to correlated error terms (Thompson 2011). Finally, the moderators were not quantitatively comparable except for the duration of refusal avoidance trainings and so a qualitative systematic review was necessary. Fifth, the literature search in this meta-analysis did not identify all relevant studies. In particular, older studies and studies from edited volumes could not be located (see online Appendix Section 4.6.4 for a listing of studies).

Both the lack of statistical performance and the lack of variation on some moderators were caused by the small number of primary research studies. Therefore, we strongly encourage researchers to conduct additional experimental primary research on training methods and especially on training content. The aim for future research should be to develop evidence-based interviewer training that can be implemented in actual field-based settings.

Acknowledgments

Special thanks to Kristen Olson and Jennifer Dykema for their feedback, which has vastly improved this chapter. They have not only identified methodological problems, but also made excellent suggestions for solutions.

Bibliography (* indicate studies included in systematic review/meta-analysis)

Alcser, K., J. Clemens, L. Holland, H. Guyer, and M. Hu. 2016. *Interviewer recruitment, selection, and training. Guidelines for best practice in cross-cultural surveys.* Ann Arbor, MI: Survey Research Center, Institute for Social Research, University of Michigan. Retrieved 03, 2018, from www.ccsg.isr.umich.edu.

American Association for Public Opinion Research (AAPOR). 2016. *Standard definitions: Final dispositions of case codes and outcome rates for surveys* (9th ed.). Oakbrook Terrace, IL: American Association for Public Opinion Research.

*Benson, M. S., and M. B. Powell. 2015. Evaluation of a comprehensive interactive training system for investigative interviewers of children. *Psychology, public policy, and law* 21(3):309–322.

*Billiet, J., and G. Loosveldt. 1988. Improvement of the quality of responses to factual survey questions by interviewer training. *Public opinion quarterly* 52(2):190–211.

Borenstein, M., L. V. Hedges, J. P. T. Higgins, and H. R. Rothstein. 2009. *Introduction to meta-analysis*. Hoboken, NJ: John Wiley and Sons.

Bosnjak, M. 2017. Mixed-mode surveys and data quality. In *Methodische Probleme von Mixed-Mode-Ansätzen in der Umfrageforschung*, ed. S. Eifler, and F. Faulbaum, 11–25. Wiesbaden: Springer VS.

*Cannell, C. F., L. Oksenberg, and J. M. Converse. 1977. Striving for response accuracy: Experiments in new interviewing techniques. *Journal of marketing research* 14(3):306–315.

*Cantor, D., B. Allen, S. J. Schneider, T. Hagerty-Heller, and A. Yuan. 2004. Testing an automated refusal avoidance training methodology. Annual Meeting of the American Association for Public Opinion Research (AAPOR), Phoenix, AZ.

*Dahlhamer, J. M., M. L. Cynamon, J. F. Gentleman, A. L. Piani, and M. J. Weiler. 2010. *Minimizing survey error through interviewer training: New procedures applied to the National Health Interview Survey (NHIS)*. Chicago, IL: Joint Statistical Meetings.

Daikeler, J., H. Silber, M. Bosnjak, A. Zabal, and S. Martin. 2017. A general interviewer training curriculum for computer-assisted personal interviews. *GESIS Survey Guidelines*, Version 1.

*Durand, C., M. E. Gagnon, C. Doucet, and E. Lacourse. 2006. An inquiry into the efficacy of a complementary training session for telephone survey interviewers. *Bulletin de Méthodologie Sociologique* 92:5–27.

Fischer, M., B. T. West, M. R. Elliott, and F. Kreuter. 2018. The impact of interviewer effects on regression coefficients. *Journal of survey statistics and methodology* 7(2):250–274.

Fowler, F. J., and T. W. Mangione. 1986. Reducing interviewer effects on health survey data. Executive summary. Report, National Center for Health Services Research and Health Care Technology. Rockville, MD: National Center for Health Services Research and Health Care Technology (DHHS/PHS).

*Fowler, F. J., and T. W. Mangione. 1990. *Standardized survey interviewing: Minimizing interviewer-related error*. Newbury Park, CA: Sage Publications.

Groves, R. M., R. B. Cialdini, and M. P. Couper. 1992. Understanding the decision to participate in a survey. *Public opinion quarterly* 56(4):475–495.

Groves, R. M., F. J. Fowler, M. P. Couper, J. M. Lepkowski, E. Singer, and R. Tourangeau. 2009. *Survey methodology*, volume 2. Hoboken, NJ: John Wiley & Sons.

Groves, R. M., and L. J. Magilavy. 1986. Measuring and explaining interviewer effects in centralized telephone surveys. *Public opinion quarterly* 50(2):251–266.

*Groves, R. M. and K. A. McGonagle. 2001. A theory-guided interviewer training protocol regarding survey participation. *Journal of official statistics* 17(2):249–265.

*Guest, L. 1954. A new training method for opinion interviewers. *Public opinion quarterly* 18(3):287–299.

ISO. 2012. Market, opinion and social research – Vocabulary and service requirements. *ISO standards 20252:2012(E)*.

Jäckle, A., P. Lynn, J. Sinibaldi, and S. Tipping. 2013. The effect of interviewer experience, attitudes, personality and skills on respondent co-operation with face-to-face surveys. *Survey research methods* 7(1):1–15.

*Karlsson, A. Ö. 2010. Recent developments in interviewer training at Statistics Iceland: Minimizing interviewer effects and reducing refusal rates. *Nordisk Statistikermøde* 1:7.

Kelly, M. 2010. Learning styles-Understanding and using learning styles. http://712educators.about.com/od/learningstyles/a/learning_styles.htm?p= (accessed February 25 2019).

Knowles, M., E. F. Holton III, and R. A. Swanson. 2005. *The adult learner: The definitive classic in adult education and human resource development* (6th ed.). Burlington, NJ: Elsevier.

Kreuter, F. 2008. Interviewer effects. In *Encyclopedia of survey research methods*, ed. P. J. Lavrakas. Thousand Oaks, CA: Sage Publications, Inc.

Lessler, J. T., J. Eyerman, and K. Wang. 2008. Interviewer training. In *International handbook of survey methodology*, E. D. de Leeuw, J. J. Hox, & D. A. Dillman (eds.), 442–460. New York: Taylor & Francis Group.

Lipsey, M. W., and D. B. Wilson. 2001. Analysis issues and strategies. In *Practical meta-analysis*, Lipsey, M. W., and D. B. Wilson (eds.), 105–128. Thousand Oaks, CA: SAGE Publications, Inc.

Loosveldt, G., K. Beullens, C. Vandenplas, H. Matsuo, L. Winstone, A. Villar, and V. Halbherr. 2014. ESS interviewer briefing: Note for national coordinators. Interviewer Manual. London, UK: ESS ERIC Headquarters.

*Mayer, T. S., and E. O'Brien. 2001. Interviewer refusal aversion training to increase survey participation. *Proceedings of the Annual Meeting of the American Statistical Association*, Atlanta, Georgia.

Means, B., Y. Toyama, R. Murphy and M. Baki. 2013. The effectiveness of online and blended learning: A meta-analysis of the empirical literature. *Teachers college record* 115(3):1–47.

*Miller, P. V., and C. F. Cannell. 1982. A study of experimental techniques for telephone interviewing. *Public opinion quarterly* 46(2):250–269.

OECD. 2014. *PIAAC technical standards and guidelines*. Organisation for Economic Co-operation and Development (OECD); https://www.oecd.org/skills/piaac/PIAAC-NPM(2014_06)PIAAC_Technical_Standards_and_Guidelines.pdf.

Olson, K., A. Kirchner, and J. Smyth. 2016. Do interviewers with high cooperation rates behave differently? Interviewer cooperation rates and interview behaviors. *Survey practice* 9(2):1–11.

O'Muircheartaigh, C., and P. Campanelli. 1998. The relative impact of interviewer effects and sample design effects on survey precision. *Journal of the royal statistical society: Series A (Statistics in Society)* 161(1):63–77.

Schaeffer, N. C. 1980. Evaluating race-of-interviewer effects in a national survey. *Sociological methods & research* 8(4):400–419.

*Schnell, R., and M. Trappman. 2006. The effect of the refusal avoidance training experiment on final disposition codes in the German ESS-2. Report, Working Paper 3/2006, Center for Quantitative Methods and Survey Research, University of Konstanz, Germany.

Singer, E., M. Frankel, and M. B. Glassman. 1983. The effect of interviewer characteristics and expectations on response. *Public opinion quarterly* 47(1):68–83.

Thompson, C. 2011. Impact of multiple endpoint dependency on homogeneity measures in meta-analysis. Unpublished thesis, Florida State University.

Thorndike, E. L. 1913. *The psychology of learning*, volume 2. New York: Teachers College, Columbia University.

Tusting, K., and D. Barton. 2003. *Models of adult learning: A literature review*. UK: NIACE.

Viterna, J. S., and D. W. Maynard. 2002. How uniform is standardization? Variation within and across survey research centers regarding protocols for interviewing. In *Standardization and tacit knowledge*, ed. D. W. Maynard, H. Houtkoop-Steenstra, N. C. Schaeffer, and J. van der Zouwen, 365–401. New York: Wiley.

West, B., and A. G. Blom. 2017. Explaining interviewer effects: A research synthesis. *Journal of survey statistics and methodology* 5(2):175–211.

Section III

Managing and Monitoring Interviewers and the Survey Process

5

Exploring the Mind of the Interviewer: Findings from Research with Interviewers to Improve the Survey Process

Robin Kaplan and Erica Yu

CONTENTS

5.1 Introduction

The survey interviewer's task in the data collection process is a complex one. Interviewers make many judgments and decisions during the process of interacting with respondents that may contribute to total survey error (West and Blom 2017). Similar to the components of the response process undertaken by respondents during the survey interview (Tourangeau, Rips, and Rasinski 2000), interviewers must: (1) understand what the respondent said; (2) compare the response with the intent of the question and/or match the response to the available response options; (3) judge whether the response was adequate; (4) decide whether or not to probe the respondent's answer if it is inadequate; and (5) record the response (e.g., Japec 2008; Ongena and Dijkstra 2007; Sander, et al. 1992).

How interviewers go about accomplishing these complex tasks can vary. For example, survey organizations typically train interviewers using some form of standardized interviewing (Fowler and Mangione 1990). Standardized approaches aim to minimize variation in the ways that interviewers ask questions, primarily by requiring interviewers to read every question as scripted and use standardized probes; this approach may reduce error. Conversational interviewing, in contrast, acknowledges the dynamic nature of the survey interview process and allows interviewers the flexibility to probe without substantially changing the question meaning, which can increase data quality (Schober and Conrad 1997; Conrad and Schober 2000; Groves and Couper 1998; West, et al. 2018). Interviewers encounter many situations in which the rules of standardization are lacking or incomplete, or they may find it difficult to adhere to standardized practices (see Olson, Smyth, and Cochran 2018, and Chapter 3 of this volume). For example, interviewers are more likely to deviate from an interview script when survey questions are repetitive, sensitive, or difficult (e.g., Haan, Ongena, and Huiskes 2013; Houtkoop-Steenstra and Houtkoop-Steenstra 2000). Minor deviations from interview scripts can occur as often as 33% of the time (Ongena and Dijkstra 2006), demonstrating that interviewers often feel a need to make changes to repair question wording.

How interviewers probe responses is a critical yet understudied aspect of the survey process, and data quality and measurement issues may arise during probing (e.g., Olson, Smyth, and Ganshert 2019). Probing involves decisions about when and how to get additional information about a response. While interviewers are typically trained to probe neutrally, probes are often conversational, emergent, and unscripted, as interviewers respond to the survey context in real-time. Because the survey process can be unpredictable, survey organizations may face challenges training interviewers to use standardized probes consistently for unexpected situations. Interviewer training may help with consistency, but training on how to handle challenging situations is more difficult when interviewers work on multiple survey topics with varying question sensitivity or difficulty.

5.1.1 Motivation

The motivation to conduct this research was to learn more about interviewers' cognitive and decision-making processes and to use this information to help support interviewers and improve interviewer training. Most research conducted directly with interviewers has been informal, such as debriefings with interviewers, limiting the ability to generalize from these findings. In addition to probing, the use of unscripted question lead-ins, such as apologizing (e.g., Dykema and Schaeffer 2005), forgiving wording (e.g., Naher and Krumpal 2012; Peter and Valkenburg 2011), and distancing oneself from the survey question or organization (e.g., Schaeffer, et al. 2008) are well-documented, but little is known about how interviewers decide to use these techniques. The current research was designed to take a more systematic approach to understanding interviewer cognition and decision-making in the field by conducting in-depth interviews with interviewers, and asking them to react to vignettes that closely represent situations that they might face during real survey interviews.

5.1.2 Research Topics

The study presented in this chapter sought to investigate the following three sets of research areas:

1) *Sensitive questions* (i.e., questions that are perceived as personal, invasive, or threatening) are typically studied from the perspective of the respondent rather than the interviewer. However, the task of asking respondents sensitive questions may also

affect the survey process. For example: Do interviewers perceive some questions as being sensitive to ask of respondents? Do interviewers ask sensitive questions differently than non-sensitive questions?

2) *Difficult questions* (i.e., questions that are cognitively burdensome; require calculations, estimation, or looking up information in records; or are effortful to answer due to insufficient knowledge or recall problems) are also often examined from the respondent's perspective. However, interviewers play an important role in motivating and assisting respondents in the process of answering difficult questions. For example: Do interviewers perceive some questions as being difficult to ask of respondents? Do interviewers ask difficult questions differently than less difficult questions? What cues do interviewers look for to determine if respondents are having difficulty responding?

3) *How do interviewers approach the process of probing respondents' answers?* For instance: What cues or features of the interaction (e.g., uncodable answers) do interviewers look for to determine when a response needs to be probed further? How do interviewers decide whether or not to probe? How do interviewers approach probing responses to sensitive or difficult questions?

5.2 Method

Two researchers conducted a total of 27 semi-structured interviews with trained interviewers. Each interview lasted 60 minutes. All interviewers were from a single organization, with representation from regional offices across the United States. All interviewers were trained in the standardized interviewing approach. All interviewers were notified that their responses would be completely anonymous and confidential, and that only the research team would have access to their responses. Interviewers were paid their usual hourly rate for participating in this research. The interviewers had a wide range of experience interviewing, from under 1 year to over 15 years. Most had experience across multiple surveys, spanning topics such as employment, health, housing, crime, and household expenditures.

The protocol (see Online Appendix 5A) started with introductions and an explanation of the purpose of the interviews. Afterward, topics ranged from how interviewers approach asking sensitive and difficult questions, deciding whether and how to probe, and general aspects of interviewer–respondent interactions. Due to time constraints, not all questions in the protocol were asked. Afterward, the researchers administered fictional vignettes (see Online Appendix 5B) depicting common interviewer–respondent interactions. The goal of using these vignettes was to get feedback on how interviewers would handle scenarios like these in the field. The vignettes varied in their level of sensitivity and difficulty (e.g., asking a respondent who was recently laid off about their employment status). Interviewers rated the sensitivity and difficulty of each vignette and then provided feedback on how they would approach each scenario in the field. The resulting qualitative data were analyzed with a grounded theory approach to identify themes and patterns and to gain a deeper understanding of the range of interviewer responses (Glaser and Strauss 2017).

5.3 Results

5.3.1 Sensitive Questions

Interviewers perceived questions about income, children, housing/assets, disability, taxes, where respondents were born, and other demographics as the most sensitive topics from the respondent's point of view. Interviewers reported that they believed respondents seem to find these questions sensitive because they are highly personal or because respondents believe they are none of the government's business. Interviewers reported using techniques such as distancing, apologizing, and repeating the question to help administer questions that they anticipate respondents will find sensitive. Several interviewers reported being able to anticipate which respondents will find income questions sensitive. In those cases, some interviewers reported skipping the scripted question and asking immediately for an income range (e.g., "I don't need to know an amount, just give me a category").

Methods for administering sensitive questions. Distancing themselves from the survey was common as many interviewers reported showing respondents their CAPI screen during in-person interviews. This seemed to accomplish two goals: (1) proving that the question is real and that they are required to ask it; and (2) allowing respondents to select a category without having to tell the interviewer the answer, thereby turning the question into a quasi-self-administered item. For example: "I ask for their best estimate, or let them pick a range from the [CAPI] screen. Once they start, they're usually pretty cooperative" and "I show the screen to respondents to show them what the question and response options are. For income questions, the screen is evidence that we aren't looking for a precise dollar value, just categories."

Interviewers also distanced themselves from the survey by simply informing respondents that they are required to read all questions as worded as mandated by the survey organization:

> "I'm required to ask all questions but you aren't required to answer;" and "[I will say to respondents] the question is written like this and it's not up to me; I am low on the totem pole. It's okay to refuse to answer, but your response is helpful data."

Several interviewers mentioned that respondents find questions sensitive because they have confidentiality concerns. Once the interviewers explain how the data are used and the penalties for breaches, more respondents become assured. For example: "Most people are concerned about their confidentiality. We explain their names are stripped from the data and they are just a statistic. Once you explain how it's used, most people don't care;" and "I tell respondents that data are confidential and represent many other households in the aggregate. I mention that there is a fine of $250,000 for breaking confidentiality."

Other interviewers make sure respondents know that they do not have to answer every question. For instance: "I say to all respondents before the interview starts that if there's something you don't want to tell me, we can just skip over it. I am the only one who will know the response, no one else will."

Other interviewers used lead-ins to questions to help ease respondents' observed hesitation: "I understand you might be uncomfortable..." and "If there are any questions that make you uncomfortable, just say 'No thank you.'"

Questions perceived as sensitive to ask. Interviewers identified questions related to sexual orientation, gender identity, disability, and religion as being sensitive for them to administer. These questions or topics can be awkward for interviewers to ask, so they sometimes modify those questions, again using distancing and apologizing lead-ins: "[For in-person interviews] it's awkward to ask 'Are you male or female?' in front of them. I add, 'I'm sorry I have to ask everything as worded.' Then they laugh and answer. It's easier to do over the phone;" and "The biological children question. Some people get offended you'd even ask it. I wouldn't answer such a question either; it's too personal. I always stress the confidentiality of the question."

Questions may also be sensitive given the context of the survey interview. There are situations where interviewers may encounter unanticipated sensitivity, and have to think in real-time about how to handle the question-asking process, as several interviewers explained: "Asking about sexual orientation and gender identity was recently added to the survey. This is sensitive when asking people from other cultures or countries, or they have religious memorabilia in the house;" "Asking about biological, adopted, stepchildren when maybe the kids are within earshot. Kids may not know they are adopted; stepchildren may be considered their own;" and "I despise asking the question about whether anyone in the home has difficulty dressing or bathing. It's awkward to ask this question to a 30-year-old woman."

Interviewers reported that they did not receive much training in asking sensitive questions. In general, their training focused on using a standardized interviewing approach, reading each question as scripted, keeping their voice modulated, not showing emotion, not leading respondents, and not giving one question any more weight than another. While interviewers reported that they are trained to make respondents feel comfortable, generally they are not given much guidance on how to do so when administering sensitive questions.

5.3.2 Difficult Questions

Difficult questions were defined as those where respondents may have trouble understanding the question, retrieving the information needed to answer the question, or forming an answer. Like sensitive questions, a general pattern was found where interviewers handled difficult questions with emergent, non-standardized probes. For example, interviewers reminded respondents that they do not have to answer each question or that they can provide rough estimates. Interviewers often tried to anticipate questions that respondents would have difficulty answering (e.g., questions that ask about detailed household expenses, Medicare and Medicaid payments, health insurance deductibles, and income before taxes). For some of these questions, it is an issue of complexity and difficulty with comprehension or taking the information from the records respondents have available and matching it to answer the survey question: "Respondents won't know the deductible. They might just know they spent $50 at the doctor's office;" "Respondents don't know the amount of their Social Security that gets deducted for Medicare. [I have to] explain this question to them;" and "Respondents don't always know or understand the amount they spend on health insurance. Some of them hand me their paystub to interpret the output for them. This section [of the survey] needs attention. Some questions are oddly worded."

One interviewer expressed that he sometimes does not know how to provide useful probes or additional information to the respondent on how to answer these questions: "Medical expenses – I'm not always sure what expenses count or go into what category. I don't know much about this myself so sometimes I am afraid to probe in case I get it wrong."

Other interviewers ran into difficulty for respondents stemming from knowledge or memory issues in answering the questions: "Respondents don't know the exact amount of their utility bills. I have to ask them to look at records."

Interviewers reported using a variety of strategies to help respondents remember information or look it up. A common method was to encourage the use of records to get the information. When it was not possible to use records to get exact answers from respondents, interviewers used a variety of unscripted probes to aid respondent recall. Strategies included providing examples and asking about dates, seasons, and events. For example, as one interviewer described: "I try to ask if the event happened near a date, a holiday, or when a child was born." A minority of interviewers expressed that using probes to stimulate recall was not helpful: "On recall questions, probing isn't always worth it. You might get more detail but it is a big ask and you don't usually get much more detail."

Identifying respondent difficulty. Interviewers reported being able to decipher verbal and nonverbal respondent behaviors that are indicative of question difficulty, such as when respondents ask them to repeat the question, pause, hesitate, or provide blank expressions. One interviewer said that she uses educational level as a cue for how difficult respondents might find the questions. Cues are easier to observe for in-person interviews than telephone interviews, but interviewers reported that long pauses in both modes often indicate difficulty. A few interviewers reported modifying the questions and their lead-ins when they observe that respondents are having difficulty with questions, e.g., "If respondents find a question difficult or unclear, I'll add examples when reading it initially." Some interviewers reported using techniques that fall on the more conversational end of the spectrum:

> Sometimes the respondent gives a vague answer, and you have to probe to get more detail to fit one of the response options. When you have to keep probing, it can sound like you're judging them. I'll change the scripted probe wording from "Anything else?" to "Is there anything else you tried that you think we should know about?"

Depending on the survey context, and whether interviewers are supposed to read the response options or use scripted probes, these techniques could potentially lead respondents to give particular answers: "When this happens, I will read a subset of the response options, and check which ones might match the respondent's answer and verify it."

Repetitive questions. Interviewers often have to ask questions that respondents perceive as repetitive. For example, the survey may contain similar questions or repeat questions over multiple waves of a survey, or respondents may spontaneously report information that answers subsequent questions, causing later scripted questions to appear repetitive. When these situations occur, some interviewers reported using a special lead-in (e.g., apologizing, tailoring, or distancing) to acknowledge that they already addressed the question earlier in the interview: "I acknowledge that I understand why the respondent would feel the questions are tedious. It helps to be sympathetic, understanding. I explain we have to ask the same questions of everyone;" and "I'll apologize, and say sorry, I have to read it. If they get annoyed, I move quickly. Give them an option not to answer. I agree with them, say I know what you mean, helps rapport."

Irrelevant questions. Another challenge that interviewers face is having to ask a question that may not apply or may not be relevant to the respondent. For example, an interviewer might have to ask about maternity or paternity leave, although the respondent does not have children. This could occur because the survey instrument did not have adequate filters to skip irrelevant questions, or when respondents spontaneously offered information signaling to the interviewer that a question may not be relevant to them. A minority of interviewers reported simply reading the question verbatim anyway, even though they were aware that it was not relevant to the respondent. However, most interviewers reported that they improved rapport with respondents by adding a lead-in to the question to acknowledge the question was irrelevant to them. For example: "This might not apply to you, but I have to ask this." Other interviewers reported directly acknowledging something the respondent said earlier as a lead-in to the question, for example: "You told me you're retired, so did you do any work for pay?"

5.3.3 Deciding Whether to Probe

Finally, interviewers were asked open-ended questions about when they decide to probe a response or not. The majority of interviewers reported that the main reason they decide to probe is usually to clarify the meaning of a question or to verify the answer a respondent has given. For example: "It depends on the question. I would not probe if they might be lying. I probe mostly to clarify the question, or reiterate/verify what they said. Sometimes people say the answer ahead of time, [so I] won't repeat the full question but will verify instead.;" and "The insurance questions are hard to answer cleanly, sometimes it's clear that an insurance payment amount is too low. Respondents want to give an answer to be helpful, but they aren't sure about the accuracy, so I'll probe when something sounds off."

Interviewers reported deciding not to probe when the respondent had limited time or was reluctant to participate in the interview, or the question topic was highly sensitive.

5.4 Vignettes

Vignettes were administered to get interviewers' feedback on a variety of scenarios that mimicked real survey interviews (see Online Appendix 5B). The first set of vignettes was based on a survey about employment. Three of the vignettes were designed to be sensitive and three were designed to be neutral. Interviewers first rated the sensitivity of each survey scenario on a three-point scale, where 1 was "not at all sensitive," 2 was "moderately sensitive," and 3 was "very sensitive," from the perspective of both the respondent answering the question and the interviewer administering the question. The second set of vignettes was based on questions used in a survey about income and program participation. Interviewers rated difficulty on a three-point scale, where 1 was "easy," 2 was "neither easy nor difficult," and 3 was "difficult," from both perspectives again. Vignettes were a mix of cognitively burdensome (e.g., out-of-pocket expenses for healthcare products) or sensitive questions (e.g., whether all of the children in the household have the same biological father or mother).

5.4.1 Sensitivity Vignettes

Table 5.1 shows the mean ratings interviewers assigned to the sensitive and neutral employment vignettes from the perspective of the respondent and interviewer. The ratings were similar within the sensitive and neutral vignette categories, and are collapsed here.

Most interviewers rated the sensitive vignettes as sensitive for respondents to answer as well as for interviewers to ask. Interestingly, interviewers rated answering the questions as more sensitive than asking them. This is likely because respondents volunteer personal information about their own lives, whereas interviewers just ask the questions. For the neutral vignettes, interviewers rated these questions as equally non-sensitive regardless

TABLE 5.1

Mean Interviewer Ratings (Standard Errors in Parentheses) Assigned to the Sensitive and Neutral Employment Vignettes by Respondent and Interviewer Perspective (n = 25)

	Employment Vignettes	
	Sensitive Vignettes	**Neutral Vignettes**
Respondent perspective (answering a question)	2.4 (0.09)	1.1 (0.05)
Interviewer perspective (asking a question)	1.9 (0.10)	1.1 (0.05)

Notes: 1 to 3-point scale; higher scores indicate greater sensitivity.

of perspective. Some interviewers reported that they would use a conversational approach to address the sensitive questions, including the use of unscripted probes, tailored question lead-ins to maintain rapport, displays of empathy, or distancing themselves from the survey: "I would try to put them at ease and emphasize they [unemployed people] are why the survey is important;" "I would apologize and say, 'I understand it's sensitive, but… [repeat question.];'" and "I would tread lightly here. Need to be careful not to lose the respondent. I wouldn't spend too much time probing."

A few interviewers noted that administering potentially sensitive questions can be difficult because they do not know the answer to a question until it has been asked. Sometimes a question is unexpectedly sensitive for a respondent that might be neutral for most respondents. For example, one interviewer who rated the items as sensitive stated, "I go into the question not knowing what the respondent's answer is going to be."

Several interviewers felt that the questions were worded in a way that made them extra sensitive, and would rephrase these questions in the field. For example, "The question about [what they're doing to look for work] is worded weirdly. It sounds like you think they're not doing anything, slacking. I would rephrase this in a more positive way, 'Have you been job hunting?'"

5.4.2 Difficulty Vignettes

Table 5.2 shows the mean interviewer ratings in terms of difficulty answering and asking questions about income and program participation across all six vignettes. Interviewers found the questions to be moderately difficult for respondents to answer, but slightly less

TABLE 5.2

Mean Difficulty Ratings (Standard Errors in Parentheses) of the Income and Program Participation Vignettes by Respondent and Interviewer Perspectives (n = 25)

	Roster (Neutral)	Support Payments (Sensitive)	Savings Account (Sensitive)	Biological Children (Sensitive)	Health Products (Difficult)	Recontact Consent (Difficult)	Overall Mean
Respondent perspective (answering a question)	1.1 (0.05)	1.9 (0.16)	1.6 (0.15)	1.9 (0.17)	2.2 (0.17)	1.4 (0.13)	1.7 (0.07)
Interviewer perspective (asking a question)	1.0 (0.04)	1.6 (0.16)	1.4 (0.14)	2.0 (0.19)	1.2 (0.10)	1.4 (0.13)	1.4 (0.06)

Notes: 1 to 3-point scale; higher scores indicate greater difficulty.

difficult for interviewers to administer, consistent with the pattern observed in the previous vignettes on employment. The question about household rosters was rated as relatively neutral for both respondents and interviewers, although for some populations household rosters are perceived as sensitive (Tourangeau, et al. 1997). The remaining vignettes were rated as easy or neither easy nor difficult, but the question about expenses on healthcare products was rated as more difficult for respondents to answer than for interviewers to ask because the information was difficult to recall and calculate.

As observed earlier, interviewers reported being attuned to the fact that they cannot always predict how respondents will answer even seemingly neutral questions. Although interviewers are aware that unexpected situations can arise during the interview process, they do not always know how to respond in those situations. In terms of the questions about support payments, savings accounts, and spending on healthcare products, interviewers reported occasionally getting negative reactions from respondents on these questions for a variety of reasons. Like questions about income, these questions can also feel invasive to respondents: "I'd be worried that the fact of making support payments is sensitive. Sometimes people don't want to disclose the [payment] amount. For most people it's easy, get refusals sometimes;" and "Most of the time these questions are fine, but it could be intrusive, like we're checking up on them."

Although these questions were seen as sensitive for some interviewers, many mentioned during the interviews that the question about support payments in particular was difficult to administer due to the length of the question and having to read a large number of response options out loud:

> "I'll get interrupted [while reading the response options]. I make the respondent wait to hear all of them because the answer is [the last response option listed] 'no payments made;'" "I would ask yes or no, line by line [for each response option]. I'd anticipate some sensitivity on this topic;" and "The question is long, and the first option should be [the most common response] 'no payments made.'"

Interviewers recognized that the question was long and making respondents listen to a set of response options that may not apply to them could be tedious. Several interviewers noted that they would modify the question to filter out people who did not make support payments, or would change it to a yes/no format while reading the response options to help the interview flow better. For other questions, interviewers mentioned that problems might occur when respondents attempted to answer a cognitively burdensome question, such as the item about out-of-pocket expenses on healthcare products. This question could pose difficulty because respondents often do not know the information needed off the top of their head, or it may require complex calculations or estimation techniques such as addition, multiplication, or guessing the amount: "I have to probe this question when the respondent says, 'I don't spend anything on that.' Then I ask if they have allergies or take Tylenol and usually the respondent says yes. You have to help them add up what they spent. I help the respondent answer with memory probes;" and "Most people don't keep track of this throughout the year. They start thinking about all they buy, and trying to think of everything that falls under that, how often they buy it. If the respondent starts listing what they have, mentions the bottle size, I try to see how long it lasts them, figure out how often they have to buy them, for instance, once per month, every few months."

The question about biological children in the household seemed to be perceived as potentially sensitive and difficult for interviewers to ask. A minority of interviewers felt that this question was not sensitive because divorce is common, and that most respondents are open and forthcoming: "I try to make them comfortable, but difficult to change. If they aren't comfortable, I might say 'There are all different households these days;'" "This is uncomfortable to ask because there is a possibility the respondent will get offended. Don't need to label the child as adopted or biological. Makes people uncomfortable. I don't know why the survey asks about it;" and "I would show the screen to the respondent."

Finally, interviewers were asked about a question in which respondents are asked for their consent to use previously recorded information during a subsequent wave of the survey to shorten the duration of future interviews. Most interviewers recognized the value of asking for this, but felt that the question wording was long and awkward to administer: "Question is lengthy, the last sentence has two sets of parentheses;" "The question is long, kind of beats around the bush. A lot of people have commented that they'd like to do the survey online instead;" and "This question needs to be streamlined, otherwise people don't know what you're asking. I would read it as scripted but if I needed to probe, I'd say, 'This will make the [next] interview shorter.'"

5.5 Summary and Conclusions

Learning directly from professional survey interviewers about how they do their jobs is a critical step in understanding how to develop realistic data collection models, improve question wording, and improve interviewer training. Most interviewers handled asking sensitive and difficult questions in similar ways, using techniques such as distancing, apologizing, and repeating the question. Despite being trained to read questions verbatim and use scripted probes (when provided) or neutral probes, interviewers reported using a variety of unscripted and non-neutral probes and other techniques when faced with sensitive or difficult questions and unexpected situations. Interviewers articulated the importance of being flexible and adapting to respondents in real-time, while maintaining rapport to help complete the interview, despite not receiving much formal training on these skills. The data from the interviews and vignettes illustrated the interview as a social interaction between the interviewers and respondents, which drives the need to maintain rapport over and above following the scripted questionnaire.

The demands of interviewing respondents (e.g., the ability to be flexible, conversational, and maintain rapport with respondents without compromising data quality) may have led to some of the inconsistencies in techniques used to handle sensitive and difficult questions and contexts that were observed across interviewers. Some interviewers noted that they make it apparent at the beginning of an interview that respondents can skip any questions they do not want to answer; some used distancing techniques; and others used non-scripted lead-ins or selective reading of response options. Most interviewers reported that they probe when respondents have comprehension or recall problems, concerns about confidentiality, or uneasiness with a particular topic. Interviewers decided not to probe when respondents had limited time, were reluctant to participate, or the question topic was highly sensitive.

Interviewers reported using a wide range of probing strategies. Some modified the question or tailored their lead-ins with information respondents previously provided. Others changed scripted probes, read a subset of response options, changed open-ended questions to closed-ended questions, and shortened questions, while a minority did little probing at all. In response to vignettes depicting realistic survey scenarios, interviewers often reported they used a conversational approach, unscripted probes, tailored question lead-ins, displays of empathy, or distanced themselves from the survey as needed to maintain rapport and complete the interview. Although these changes to the interview script are well-intentioned, they may affect measurement error and survey estimates. The vignettes were a useful tool to understand how interviewers react to unexpected survey contexts, including sensitive or difficult interactions, and shed light on interviewers' decision-making processes when facing these challenges.

5.5.1 Recommendations for Interviewer Training

This research identified several ways that survey organizations could improve support and training for interviewers. Based on these findings, it is recommended that survey organizations focus more training on topics such as probing sensitive or difficult questions. Training materials should help interviewers understand why a question is included in the survey and how to effectively explain complex concepts to respondents. Many interviewers expressed frustration with not only getting respondents to participate in the first place, but then having to administer questions that they themselves find uncomfortable or lack confidence in asking. It is also recommended that survey organizations include interviewers in final questionnaire decisions and throughout the questionnaire development process. Additionally, in the absence of understanding the value and purpose of survey questions, interviewers may alter the meaning of the original question or probe in non-standardized ways, contributing to measurement error. Training on the purpose of each survey question and how to effectively handle sensitive or difficult survey questions is likely to improve how interviewers administer these questions.

5.5.2 Recommendations for Future Research

Conducting research directly with interviewers is critical to improving our understanding of the complex nature of interviewer–respondent interactions. Systematic research on interviewers' decision-making, cognitive processes, and probing in sensitive and difficult survey contexts in particular is needed in the survey methods literature. New technologies such as CARI recordings make this type of research possible and enable organizations to see how interviewers actually administer questions in the field, which can help organizations identify and improve problematic questions.

Many of the themes identified in this research related to inconsistent probing across interviewers. This is problematic, as interviewer effects are often tied to the rate at which respondents provide inadequate answers that require probing (Mangione, Fowler, and Louis 1992; West and Blom 2017). Future research should investigate when and why interviewers deviate from the script so that survey organizations can provide interviewers with the tools to respond to these concerns consistently and in a manner that can be measured and monitored. For instance, conducting research on the effectiveness of building scripted probes directly into survey instruments could prove beneficial. The use of scripted probes has the potential to help better standardize survey interviews, increase data quality, and reduce measurement error. It is also recommended that researchers make use of vignettes for evaluating questions that may be problematic. Question pretesting is often focused solely on potential problems with respondent processing, but question pretesting with interviewers using vignettes could improve survey design – including question wording and probes. Investments in interviewer training and research conducted directly with survey interviewers are likely to have a positive impact on survey administration in the future, and help to reduce the total survey error associated with sensitive and difficult survey questions.

References

Conrad, F. G., and M. F. Schober. 2000. Clarifying question meaning in a household telephone survey. *Public opinion quarterly* 64(1):1–28.

Dykema, J., and N. C. Schaeffer. 2005. An investigation of the impact of departures from standardized interviewing on response errors in self-reports about child support and other family-related variables. *Paper Presented at the Annual Meeting of the American Association for Public Opinion Research*, Miami, FL.

Fowler, F. J., and T. W. Mangione. 1990. *Standardized survey interviewing: Minimizing interviewer-related error*. Newbury Park, CA: Sage Publications.

Glaser, B. G., and A. L. Strauss. 2017. *Discovery of grounded theory: Strategies for qualitative research*. New York, NY: Routledge, CRC Press/Taylor & Francis.

Groves, R. M., and M. P. Couper. 1998. *Nonresponse in household interview surveys*. New York, NY: John Wiley & Sons.

Haan, M., Y. Ongena, and M. Huiskes. 2013. Interviewers' questions: Rewording not always a bad thing. In *Interviewers' deviations in surveys: Impact, reasons, detection and prevention*, ed. P. Winker, N. Menold, and R. Porst, 173–193. Frankfurt on the Main: Peter Lang Academic Research.

Houtkoop-Steenstra, H., and J. P. Houtkoop-Steenstra. 2000. *Interaction and the standardized survey interview: The living questionnaire*. Cambridge, UK: Cambridge University Press.

Japec, L. 2008. Interviewer error and interviewer burden. In *Advances in telephone survey methodology*, ed. E. de Leeuw, L. Japec, P. J. Lavrakas, M. W. Link, and R. L. Sangster, 185–211. Hoboken, NJ: John Wiley & Sons.

Mangione, T. W., F. J. Fowler, and T. A. Louis. 1992. Question characteristics and interviewer effects. *Journal of official statistics* 8:293–293.

Näher, A. F., and I. Krumpal. 2012. Asking sensitive questions: The impact of forgiving wording and question context on social desirability bias. *Quality & quantity* 46(5):1601–1616.

Olson, K., J. D. Smyth, and B. Cochran. 2018. Item location, the interviewer –respondent interaction, and responses to battery questions in telephone surveys. *Sociological methodology* 48(1):225–268.

Olson, K., J. D. Smyth, and A. Ganshert. 2019. The effects of respondent and question characteristics on respondent answering behaviors in telephone interviews. *Journal of survey statistics and methodology* 7(2):275–308.

Ongena, Y. P., and W. Dijkstra. 2006. Methods of behavior coding of survey interviews. *Journal of official statistics* 22:419–451.

Ongena, Y. P., and W. Dijkstra. 2007. A model of cognitive processes and conversational principles in survey interview interaction. *Applied cognitive psychology: The official journal of the society for applied research in memory and cognition* 21(2):145–163.

Peter, J., and P. M. Valkenburg. 2011. The impact of "forgiving" introductions on the reporting of sensitive behavior in surveys: The role of social desirability response style and developmental status. *Public opinion quarterly* 75(4):779–787.

Sander, J. E., F. G. Conrad, P. A. Mullin, and D. J. Herrmann. 1992. Cognitive modelling of the survey interview. In *Proceedings of the Joint Statistical Meetings, Survey Research Methods Section*, 818–823.

Schaeffer, N. C., J. Dykema, D. Garbarski, and D. W. Maynard. 2008. Verbal and paralinguistic behaviors in cognitive assessments in a survey interview. In *Proceedings of the Joint Statistical Meetings, Survey Research Methods Section*, 4344–4351.

Schober, M. F., and F. G. Conrad. 1997. Does conversational interviewing reduce survey measurement error? *Public opinion quarterly* 61:576–602.

Tourangeau, R., L. J. Rips, and K. Rasinski. 2000. *The psychology of survey response.* Cambridge, UK: Cambridge University Press.

Tourangeau, R., G. Shapiro, A. Kearney, and L. Ernst. 1997. Who lives here? Survey undercoverage and household roster questions. *Journal of official statistics* 13(1):1–18.

West, B. T., and A. G. Blom. 2017. Explaining interviewer effects: A research synthesis. *Journal of survey statistics and methodology* 5(2):175–211.

West, B. T., F. G. Conrad, F. Kreuter, and F. Mittereder. 2018. Can conversational interviewing improve survey response quality without increasing interviewer effects? *Journal of the royal statistical society: series A (statistics in society)* 1:181–203.

6

Behavior Change Techniques for Reducing Interviewer Contributions to Total Survey Error

Brad Edwards, Hanyu Sun, and Ryan Hubbard

CONTENTS

6.1 Introduction

In the total survey error (TSE) paradigm, nonsampling errors can be difficult to quantify, especially errors that occur in the data collection phase of face-to-face surveys. It may seem strange to focus on the face-to-face mode because it is so expensive, and therefore has lost much ground to telephone (in the 1960s through the 1990s) and (more recently) web modes (Groves 2011). However, it is the only way to collect some kinds of survey data, particularly for official government surveys, surveys that collect certain kinds of biomarker or environmental data, and surveys in cultures with no written language.

Field interviewers play "dual roles as recruiters and data collectors" (West et al. 2018, 335), and are therefore potential contributors to both nonresponse error and measurement error. Recent advances in technology, paradata, and performance dashboards offer an opportunity to observe interviewer effects almost as the interviewer engages in the behavior that produces the effect, and to intervene quickly to curtail undesired behaviors. For example, improvements in computer-assisted recorded interviewing (CARI) in the previous decade have greatly increased the efficiency of listening to selected questions across all field interviewers working on a survey. Edwards, Maitland, and Connor (2017) report on an experimental program using rapid feedback of CARI coding results to improve interviewers' question-asking behavior. Mohadjer and Edwards (2018) describe a system for visualizing quality metrics and displaying alerts to inform field supervisors of anomalies (such as very short interviews) detected in data transmitted overnight. These features allow supervisors to quickly investigate, intervene, and correct interviewer behavior that departs from the data collection protocol. From the interviewer's perspective, these interactions can be viewed as a form of experiential learning, consistent with the literature on best practices in adult learning (Kolb, Boyatzis, and Mainemelis 2001). From the survey manager's perspective, they can be important features in a continuous quality improvement program. We build on these initiatives to focus on specific areas where interviewer error can be a major contributor to TSE with major impact on key survey statistics.

6.1.1 Review of Relevant Literature

In the development of survey research as a social science discipline, standardized interviewing looms nearly as large as probability sampling (Fowler and Mangione 1990). Asking all sampled individuals the same question was a major leap forward over interviewers having their own conversations with respondents. Asking all questions exactly as worded is common practice in many telephone centers. It is much less frequently observed in field settings, which are unpredictable and not easily monitored. Interviewers and respondents are social creatures and their conversations – even those as structured as the standardized survey interview – are complex interactions that are difficult to confine to words on a page or a computer screen (Schaeffer 1991). A nuanced view of the interviewer's role has been incorporated in the TSE paradigm, beginning in the late 1970s (Groves 1987; Groves and Lyberg 2010). In TSE, interviewer behavior is only one of a number of error sources, and the sources interact – a decrease in one can lead to an increase in another. For example, Tourangeau, Kreuter, and Eckman (2012) demonstrate the tradeoff between nonresponse error and coverage error in screening for eligibility by age. The errors could be an effect of interviewer behavior, but their findings suggest respondents are the source.

Interviewer training improves skills and has the potential for reducing interviewer effects in TSE. Ackermann-Piek (2019) maps training content for three multinational surveys onto the TSE view of error sources. Best practices for training interviewers are based in part on adult learning theory, which holds that adults learn differently than children. Key tenets are just-in-time content (knowledge or skills they can apply to work activities immediately at hand), peer learning (learners interacting with each other to achieve objectives), and close-to-real-world practice (learners performing activities that are similar to the work required on the job) (Knowles 1980; Knowles, Horton, and Swanson 2005). It has been difficult to incorporate these core concepts into interviewer training. Classroom training is typically staged as closely as possible to the beginning of data collection, and often attempts to train interviewers for rare situations they may encounter months later if at all. Current training often makes minimal use of peer learning or close-to-real-world practice.

Interviewers working in a call center benefit from supervisor feedback in real time. This can be an important component of a continuous training program. It corrects undesirable behavior immediately, and reinforces adherence to standard operating procedures. In contrast, field interviewers typically work in a disconnected mode. They work independently and may not interact with supervisors more than several times a week, through phone calls or emails. Paradoxically, their work is often more challenging than work in a call center. Interviewing face-to-face enables interviewers to communicate with respondents much more fluidly than is possible through voice alone. This contributes to greater success in gaining cooperation and in maintaining rapport in the interview, enabling much longer interviews and collection of biomarker and environmental data efficiently with a relatively high degree of compliance. However, monitoring quality and improving performance in field conditions is difficult.

Smart phones and near-universal internet coverage (at least in the U.S. and Europe) can support constant communication between field interviewers and supervisors, and allow supervisors to observe many aspects of the interviewers' behavior (e.g., doorstop documentation of contact attempts; travel from one respondent's home to another). Quality issues can be visualized for supervisors on dashboards and a system of alerts can inform them overnight about problems that require immediate attention (Mohadjer and Edwards 2018). These tools have the potential to bring field operations under much more rigorous control (Edwards, Maitland, and Connor 2017). Edwards, Maitland, and Connor (2017) reported results from an experiment using CARI behavior coding and rapid feedback (both written and verbal), showing improvements in general interview quality. In this chapter we investigate rapid feedback on specific questions as a tool for on-the-job training of field interviewers that can impact key survey statistics.

6.2 Data and Methods

6.2.1 The Medical Expenditure Panel Survey (MEPS)

This chapter uses data from the Medical Expenditure Panel Survey, a longitudinal survey that produces annual public use files on the use and costs of health care, health insurance coverage, and a host of other data on U.S. households (https://meps.ahrq.gov/mepsweb/). The sample is drawn from households that completed the National Health Interview Survey (NHIS) the previous year. Five in-person data collection visits using a computer-assisted personal interviewing (CAPI) system occur with each panel over a 2.5-year period in an overlapping panel design. Data collection is divided into two phases. The first half of each year includes the first-time-in-MEPS interviews with the newest sample panel; third interviews with the previous year's panel; and final interviews with the panel that is exiting MEPS. The second half of the year includes the second and fourth interviews with the remaining panels. Data for this chapter are from work in fall 2018. About 400 field interviewers conduct more than 25,000 interviews with a household informant each year. Answers to most questions are combined with others to create about 4,000 constructed variables for annual estimates. Key survey statistics include the number of hospital stays, medical provider visits for ambulatory care, and prescribed medicines dispensed each year. MEPS is sponsored by the Agency for Healthcare Research and Quality (AHRQ) in the U.S. Department of Health and Human Services (HHS).

To understand the magnitude and rich detail of these data, consider prescribed medicines for older people. MEPS reports that just over 90% of people aged 65 and older obtained prescribed medicines in 2016. The mean annual expenditure for these people was $3,289. This key statistic is quite sensitive to data from outliers and thus puts a premium on high quality interviewing. (See Table A6A.1 in Appendix 6A in the online supplementary materials for more detailed MEPS data on prescribed medicines.)

In 2018 MEPS implemented a major upgrade of the CAPI system, simplifying some questionnaire sections and entry mechanisms. Interviewers were trained on the new system in January 2018. Researchers were keen to observe how the new system was operating in the field, and to determine whether any corrections might be needed.

6.2.2 Two Critical Question Series: The Calendar Series and Provider Probes Series

Two question series were of particular interest because they were asked in all interviews, they were always recorded in CARI (almost all respondents gave consent to record), and they were critical building blocks for producing data on the use and cost of health care services; key MEPS statistics were: (1) use of calendar aids and other records of medical care, and (2) provider probes (filter questions that prompt the respondent to recall services from various types of medical providers). A striking characteristic of health care surveys is the large degree of underreporting: respondents forget events, or remember events that occurred before the interview reference period (telescoping) (see Miller and Mathiowetz, Chapter 2, this volume).

Increased levels of event reporting are associated with the presence of records (Kashihara and Wobus 2006). The calendar series asks if various sorts of records are available in the interview (e.g., a calendar with entries for medical visits, insurance statements, a patient portal, prescription records or bottles, etc.), and who in the household was associated with each record type. The CAPI screen presents the entry area for these items as a grid, with each household member listed on a row and each record type as a column header. Interviewers can enter answers in any order, by person or by record type (see Online Appendix 6B). The design objective is to encourage respondents to bring records for all family members into the interview and to structure the questioning so that the records can be incorporated into the interview in any order. The grid allows the interviewer to follow the respondent's lead (a respondent-driven path). Grids that allow movement from any cell to any other cell are one of the most difficult item types for interviewers to navigate in a standardized manner because they require the interviewer (rather than the CAPI instrument) to decide where to go next (see Edwards, Schneider, and Brick 2008; Sperry et al. 1998). As part of the technical upgrade, interview skip patterns were revised to take advantage of record types mentioned in this grid, streamlining the interview for respondents with records.

The provider probes are a series of 15 questions about various kinds of health care providers. They were re-ordered in the technical upgrade to begin with three that accounted for the highest expenditures. For example, the first provider probe in the 2018 Round 1 interview read: "since January 1, 2018, were you [...] admitted to the hospital?" (see Online Appendix 6B). If the answer was yes, the interviewer determined who was admitted, collected details about the event, and administered probes about emergency room visits, outpatient visits, and other visits and events.

6.2.3 Behavior Coding of Audio-Recordings

Audio-recordings of the calendar series and the provider probes series in 212 interviews were reviewed by two behavior coders, using Westat's CARIcode system. In CARIcode, the coder is able to call up a specific interviewer or question in an interview. A coding screen presents the question as it appeared to the interviewer, the coding form, a button to play the recording, and some paradata items that might be associated with the quality of the interview, such as interview length (see CARIcode screenshots in Appendix 6C in the online supplementary materials). Project-specific training was conducted to ensure that the two coders understood the coding scheme and could apply the behavior codes consistently. Coders evaluated the overall quality of the interview and of each instance of asking the calendar series and the provider probes. The inter-coder agreement rate was .82. The coding scheme for the calendar series and provider probes are presented in Appendix 6D in the online supplementary materials.

The rapid feedback process is diagrammed in Appendix 6E in the online supplementary materials. Following the procedures described by Edwards, Maitland, and Connor (2017), verbal and written feedback was provided to the interviewer quickly (ideally within 72 hours of the interview). The feedback process included positive as well as negative points. The next interview conducted by the interviewer was coded, resulting in each interviewer having a pair of interviews in the data set, one just before and one just after feedback. Because the process was implemented in late fall, only a subset (112) of the MEPS interviewers were available to participate in the study, resulting in 224 interviews in the data set. Data about the feedback interaction was captured (such as whether the interviewer agreed with the feedback or asked for clarification). Selected interviewer characteristics were also available, such as years of MEPS experience. We hypothesized that more desirable behavior (more consistent with the protocol) would be observed after feedback, both for overall interview quality and for each question series.

Coders reviewed more than 3,000 instances (i.e., question series) of interviewer question-asking behavior in the first selected interview for each interviewer. For each instance, the coders assessed behavior using a three-point scale indicating whether the interviewer: (1) followed protocol exactly; (2) departed from exact protocol but maintained the series meaning; or (3) did not maintain the series meaning. For the calendar series, the exact protocol was to follow the order indicated by the respondent. For the provider probes series, the exact protocol was to ask each question verbatim in the order presented by the CAPI screens. Field directors gave interviewers verbal and written feedback based on the coding. Coders reviewed another 2,200 instances in the interviews conducted right after feedback.

6.2.4 Supervisor Alerts

In addition to the CARIcode data, we reviewed data from alerts generated for supervisors to investigate suspicious interviewer behavior. The alerts were displayed on a dashboard, which also displayed useful production and cost metrics by region and interviewer and offered a simplified way to drill down into interviewer details. The alerts included four operational issues representing outlier behaviors: CARI consent refusal, whether the time of day the interview was completed exceeded boundaries, excessive contact attempts, and cases with high potential to yield completed interviews not worked during reporting period. Alerts also flagged specific data quality concerns such as unusual event date patterns and low record use. (Low record use could reflect interviewers failing to encourage respondents to produce records or failing to follow the calendar series protocol.) Alerts were generated early each morning from data transmitted to the home office overnight, and displayed on the supervisor's dashboard. The supervisor was expected to act on the alert within 24 hours, and might begin by reviewing paradata from various sources: records of contact attempts, CARI recordings, geographic information system (GIS) data, etc. If the anomaly could not be explained by reviewing paradata, the supervisor discussed the issue with the interviewer, determined whether there was a problem, diagnosed it, and took corrective action with the interviewer. Actions included explaining proper protocol or referring the interviewer to other training material. The supervisor documented alert actions and status on the dashboard. Reviewing alerts with interviewers can be viewed as another form of rapid feedback and on-the-job training. We examined alert frequency by type and by interviewer over the course of the field period. We hypothesized (1) alerts would diminish over time and (2) most interviewers who generated an alert would not continue to perform in ways that generated alerts.

6.3 Results

We review the results of the behavior coding and rapid feedback and then summarize the results for the supervisor alert system.

6.3.1 Behavior Coding and Rapid Feedback

Table 6.1 shows the rapid feedback results at the level of the question series. Interviewers maintained the meaning of the questions but did not follow the protocol exactly in the majority of instances in which the two series of questions were asked (n = 5,259), both before and after feedback. Question-asking behavior that followed the protocol exactly increased from 33.4% before feedback to 43.4% after feedback. Behavior that failed to maintain the question meaning decreased from 9.8% before feedback to 3.7% after feedback. The same direction and magnitude of results were observed for both question series. However, behavior differed somewhat by question series type. More desirable interviewer behavior was observed in the provider probes than in the calendar series, both before and after feedback.

After each feedback session the field director recorded whether the interviewer asked for clarification (a case-level dichotomous variable). Table 6.2 shows the interviewers' behavior by question series before and after feedback, broken out by whether the interviewer asked for clarification. As expected, requests for clarification were more frequent when the interviewer had not followed protocol exactly before feedback. For the calendar series, in more than 14% of the question series in which clarification was sought, the interviewer had not maintained the series' meaning in the interview conducted before feedback. It is striking that clarification was requested in none of the question series in which the interviewer had followed the calendar series protocol exactly before feedback. After feedback sessions

TABLE 6.1

Question-Asking Results Before and After Feedback by Question Series

Interviewer Behavior	Calendar Series		Provider Probes		Both Series Combined	
	Before	After	Before	After	Before	After
Followed protocol exactly	18.6%	27.9%	43.3%	51.7%	33.4%	43.4%
Maintained meaning but did not follow protocol exactly	68.9%	65.1%	48.7%	46.4%	56.8%	52.9%
Did not maintain meaning	12.5%	7.0%	8.0%	2.0%	9.8%	3.7%
Total	100.0%	100.0%	100.0%	100.0%	100.0%	100.0%
N	1,240	759	1,832	1,428	3,072	2,187

TABLE 6.2

Interviewer Behavior Before and After Feedback, with and without Request for Clarification, by Question Series

	Calendar Series				Provider Probes			
	Clarification		No Clarification		Clarification		No Clarification	
Interviewer Behavior	Before	After	Before	After	Before	After	Before	After
Followed protocol exactly	0.0%	12.7%	34.4%	44.4%	36.4%	41.9%	50.9%	53.4%
Maintained meaning but did not follow protocol exactly	85.6%	77.6%	55.6%	52.6%	55.3%	55.9%	43.4%	44.5%
Did not maintain meaning	14.4%	9.7%	10.0%	3.0%	8.3%	2.2%	5.7%	2.1%
Total	100%	100%	100%	100%	100%	100%	100%	100%
N	548	330	581	331	734	454	970	678

Note: Whether the interviewer asked for clarification during the feedback session was recorded for 88% of the question series.

with clarification requests, 12.7% of the calendar series had question-asking behavior that followed the protocol exactly, a considerable increase from zero but still far below the level observed in sessions with no requests for clarification.

Similar results were found for behavior in the provider probes, but exact protocol was followed much more frequently than in the calendar series, and failure to maintain the series meaning was exhibited less often. For provider probes in sessions with clarification requests, 36.4% had question-asking behavior that followed the protocol exactly before feedback, increasing to 41.9% after feedback; 8.3% failed to maintain the meaning before feedback, decreasing to only 2.2% after.

We estimated multilevel multinomial logistic regressions to examine the role that asking for clarification during feedback might have played (see Table A6F.1 in Appendix 6F in the online materials). Figure 6.1 shows the predicted probability of the interviewer following protocol exactly versus the probability of not maintaining the meaning, taking into account the design effect of the clustering by interviewer. The first set of bars is before feedback and the second set after feedback. The darker bars are interviewers who asked for clarification; the lighter bars are interviewers who did not ask for clarification. (It should be noted that the interviewer interclass correlation coefficients are large due to the within-subject design.)

As expected, interviewers who did not maintain meaning in the first interview and asked for clarification during feedback were more likely to read verbatim after feedback. However, interviewers who did not ask for clarification showed the opposite effect: they were less likely to ask questions verbatim after feedback.

Figure 6.2 shows the results by question series. It is only in the provider probes that we see that interviewers who did not maintain meaning were more likely to read the questions verbatim after feedback if they asked for clarification. On the flexible calendar grid, interviewers who did not maintain the meaning were *less* likely to follow the protocol exactly, even slightly less likely when they asked for clarification. This relationship held for the predicted probability of interviewers making minor working changes after feedback (data not shown) – it was more likely only if the interviewer asked for clarification.

We added interviewer experience, both as a continuous variable (years of experience on MEPS) and as a dichotomous variable (less than one year versus one year or more) and found it had no effect on the results (see Table A6G.1 in Appendix 6G in the online materials).

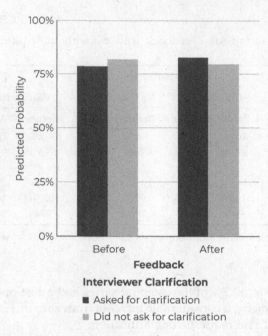

FIGURE 6.1
Predicted probability of following protocol exactly (versus not maintaining meaning) before and after feedback by whether the interviewer asked for clarification during feedback.

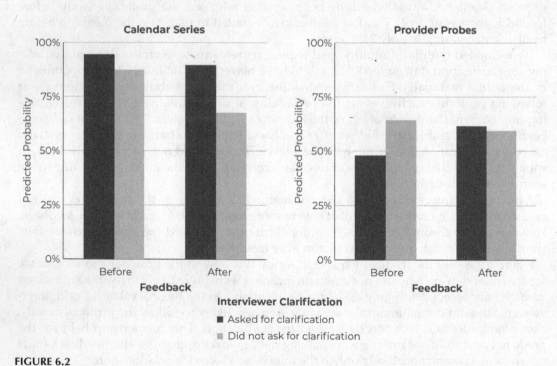

FIGURE 6.2
Predicted probability of following protocol exactly (versus not maintaining meaning) before and after feedback by whether the interviewer asked for clarification during feedback, by question series type.

6.3.2 Supervisor Alert System

Supervisor alerts, specifically geared toward identifying issues that influence data quality during the interview, were issued throughout the fall field period from a base of 7,361 interviews. The project has employed a number of operational alerts for years related to efficiency and production with known patterns and tested retraining approaches. While effective, they do not directly influence the collection of key data elements. By contrast, the newly instituted data quality alerts offer a novel, rapid approach to identifying and addressing within-interview behavior that more directly influences key study estimates. The focus specifically on data quality alerts reflects the evolution of this system from monitoring production to assessing data collection. One data-quality alert type dominated the process: record usage for medical events (see Table A6H.1 in Appendix 6H in the online materials). Almost 2,000 alerts (84.4%) were generated because answers to the calendar series indicated the respondent had no records yet reported events later in the interview. The second most common alert, the lack of records for prescribed medicines for those aged 65 and over, occurred less than 250 times (10.4% of alerts). Alerts for hospital stays that started and ended on the same day occurred in the data at less than half that rate, and the fourth type of anomaly (respondents younger than 18) was negligible.

MEPS data collection is not spread evenly over the fall months. Rather, all cases are assigned at the beginning and more interviews are completed in August and September than in October and November. Figure 6.3 shows the distribution of alerts by month, adjusted to reflect the number of interviews each month. More alerts were generated in August even when adjusted for caseload, and the trend is toward fewer alerts per month over the course of the fall. Figure 6.4 shows the distribution of alert volume by interviewer and by alert type. All interviewers were associated with at least one alert. For a given alert, many interviewers had only one, and many others only two. This pattern holds for most alerts, including no prescribed medicine records for the 65+ and hospital stays that started and ended on the same day.

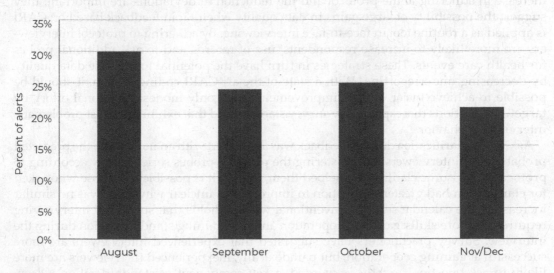

FIGURE 6.3
Caseload-adjusted alert distribution over field period.

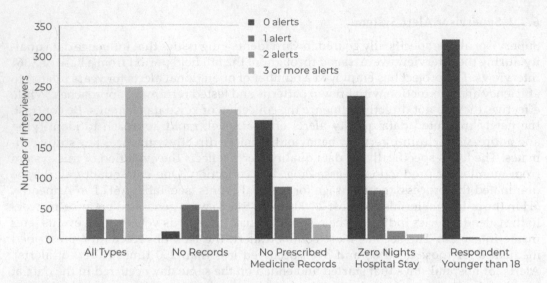

FIGURE 6.4
Alert occurrence among interviews by alert type.

6.4 Discussion

6.4.1 Behavior Coding and Rapid Feedback

With rapid verbal and written feedback of the behavior coding results from CARI, interviewer question-asking behavior improved. The behavior changes are consistent with the rapid feedback results for general interview quality reported by Edwards, Maitland, and Conner (2017). Following the protocol exactly for two series of questions increased from 33.4% before feedback to 43.4% afterward. From a total survey error perspective, the increase in adhering to the protocol and the reduction in deviations are important: they suggest the possibility of large gains in data quality when rapid feedback based on CARI is applied as a routine tool in face-to-face interviewing. By adhering to protocol interviewers are more likely to increase respondents' use of records and to ask additional probes for health care events. These strategies in turn have the potential to increase data quality by decreasing underreporting. With a state-of-the-art CARI coding system, it should be possible to achieve major quality improvements with only modest additional effort, by targeting questions that capture key survey statistics and that exhibit the most problematic interviewer behavior.

Asking for clarification during feedback was associated with an increase in the predicted probability of interviewers administering the provider probes series strictly according to protocol. We assume clarification helped them, though it is possible that those who asked for clarification had greater motivation to improve. It is unclear why there was no similar increase in the calendar series. Conventional wisdom holds that successful interviewing requires two core skills: gaining cooperation and maintaining standardization during the interview. Survey practitioners have suggested that experienced interviewers are more successful at gaining cooperation, but paradoxically, inexperienced interviewers are more likely to ask questions exactly as worded, a key component of standardization (Olson, Kirchner, and Smyth 2016). New interviewers may have learned strict question-asking behavior in their initial training, but with experience they learn to cut corners to shorten interviews or maintain rapport. However, interviewers' level of experience was not related to the behavior we observed in these two question series.

It is not surprising that failure to maintain question meaning was greater in the calendar series than in the provider probes. As noted earlier, the calendar series format is more difficult to navigate and requires more cognitive skill to follow correct protocol. We also speculate that some interviewers may perceive the calendar series as less critical to the survey's key statistics. If that is the case, some interviewers may judge that correct protocol can be abandoned for questions like those in this series in favor of speeding up the interview or maintaining rapport with the respondent (Hubbard, Edwards, and Maitland 2016).

6.4.2 Supervisor Alert System

Alerts can be considered another form of rapid feedback to interviewers, but one that draws from data as well as paradata, at least for MEPS. The overwhelmingly dominant alert type – no use of records for medical events – was triggered so often that it became obvious early in the field period that it was not well-suited to the purpose of flagging incorrect interviewer behavior. Many respondents do not keep records of medical events, and thus the respondent's behavior is not fully under the interviewer's control.

The second most frequent alert – no use of records for prescribed medicine use by the elderly – was perhaps more appropriate because most respondents do have such records; for example, many medications for the elderly are for chronic conditions and are therefore current and available when the MEPS interview occurs. Arguably though, this alert and the one for all medical records might be more effective at the interviewer level rather than the *interview* level, flagging a pattern of no use of records across a number of interviews.

Despite these caveats, the alert system and the corresponding retraining program appear to have been effective in reducing the frequency of these quality-related data anomalies over the course of the field period as evidenced by the low recurrence rate. The other two data alerts were triggered infrequently, and the underlying problems they attempted to address could be classified better as issues in the instrument design rather than in interviewer behavior. Instrument changes in the following year eliminated them.

6.5 Conclusion

With its mix of public and private insurance, its fragmentation of medical providers and problems with coordination of care, and its large health care disparities, the U.S. health care system is very complex. A probability-based face-to-face survey provides the highest quality and richest set of metrics needed for investigating and understanding health care utilization and costs at the person and family levels and for modeling the effects of changes in policy, but such a survey requires skilled interviewers with the capability to collect complex data in the home.

In an earlier section of this chapter, we used prescribed medicine utilization and expenditures to illustrate key MEPS statistics, and discussed the effect of high utilizers. Interviewer effects on prescribed medicine data quality were targeted with rapid feedback of CARIcode data from the calendar series and the provider probes, and with overnight alerts to supervisors about interviews with elderly that reported no prescribed medicines obtained during the reference period. We have shown that these techniques can improve interviewers' adherence to correct protocol. One of the biggest problems in surveys of health care utilization is respondent error, particularly the underreporting of events. This error can be compounded by interviewer error. On MEPS, other such targeted efforts in interviewer classroom training have increased event reporting, a positive effect that has persisted for years. Rapid feedback from CARIcode and alerts about data anomalies is another form of interviewer training that can pinpoint problematic behavior when memories are fresh and there is incontrovertible evidence. The combination of interviewer and respondent error may be a large component of total survey error in the MEPS prescribed medicine estimates. If the measures undertaken in fall 2018 succeed in reducing the interviewer's contribution to total survey error, the estimates for total prescribed medicine expenditures for the elderly might increase by billions of dollars.

References

Edwards, B., A. Maitland, and S. Connor. 2017. Measurement error in survey operations management: Detection, quantification, visualization, and reduction. In *Total survey error in practice*, ed. P. P. Biemer, E. de Leeuw, S. Eckman, B. Edwards, F. Kreuter, L. E. Lyberg, N. C. Tucker, and B. T. West, 255–277. Hoboken, NJ: John Wiley & Sons.

Edwards, B., S. Schneider, and P. D. Brick. 2008. Visual elements of questionnaire design: Experiments with a CATI establishment survey. In *Advances in telephone survey methodology*, ed. J. Lepkowski, N. C. Tucker, J. M. Brick, E. D. de Leeuw, L. Japec, P. J. Lavrakas, M. W. Link, and R. L. Sangster, 276–296. New York: John Wiley & Sons.

Fowler Jr., F. J., and T. W. Mangione. 1990. *Standardized survey interviewing: Minimizing interviewer-related error*. Vol. 18. Newbury Park, CA: Sage.

Groves, R. M. 1987. Research on survey data quality. *Public opinion quarterly* 51:S156–S172.

Groves, R. M. 2011. Three eras of survey research. *Public opinion quarterly* 75(5):861–871.

Groves, R. M., and L. Lyberg. 2010. Total survey error: Past, present, and future. *Public opinion quarterly* 74(5):849–879.

Hubbard, R., B. Edwards, and A. Maitland. 2016. The use of CARI and feedback to improve field interviewer performance. *Paper Presented at the Annual Meeting of the American Association for Public Opinion Research, Austin, TX.*

Kashihara, D. and D. Wobus. 2006. Accuracy of household-reported expenditure data in the Medical Expenditure Panel Survey. *Proceedings of the American Statistical Association*, 3193–3200. Alexandria, VA: American Statistical Association.

Knowles, M. 1980. *The modern practice of adult education: From pedagogy to androgogy*. Wilton, CT: Association Press.

Knowles, M., E. F. Holton III, and R. A. Swanson. 2005. *The adult learner: The definitive classic in adult education and human resource development* (6th ed.). Burlington, NJ: Elsevier.

Kolb, D. A., R. E. Boyatzis, and C. Mainemelis. 2001. Experiential learning theory: Previous research and new directions. *Perspectives on thinking, learning, and cognitive styles* 1(8):227–247.

Medical Expenditure Panel Survey, Agency for Healthcare Research and Quality. Accessed 6/8/19 at https://meps.ahrq.gov/mepsweb/.

Mohadjer, L., and B. Edwards. 2018. Paradata and dashboards in PIAAC. *Quality assurance in education* 26(2):263–277.

Olson, K., A. Kirchner, and J. Smyth. 2016. Do interviewers with high cooperation rates behave differently? Interviewer cooperation rates and interview behaviors. *Survey practice* 9(2):1–11.

Schaeffer, N. C. 1991. Conversation with a purpose—or conversation? Interaction in the standardized interview. In *Measurement errors in surveys*, ed. P. P. Biemer, R. M. Groves, L. E. Lyberg, N. A. Mathiowetz, and S. Sudman, 367–391. New York: John Wiley & Sons, Inc.

Silber, H. 2019. Interviewer training programs of multinational survey programs mapped to the total survey error. *Paper* Presented *at the* Annual conference *of the World Association for Public Opinion Research*, Toronto, Ontario.

Sperry, S., B. Edwards, R. Dulaney, and D. E. B. Potter. 1998. Evaluating interviewer use of CAPI navigation features. In *Computer assisted survey information collection*, ed. M. Couper, R. Baker, J. Bethlehem, C. Clark, J. Martin, W. L. Nicholls II, and J. M. O'Reilly, 351–365. Hoboken, NJ: Wiley.

Tourangeau, R., F. Kreuter, and S. Eckman. 2012. Motivated underreporting in screening interviews. *Public opinion quarterly* 76(3):453–469.

West, B., F. G. Conrad, F. Kreuter, and F. Mittereder. 2018. Nonresponse and measurement error variance among interviewers in standardized and conversational interviewing. *Journal of survey statistics and methodology* 6(3):335–359.

7

Statistical Identification of Fraudulent Interviews in Surveys: Improving Interviewer Controls

Silvia Schwanhäuser, Joseph W. Sakshaug, Yuliya Kosyakova, and Frauke Kreuter

CONTENTS

7.1 Introduction

Survey data are important for establishing new insights in many disciplines such as sociology, economics, and others. However, various sources of error, summarized in the Total Survey Error (TSE) framework, can undermine survey data, with interviewers being one important source (Groves 2005). Interviewers can deviate from their instructions and, in the most blatant case, falsify or manipulate the data. Although empirical evidence suggests that interviewer falsification is a rare event (Blasius and Friedrichs 2012), even small amounts of undetected falsification can lead to substantial bias in multivariate analyses (Schräpler and Wagner 2005). Therefore, falsification detection strategies such as random re-contacting procedures are crucially important for optimizing data quality. Statistical identification methods, which identify suspicious patterns in the data to reveal potential falsifiers, are used less often. Yet a number of such methods have been developed for detecting a wide variety of falsification types, such as duplicates and complete or partial falsification, in a cost-effective way. This chapter contributes to the literature by providing a broad overview of statistical methods for identifying interviewer falsification and demonstrating promising statistical identification strategies using data from a large-scale refugee survey in Germany that includes confirmed falsifications.

7.2 Interviewer Falsification – An Overview

7.2.1 Forms of Falsification

The American Association for Public Opinion Research defines *interviewer falsification* as "the intentional departure from the designed interviewer guidelines or instructions, unreported by the interviewer which could result in the contamination of data" (AAPOR 2003: 1). There are multiple forms of falsification. The most blatant form, complete falsification, occurs when no interview is conducted and instead the interviewer provides fictitious data (Schreiner, Pennie, and Newbrough 1988). An attenuated form is the partial falsification of interviews where interviewers conduct "short interviews," meaning that some parts of the questionnaire contain real data provided by the respondent while other parts contain fictitious data. Particularly long or difficult parts of the questionnaire are more prone to this type of falsification (Biemer and Stokes 1989). In addition to providing fictitious data, other falsification forms include interviewers deviating from respondent selection rules and interviewing the wrong person (AAPOR 2003; Schreiner, Pennie, and Newbrough. 1988), misclassifying persons or addresses as ineligible cases, deviating from the intended interview mode (Biemer and Stokes 1989; Schreiner, Pennie, and Newbrough), or incorrectly entering answers to manipulate questions that trigger skip patterns and thus shorten the interview (AAPOR 2003; Kosyakova, Shopek, and Eckman 2015; Schnell 2012; Tourangeau, Kreuter, and Eckman 2012). In addition, duplicate response patterns that are highly unlikely to be attributed to respondents represent a special form of falsification because they can be caused by either the interviewer or other survey staff (Sarracino and Mikucka 2016; Slomczynski, Powalko, and Krauze 2017).

7.2.2 Frequency of Falsification

Overall, the proportion of falsified interviews is low and falsifiers usually represent a minority of interviewers. For example, the percentage of detected falsifications (any of the above-mentioned types) in the US Current Population Survey was 0.4% per month (Schreiner, Pennie, and Newbrough, 1988); in the New York City Housing

Vacancy Survey the share was comparatively high at 6.5% (Schreiner, Pennie, and Newbrough, 1988); and in the German Socio-Economic Panel it was between 0.1% and 2.1% (Schräpler and Wagner 2005). Although studies with high rates of interview falsification appear occasionally (Hyman et al. 1954; Turner et al. 2002), rates between 3% and 5% can be regarded as realistic (Biemer and Stokes 1989). Higher rates have been found in studies employing only a small interviewer staff (Bredl, Storfinger, and Menold 2013). In panel surveys, falsification rates are often lower and partial rather than complete falsification is more likely to occur (Blasius and Friedrichs 2012; Schräpler and Wagner 2005).

7.2.3 Reasons for Falsification

The motivation to falsify data primarily arises from conditions or situations that discourage interviewers from fulfilling their roles adequately. A difficult questionnaire, administrative factors related to the interviewer's employment, or other external factors like the survey location are typical reasons for interviewers becoming discouraged and thus increasing the chance they falsify data (Biemer and Stokes 1989; Crespi 1945; Winker 2016). These interview conditions can be divided into intrinsic and extrinsic factors (Gwartney 2013). Intrinsic factors are partly under the control of the researcher and include the sampling design (e.g. difficult selection rules), the survey instrument (e.g. poorly designed questionnaires, programming errors), the survey institute (e.g. high workloads, poorly communicated standards), or the respondent (e.g. difficult interviewees) (Gwartney 2013). Extrinsic factors include the interview location, the interviewer's personal situation, how interviewers are paid, or whether they know about quality control and monitoring procedures (Gwartney 2013; Koczela et al. 2015). Many factors can be taken into account during the process of designing the survey to minimize interviewer burden and the overall falsification likelihood (Crespi 1945; Biemer and Stokes 1989; Blasius and Friedrichs 2012). Nevertheless, the remaining risk of falsifications calls for quality control procedures.

7.2.4 Effects of Falsification on Data Quality

Since falsifications are systematic deviations (Gwartney 2013), even small proportions of falsified data may introduce bias, resulting in misleading inferences (Schnell 1991, 2012; Schräpler and Wagner 2005). Descriptive statistics, such as means and variances, may be only slightly distorted: the bias in a mean cannot exceed the share of falsified records. Yet larger distortions are evident for multivariate statistics since falsifiers are unlikely to reproduce complex multidimensional relationships between variables (Reuband 1990; Schnell 1991; Schräpler and Wagner 2005).

7.3 Non-Statistical Identification Strategies

In practice, non-statistical strategies for detecting falsifications are usually part of standard quality control methods. Some approaches are used during the field period, while others are used after all interviews have been conducted.

7.3.1 Re-contact

The most frequently used method for identifying falsifications is the re-contact method, also referred to as the re-interview method or verification method (AAPOR 2003; Biemer and Stokes 1989; Crespi 1945; Hauck 1969; Schnell 2012). During the verification process, some or, in rare cases, all respondents are re-contacted after the interview to verify whether the interview actually took place. AAPOR recommends asking about (1) the household composition, (2) interview mode, (3) interview duration, (4) incentives, (5) use of computers during the survey, and (6) core questionnaire topics or item batteries (AAPOR 2003). The re-interview data may be collected using postal mail, telephone, or personal re-interviews (AAPOR 2003; Koch 1995). The postal mode may be the most cost-effective one, but it can be error-prone, due to low response rates (Harrison and Krauss 2002; Menold and Kemper 2014) and respondent recall difficulties, leading to potentially unjust suspicion against interviewers (AAPOR 2003; Hauck 1969; Koch 1995). Most valid information is obtained through personal re-interviews, but they are time-consuming and costly. Telephone re-interviews are fast and cost-efficient, but with the drawback of low response rates or missing telephone numbers (AAPOR 2003; Gwartney 2013). For cost reasons, often only a subset of interviews (5–15%, rarely more than 25%) are verified (AAPOR 2003; Koch 1995; Schnell 2012). In the past, interviews were selected at random, which – due to the small number of verified interviews and few falsifications – made the re-interview procedure very inefficient. So-called "focused re-interviews" oversample cases that are likely to be falsifications based on a falsification probability model (Biemer and Stokes 1989; Li et al. 2009), which improves the efficiency of the re-contact method (Bredl, Storfinger, and Menold 2013). However, there is no consensus on which model inputs to use; it often depends on the specific survey and available information.

7.3.2 Monitoring

Interview monitoring, mainly applied in telephone surveys (AAPOR 2003; Gwartney 2013), is another commonly used procedure for identifying falsifiers, which also serves as a deterrent to interviewers. During the so-called "silent monitoring," supervisors listen to the interview and observe the interviewer's screen in real-time without the interviewer's knowledge. Interviewers can either be selected for monitoring at random or in a focused way, depending on characteristics like the number of interviews, interview duration, response or cooperation rates, and the number of contact attempts (AAPOR 2003; Gwartney 2013). For face-to-face interviews, monitoring takes place via audio recording of the complete interview or randomly generated short recordings via built-in microphones or cameras (Koczela et al. 2015). Modern techniques also allow GPS monitoring to verify an interviewer's visit at the correct location (Finn and Ranchhod 2017; Murphy et al. 2004; Winker 2016).

7.3.3 Validation with Administrative Data

In addition, respondent data – such as age and gender – can be validated using administrative data, e.g. from a population register, to identify falsifications and interviews with the wrong person (Koch 1995; Schnell 2012). However, validation is only possible if the administrative data can be linked to the fielded sample, such as in a register-based survey. The method also assumes that information provided by the respondent or administrative data are correct and that differences are not due to measurement errors.

7.4 Statistical Identification Strategies

As described above, non-statistical methods have a number of drawbacks and are often prone to errors. To overcome some of these problems, researchers have increasingly turned to statistical identification approaches, which try to distinguish between falsified and real interviews solely on the basis of the data. These methods have several strengths: First, they can be combined with non-statistical approaches to better optimize the selection of suspicious cases for re-interviews or monitoring. Second, such approaches are comparatively inexpensive, making the falsification identification process more cost-efficient. Third, the increasing use of computer-assisted survey instruments provides additional inputs that can allow for real-time statistical controls after each interview. In addition, these approaches can uncover various forms of interview falsification, such as difficult-to-identify partial falsifications, and thus compensate for weaknesses of traditional identification approaches.

7.4.1 Identification of Complete Falsifications

7.4.1.1 Falsification Indicators

Several approaches deal with the identification of complete falsifications, following the shared idea to identify systematic differences between real and falsified data, measured by so-called "falsification indicators." When falsifying data, interviewers typically behave systematically different from how real respondents behave when answering questions, resulting in different patterns in the data. This is caused by the aims of falsifiers: to maximize their monetary benefit and reduce their time expenditure and effort without being detected (Menold et al. 2013). For example, research indicates that falsifiers demonstrate more straight-lining, but fewer illogical response patterns compared to actual respondents (Menold et al. 2013; Murphy et al. 2004; Porras and English 2004; Winker et al. 2015). A falsification indicator is defined as a measure of such patterns that could suggest falsification behavior rather than respondent answering behaviors. Falsification indicators are very similar to respondent-related quality indicators such as measures of non-differentiation or straight-lining except that they are aggregated to the interviewer-level, creating an overall quality measure across the interviewer's workload, so that suspicious (or "at risk") interviewers (rather than respondents) are flagged (Hood and Bushery 1997).

Previous "ex-post"* studies have demonstrated the effectiveness of various falsification indicators (Bredl, Winker, and Kotschau 2012; Haas and Winker 2016; Kemper and Menold 2014; Li et al. 2009; Menold et al. 2013; Menold and Kemper 2014; Murphy et al. 2004; Porras and English 2004; Schräpler and Wagner 2005). In addition, "ex-ante"† studies have demonstrated the effectiveness of these indicators in various surveys (Bushery et al. 1999; Hood and Bushery 1997; Koch 1995; Stokes and Jones 1989; Turner et al. 2002). These studies utilize different data sources, for example, indicators from *paradata* make use of automatically generated information (i.e. time stamps), whereas the actual *response data* can be used to generate *formal indicators* or *content-related indicators* (Menold et al. 2013). Formal indicators measure suspicious answering behaviors, such as the straight-lining behavior mentioned

---

* "Ex-post" studies check the effectiveness of different indicators using data with known falsifications. This makes it possible to examine the discriminatory power of indicators (Bredl, Storfinger, and Menold 2013).
† "Ex-ante" studies apply indicator-based identification approaches during the survey or field control to identify possible falsifiers (Bredl, Storfinger, and Menold 2013).

above, while content-related indicators are based on the answer distributions, such as notable overestimation of specific answers. This can be extended to *longitudinal data indicators*, which utilize panel data in longitudinal surveys (i.e. comparison of answers between waves). Detailed examples of the different indicators are provided below.

Formal indicators from response data. Previous research found that falsifiers show a lower variance between answers of their fictitious respondents (Schäfer, Schrapler, and Muller 2004), because they fail to produce the full diversity of answers that would be observed for real respondents. This is caused by specific response behaviors: First, falsifiers often choose the middle category but rarely choose extreme values in ordinal response scales to avoid inconsistencies and noticeable extreme values (Porras and English 2004; Storfinger and Winker 2013), which can be measured by the frequency of choosing the middle category or extreme values within interviews (Bredl, Winker, and Kotschau 2012; Kemper and Menold 2014; Kosyakova et al. 2019; Menold et al. 2013; Menold and Kemper 2014). Second, falsifiers tend to answer all questions instead of selecting missing categories such as "don't know," resulting in unusually low item non-response rates, which can be used as a falsification indicator (Bredl, Winker, and Kotschau 2012; Kemper and Menold 2014; Kosyakova et al. 2019; Menold et al. 2013; Menold and Kemper 2014). Third, interviewers tend to use stereotypes (Reuband 1990; Schnell 2012), i.e. preconceived notions about subgroups. By using "typical" responses, which, for example, could be provided by a well-integrated refugee, falsifiers produce more similar answers (Menold et al. 2013) within this group. This can be determined using a variability test, i.e. measuring variance as an indicator between all or some items. The variance between all items of a single questionnaire is aggregated on the interviewer-level and compared with the expected variance resulting from the entire data set (Porras and English 2004; Schäfer, Schrapler, and Muller 2004). Similarly, the standard deviation of responses in item batteries is often used to indicate the lack of differentiation or diversity of answers ("non-differentiation") (Kosyakova et al. 2019; Menold et al. 2013; Winker et al. 2015). Studies on respondent answering behavior have found that, in order to reduce cognitive effort, respondents may give answers in agreement to questions or statements of opinion (acquiescence) (Messick 1966), which could result in inconsistent answers (i.e. divergent answers concerning distinct parts of one topic). Falsifiers tend to provide acquiescent responses less frequently (Kemper and Menold 2014; Kosyakova et al. 2019; Menold et al. 2013; Menold and Kemper 2014; Winker et al. 2015). Due to the different presentations of response options, primacy (visual presentation) or recency effects (aural presentation) may arise (Krosnick and Alwin 1987). Since actual respondents hear the response options read aloud by the interviewer, and falsifiers read the questions themselves, falsified data are less likely to exhibit recency effects and more likely exhibit primacy effects (Kosyakova et al. 2019; Menold et al. 2013; Menold and Kemper 2014; Winker et al. 2015). Furthermore, actual respondents tend to round answers in open numeric questions like income (Blair and Burton 1987; Hanisch 2005). Since rounded numbers – in the opinion of falsifiers – may raise suspicion, falsifiers avoid rounding, resulting in a lower share of rounded numbers (Kosyakova et al. 2019; Menold et al. 2013; Menold and Kemper 2014). As mentioned above, falsifiers try to save time and effort while at the same time remaining undetected. To achieve these aims, answers to filter questions can be easily manipulated by choosing an answer that avoids follow-up questions (Hood and Bushery 1997), measured by the frequency of answers that avoid follow-up questions or triggering-rates (Hood and Bushery 1997; Kosyakova, Shopek, and Eckman 2015; Kosyakova et al. 2019; Menold et al. 2013). Finally, falsifiers are also less likely to select the option "Other, specify" in semi-open questions. Closed answer options provide an easy and legitimate response. Therefore, the frequency of open-ended or semi-open-ended answers can be used as an indicator of interviewer falsification (Bredl, Winker, and Kotschau 2012; Kosyakova et al. 2019; Menold et al. 2013; Menold and Kemper 2014; Winker et al. 2015).

Content-related indicators from response data. A frequently used approach that is well suited for the analysis of monetary variables (e.g. income) is "Benford's Law" (Benford 1938). It is based on the observation that for naturally occurring numbers (also for some numerical data in surveys) the initial digit follows a logarithmic probability distribution, the so-called Benford distribution (Schäfer, Schrapler, and Muller 2004). The fit to this distribution, or partially modified forms, can be used as a content-related indicator (Bredl, Winker, and Kotschau 2012; Porras and English 2004; Schäfer et al. 2004; Schräpler and Wagner 2005; Swanson et al. 2003). For further information on the assumptions of Benford's Law, see Goodman (2016). Another content-related challenge for falsifiers is correctly estimating the frequency of rare or sensitive attributes; falsifiers often lack information about the real distribution of these attributes in the population. Hood and Bushery (1997), for example, compared the percentage of buildings classified as vacant or occupied, persons classified as non-ethnic or ethnic minorities, and households classified as single or no-children households by each interviewer to known distributions from Census data to detect falsifiers. Additional studies have shown that falsifiers strongly overestimate the frequency of sexual behavior and number of lifetime sex partners (Turner et al. 2002) and underestimate drug use frequency (Murphy et al. 2004) and political participation (Menold et al. 2013). A further indicator is the rate of implausible or unusual response combinations (Murphy et al. 2004; Porras and English 2004). A special-case content-related indicator is the use of control questions. Here, respondents are provided with both real and fictitious response categories; to minimize burden, falsifiers are expected to randomly choose answers and thus their set of cases is expected to have a higher rate of selection of the fictitious categories than non-falsifying interviewers' sets of cases in which respondents would detect and avoid fictitious categories. Examples include questions about print media reading habits that include both actual names of print media (e.g. *New York Times*) and fictitious names (e.g. The Floppy Disk Gazette) (Menold et al. 2013; Menold and Kemper 2014) and the "Vocabulary and Overclaiming Test" (VOC-T), which includes real and fictitious vocabulary words (Ziegler, Kemper, and Rammstedt 2013). Although it may be the case that respondents select fictitious terms on their own, the rate of this behavior is expected to be higher in the caseload of falsifiers than in the caseload of non-falsifiers (Menold et al. 2013; Menold and Kemper 2014; Winker et al. 2015).

Indicators from longitudinal data. Longitudinal data permit the calculation of correlations between repeatedly collected items across waves. Low correlations and unusual changes or fluctuations away from an otherwise stable base value can indicate falsification because of the falsifier's inability to estimate previously given answers. Such stable items include life satisfaction scales (Schräpler and Wagner 2005).

Indicators from paradata. Paradata such as automatically recorded interview durations (calculated from time stamps) or interview dates are useful falsification indicators since the interviewer cannot easily manipulate them. Interviews or questions with short or long durations are considered suspicious, as are an unusually large number of interviews completed within a relatively short period of time (Bushery et al. 1999; Hood and Bushery 1997; Kosyakova et al. 2019; Li et al. 2009; Murphy et al. 2004). In addition, cooperation and contact rates, rates of eligible and non-eligible sample households, as well as rates of telephone number or e-mail address provision can serve as falsification indicators (Hood and Bushery 1997; Kosyakova et al. 2019; Stokes and Jones 1989; Turner et al. 2002).

7.4.1.2 Multivariate Analysis of Falsification Indicators

Although single indicators can be analyzed, stronger evidence of falsifying behavior is provided by analyzing several indicators jointly, for example, analyzing multiple indicators to identify a group of suspicious interviewers via cluster analyses (Bredl, Winker, and Kotschau 2012; Menold et al. 2013; Storfinger and Winker 2013). This idea was first introduced by Bredl et al. (2012) and applied successfully by others (Haas and Winker 2014; Haas and Winker 2016; Kosyakova et al. 2019; Menold et al. 2013; Storfinger and Winker 2013). Cluster analysis divides a larger group (in this case all interviewers) into smaller homogeneous subgroups or clusters (in this case, suspicious and non-suspicious interviewers). The subdivision is performed using grouping characteristics (here, falsification indicators) to discriminate between interviewer types.

7.4.2 Identification of Partial Falsifications

Another method for identifying partial falsification is by identifying similar response patterns across ordinal-scaled item batteries (Blasius and Thiessen 2012; Blasius and Thiessen 2013; Blasius and Thiessen 2015). Principal Component Analysis (PCA) as well as Categorical Principal Component Analysis (CatPCA) can be applied to identify similar response patterns based on the factor scores of all interviews. Suspicious interviewers are thought to produce identical response patterns particularly frequently (Blasius and Thiessen 2013). Again, one inference is that similar response patterns result from a lower variance in falsified data (Blasius and Thiessen 2013; Blasius and Thiessen 2015). An assumption that is required here but seldom discussed is the identical and independent distribution of answers across items. Often item batteries cover related topics, so that answers depend on each other and specific patterns of answers occur more frequently in the population absent any falsification. Nevertheless, if an interviewer is responsible for an unusually large number of similar response patterns, this may indicate a potential risk of falsification.

7.4.3 Identification of Duplicate Records

Duplicate records can heavily influence survey data (Koczela et al. 2015; Kuriakose and Robbins 2016), for example, by biasing point estimates (e.g. regression coefficients) upward (or downward) and affecting variance estimates leading to erroneous tests of significance. Published accounts of high incidence of duplicates in survey data sets (Kuriakose and Robbins 2016; Slomczynski, Powalko, and Krauze 2017) have resulted in recommendations for the use of duplicate analyses. A simple form of duplicate analysis that checks a data set for completely identical data rows (including missing values) is implemented in standard statistical software (e.g. Stata®, SAS®, SPSS®, among others) (Koczela et al. 2015; Kuriakose and Robbins 2016; Slomczynski, Powalko, and Krauze 2017). So-called "near duplicates" can be particularly problematic since even a change in a single value is sufficient to produce an undetectable duplicate (Kuriakose and Robbins 2016). For this reason, "high-matching" methods were developed to identify data with an unusually high correspondence (between 85% and 99%) of response values (Koczela et al. 2015; Kuriakose and Robbins 2016). However, there is a strong risk that real data will be identified as near duplicates by coincidence, producing falsely suspected cases. High-matching methods are highly sensitive to various characteristics of a survey (e.g. number of questions, number of respondents, homogeneous subgroups), and thus are not generally applicable to every data set (Simmons et al. 2016). Whether "high-matching" methods should be used requires careful consideration of these factors.

7.5 Data and Sample

To demonstrate the methods mentioned above, we use data from a German household panel study, the "IAB-BAMF-SOEP Survey of Refugees in Germany," launched in 2016 (Brücker et al. 2017). The target population was drawn from the German Central Register of Foreigners (AZR) and included asylum-seekers and refugees who arrived in Germany between January 2013 and January 2016 and adult household members (Kroh et al. 2017). The first wave yielded 4,816 respondents in 3,554 households – which corresponds to a household-level response rate of 48.84% (Kroh et al. 2017) – interviewed by 98 interviewers using computer-assisted personal interviewing (CAPI). Brief interviews were conducted with every head of the household (household interviews). More detailed interviews (person interviews) were conducted with adults in those households, including asylum-seekers and refugees, originating from various home countries, i.e. Syria (40%), Afghanistan (10%), Iraq (8%), West Balkans (7%), and Eritrea/Somalia (7%) (Kroh et al., 2017). Questionnaires were provided in seven languages (Arabic, English, Farsi/Dari, German, Kurmanji, Pashtu, and Urdu) complemented through audio files containing recordings of the questions in different languages and an interpreter hotline. During the re-contact quality checks, a first case of interviewer falsification (complete falsifications) by one interviewer (hereafter referred to as "F1") was detected (IAB, 2017). F1 accounted for 6% of all person (n = 289) and household interviews (n = 217). Analysis of these data using statistical methods not only confirmed the first falsification case but also identified two further deviating interviewers (Kosyakova et al. 2019). The second interviewer ("F2") had conducted 46 person-level interviews, and the third interviewer ("F3") had conducted 16 person-level interviews. Data collected by these interviewers were excluded from the official data release (v34). We therefore use this high-profile case to demonstrate the effectiveness of different statistical identification tools.

7.6 Findings

7.6.1 Identification of Complete Falsifications

To illustrate the statistical identification tools discussed above, we consider the following indicators: acquiescent response style, extreme response style, interview duration, middle response style, recency effects, semi-open responses, and stereotyping. All indicator values were standardized such that high positive values indicate suspicious patterns. A list of these indicators, the operationalization, and their mean and standard deviation can be found in Table 7.1. These specific indicators represent a range of possible indicators; e.g. interviewer duration is based on paradata, acquiescent response style on agreement questions, stereotyping on attitudinal items, and recency effects on unordered answer option lists.

TABLE 7.1

List of Used Falsification Indicators on the Interviewer-Level

Indicator	Description	Assumed Direction of Falsifiers	Mean	SD
Acquiescent response style	Share of positive answers ("Agree/Strongly Agree"), independent of question content	Lower share of positive answers, independent of question content for falsifiers	−0.16	1.12
Extreme responses	Share of extreme responses to rating scales	Lower share of extreme responses to rating scales for falsifiers	0.07	0.81
Interview duration	Duration of completed interviews in minutes	Shorter duration of completed interviews for falsifiers	−0.24	1.00
Middle category responses	Share of the middle response in rating scales	Higher share of middle responses on rating scales for falsifiers	0.10	0.74
Recency effects	Share of choosing the last two categories in non-ordered answer option lists	Lower share of choosing the last two categories in non-ordered answer option lists for falsifiers	−0.02	1.02
Semi-open responses	Share of responses to "other" in semi-open-ended question	Lower share of responses to "other" in semi-open-ended question for falsifiers	−0.10	1.48
Stereotyping	Strength of stereotypical response to attitudinal items measured on the basis of Cronbach's alpha	Higher strength of stereotypical response to attitudinal items for falsifiers	0.24	1.02

Source: IAB-BAMF-SOEP Survey of Refugees in Germany, 2016, own calculations.

We illustrate combining the indicators via cluster analysis, using a simple agglomerative hierarchical cluster algorithm (Single-Linkage) with a Euclidean distance measure. The Single-Linkage algorithm is particularly useful in this context, since similar observations are fused first (Kaufman and Rousseeuw 2009) and therefore the most outlying interviewers are separated from the unsuspicious ones. The optimal clustering solution was determined considering the Duda–Hart index (Duda and Hart 1973; Milligan and Cooper 1985). We excluded seven interviewers who had fewer than five interviews because for these interviewers the aggregated indicators are more likely to reflect respondent-level influences. Therefore, results are presented for 91 interviewers.

Table 7.2 includes the clusters that were identified via Single-Linkage clustering. The algorithm created one cluster containing unsuspicious interviewers (cluster 1) and five further clusters (clusters 2–6) containing seven outlying (suspicious) interviewers.

TABLE 7.2

List of Suspicious Interviewers According to Single-Linkage Clustering

Clusters	Number of Interviewers Per Cluster	Interviewer IDs (Number of Person Interviews)
Cluster 1 (unsuspicious interviewers)	84	–
Cluster 2	1	**F2 (46)**
Cluster 3	1	I58 (8)
Cluster 4	2	I73 (25), I77 (5)
Cluster 5	1	I36 (10)
Cluster 6	2	**F1 (289), F3 (16)**

Source: IAB-BAMF-SOEP Survey of Refugees in Germany, 2016, own calculations.
Note: Falsifying interviewers are highlighted in boldface type.

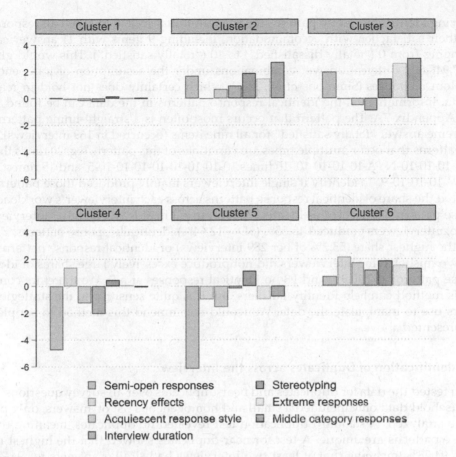

FIGURE 7.1
Mean Indicator Values per Cluster. *Source:* IAB-BAMF-SOEP Survey of Refugees in Germany, 2016, own calculations.

All three misbehaving interviewers are classified as outliers and four further interviewers are falsely suspected. Figure 7.1 shows the mean indicator values for every cluster. The largest and unsuspicious interviewer group shows no clear direction for any of the indicators. On average, all of these interviewers produced unsuspicious values. On the other hand, F2 shows suspicious indicator values for all indicators except for recency effects. F1 and F3 were grouped into one cluster, since all indicators point in the suspicious direction. I58 in cluster 3 shows some suspicious and some unsuspicious indicator values. In addition, this interviewer has a small number of interviews and quality controls did not confirm the suspicion. The other three outliers in clusters 4 and 5 do not show any systematic suspicious patterns. The only remarkable pattern is the low value on semi-open responses indicating that those interviewers, on average, had more open answers to semi-open questions. Therefore, the algorithm classifies them as outlier but not in the sense of a falsifier.

7.6.2 Identification of Partial Falsifications

Although there are no known partial falsifications included in the data, we still aim to demonstrate the previously described method of using similar (or identical) response patterns across item batteries to identify partial falsifications. We demonstrate the method for identifying identical response patterns for one attitudinal scale that asked all respondents about their satisfaction with accommodation, including 9 items with 11 answer categories ranging from 0 ("totally dissatisfied") to 10 ("totally satisfied"). This would give 11^9 = 2,357,947,691 different answer combinations under the assumption of independence, or random responses (Simmons et al. 2016), which certainly does not hold in real survey data. Information on the identical response patterns in the data can be found in the online Appendix 7A. The pattern that occurs most often is a straight-lining pattern with the extreme answer "totally satisfied" for all nine items (occurred in 169 interviews). Most other patterns that occur multiple times are variations of this pattern, e.g. 20 times the pattern 10-10-10-10-N/A-10-10-10-10, 16 times 10-10-10-10-10-10-10-10-5, and 15 times 10-10-10-10-10-10-10-10-9. To identify if single interviewers mainly produced those patterns, we calculated the share of identical response patterns across each interviewer's workload. The response pattern was labeled as identical if it occurred in at least one other interview. On average, interviewers produced 15.55% (SD = 12.64%) identical response patterns. F1 produced the highest share (53.3% of her 289 interviews) of identical response patterns. The other two misbehaving interviewers did not produce excessively large shares of identical response patterns (F2 10.9% and F3 no identical responses at all). We therefore conclude that this method can help identify falsifiers, but it is quite sensitive to the strategies that falsifiers use to manipulate the data. We would recommend this method to supplement those presented above.

7.6.3 Identification of Duplicates across the Interview

We also tested the data for duplicates and near-duplicates over all survey questions. Since the household data only included a small and homogeneous list of answers, only person data are analyzed. A standard duplication test revealed no duplicates, meaning that all response patterns are unique. A test for near-duplicates showed that the highest match score is 94.73%, meaning that at least two interviews had similar answers for 94.73% of all questions. The mean lies at 80.23%, the median at 79.90%. Taking a closer look at interviews above the 99th percentile, it is noticeable that those near-duplicates can be attributed to only a few interviewers. Ten interviewers produced 56 interviews that had very similar responses. F1 produced 25% of these interviews. To examine this in more depth, we calculated the mean near-duplicate rate for every interviewer, i.e. how many similar answers an interviewer showed on average. In particular, three interviewers show suspiciously high values: F1, F2, and a third interviewer who was not confirmed as a falsifier. Again, F3 did not show any suspicious values. We therefore conclude that near-duplicates can help to identify falsifiers, but they should still be used with caution. The data need to meet many assumptions and, in our case, this method did not identify all misbehaving interviewers.

7.7 Outlook and Discussion

Statistical identification methods have been demonstrated to be effective and are becoming increasingly popular tools for identifying falsified interviews in surveys. Through the combined use of different analysis tools, it is possible to create comprehensive measures that can identify various types of falsification and can be particularly useful for improving and supporting traditional quality control procedures. They can be automated to flag suspicious interviewers early in the field period for more extensive quality control procedures. However, so far no empirical threshold is defined on how early results become reliable, which should be addressed in future research. Although various statistical methods for detecting falsification exist, research into new and improved methods is still ongoing. Unfortunately, appropriate data for investigating falsification and testing and evaluating new and existing detection methods are often not made available to researchers. This is due to the small number of falsified interviews, but also because falsified data are usually removed from the published data upon detection. Due to the cooperativeness of our survey institute, we were able to demonstrate the usefulness of the different statistical tools for a high-profile case of interviewer falsification. By routinely making such data available to the research community, survey institutes would make a great contribution to the discussion about interviewer falsification and facilitate the development and improvement of quality control methods.

Acknowledgments

Financial support from the Charles Cannell Fund in Survey Methodology is gratefully acknowledged.

References

AAPOR. 2003. Interviewer falsification in survey research: Current best methods for prevention, detection and repair of its effects. https://www.aapor.org/AAPOR_Main/media/MainSiteFiles/falsification.pdf (accessed January 25, 2019).

Benford, F. 1938. The law of anomalous numbers. *Proceedings of the American philosophical society*, 551–72.

Biemer, P. P., and S. L. Stokes. 1989. The optimal design of quality control samples to detect interviewer cheating. *Journal of official statistics* 5(1):23–39.

Blair, E., and S. Burton. 1987. Cognitive processes used by survey respondents to answer behavioral frequency questions. *Journal of consumer research* 14(2):280–88.

Blasius, J., and J. Friedrichs. 2012. Faked interviews. In *Methods, theories, and empirical applications in the social sciences*, ed. S. Salzborn, E. Davidov, and J. Reinecke, 49–56. VS Verlag für Sozialwissenschaften.

Blasius, J., and V. Thiessen. 2012. *Assessing the quality of survey data*. London: Sage.

Blasius, J., and V. Thiessen. 2013. Detecting poorly conducted interviews. In *Interviewers' deviations in surveys*, ed. P. Winker, N. Menold, and R. Porst, 67–88. Frankfurt am Main: Internationaler Verlag der Wissenschaften.

Blasius, J., and V. Thiessen. 2015. Should we trust survey data? Assessing response simplification and data fabrication. *Social science research* 52:479–93.

Bredl, S., N. Storfinger, and N. Menold. 2013. A literature review of methods to detect fabricated survey data. In *Interviewers' deviations in surveys*, ed. P. Winker, N. Menold, and R. Porst, 3–24. Frankfurt am Main: Internationaler Verlag der Wissenschaften.

Bredl, S., P. Winker, and K. Kötschau. 2012. A statistical approach to detect interviewer falsification of survey data. *Survey methodology* 38:1–10.

Brücker, H., N. Rother, and J. Schupp. 2017. *IAB-BAMF-SOEP-Befragung von Geflüchteten 2016: Studiendesign, Feldergebnisse sowie Analysen zu schulischer wie beruflicher Qualifikation, Sprachkenntnissen sowie kognitiven Potenzialen*. Berlin: DIW Berlin, German Institute for Economic Research. http://doku.iab.de/forschungsbericht/2017/fb1317.pdf (accessed September 20, 2019).

Bushery, J. M., J. W. Reichert, K. A. Albright, and J. C. Rossiter. 1999. Using date and time stamps to detect interviewer falsification. *Proceedings of the survey research methods section*, 316–20.

Crespi, L. P. 1945. The cheater problem in polling. *Public opinion quarterly* 9(4):431–45.

Duda, R. O., and P. Hart. 1973. *Pattern classification and scene analysis*. New York: Wiley.

Finn, A., and V. Ranchhod. 2017. Genuine fakes: The prevalence and implications of data fabrication in a large South African survey. *The World Bank economic review* 31(1):129–57.

Goodman, W. 2016. The promises and pitfalls of Benford's law. *Significance* 13(3):38–41.

Groves, R. M. 2005. *Survey errors and survey costs*. New York: Wiley.

Gwartney, P. A. 2013. Mischief versus mistakes: Motivating interviewers to not deviate. In *Interviewers' deviations in surveys*, ed. P. Winker, N. Menold, and R. Porst, 195–215. Frankfurt am Main: Internationaler Verlag der Wissenschaften.

de Haas, S., and P. Winker. 2014. Identification of partial falsifications in survey data. *Statistical journal of the IAOS* 30(3):271–81.

de Haas, S., and P. Winker. 2016. Detecting fraudulent interviewers by improved clustering methods – The case of falsifications of answers to parts of a questionnaire. *Journal of official statistics* 32(3):643–60.

Hanisch, J. U. 2005. Rounded responses to income questions. *Allgemeines statistisches archiv* 89(1):39–48.

Harrison, D. E., and S. I. Krauss. 2002. Interviewer cheating: Implications for research on entrepreneurship in Africa. *Journal of developmental entrepreneurship* 7(3):319–30.

Hauck, M. 1969. Is survey postcard verification effective? *Public opinion quarterly* 33(1):117–20.

Hood, C. C., and J. M. Bushery. 1997. Getting more bang from the reinterview buck: Identifying "at risk" interviewers. *Proceedings of the survey research method section*, 820–24.

Hyman, H. H., W. J. Cobb, J. J. Feldman, C. W. Hart, and C. H. Stember. 1954. *Interviewing in social research*. Chicago, IL: Chicago Press.

Institut für Arbeitsmarkt- und Berufsforschung (IAB). 2017. Revidierter Datensatz der IAB-BAMF-SOEP-Befragung von Geflüchteten. Nuremberg. http://doku.iab.de/grauepap/2017/Revidierter_Datensatz_der_IAB-BAMF-SOEP-Befragung.pdf (accessed September 20, 2019).

Kaufman, L., and P. J. Rousseeuw. 2009. *Finding groups in data: An introduction to cluster analysis*. Hoboken, NJ: John Wiley & Sons, Inc.

Kemper, C. J., and N. Menold. 2014. Nuisance or remedy? The utility of stylistic responding as an indicator of data fabrication in surveys. *Methodology* 10:92–99.

Koch, A. 1995. Gefälschte interviews: Ergebnisse der interviewerkontrolle beim ALLBUS 1994. *ZUMA Nachrichten* 19(36):89–105.

Koczela, S., C. Furlong, J. McCarthy, and A. Mushtaq. 2015. Curbstoning and beyond: Confronting data fabrication in survey research. *Statistical journal of the IAOS* 31(3):413–22.

Kosyakova, Y., L. Olbrich, J. Sakshaug, and S. Schwanhäuser. 2019. Identification of interviewer falsification in the IAB-BAMF-SOEP Survey of Refugees in Germany. http://doku.iab.de/fdz/reporte/2019/MR_02-19_EN.pdf. (accessed April 10, 2019).

Kosyakova, Y., J. Skopek, and S. Eckman. 2015. Do interviewers manipulate responses to filter questions? Evidence from a multilevel approach. *International journal of public opinion research* 27(3):417–31.

Kroh, M., Kuhne, S., Jacobsen, J., Siegert, M., and Siegers, R. 2017. Sampling, nonresponse, and integrated weighting of the 2016 IAB-BAMF-SOEP survey of refugees (M3/M4) – revised version. SOEP Survey Papers 477: Series C. Berlin: DIW/SOEP.

Krosnick, J. A., and D. F. Alwin. 1987. An evaluation of a cognitive theory of response-order effects in survey measurement. *Public opinion quarterly* 51(2):201–19.

Kuriakose, Noble, and Michael Robbins. 2016. Don't get duped: Fraud through duplication in public opinion surveys. *Statistical journal of the IAOS* 32(3):283–91.

Li, J., J. M. Brick, B. Tran, and P. Singer. 2009. Using statistical models for sample design of a reinterview program. *Proceedings of the survey research method section*, 4681–95.

Menold, N., and C. J. Kemper. 2014. How do real and falsified data differ? Psychology of survey response as a source of falsification indicators in face-to-face surveys. *International journal of public opinion research* 26(1):41–65.

Menold, N., P. Winker, N. Storfinger, and C. J. Kemper. 2013. A method for ex-post identification of falsification in survey data. In *Interviewers' deviations in surveys*, ed. P. Winker, N. Menold, and R. Porst, 25–47. Frankfurt am Main: Internationaler Verlag der Wissenschaften.

Messick, S. 1966. The psychology of acquiescence: An interpretation of research evidence. *ETS research bulletin series* (1):i–44.

Milligan, G. W., and M. C. Cooper. 1985. An examination of procedures for determining the number of clusters in a data set. *Psychometrika* 50(2):159–79.

Murphy, J., R. Baxter, J. Eyerman, D. Cunningham, and J. Kennet. 2004. A system for detecting interviewer falsification. *Proceedings of statistics Canada symposium*, 2005.

Porras, J., and N. English. 2004. Data-driven approaches to identifying interviewer data falsification: The case of health surveys. *Proceedings of the survey research method section*, 4223–8.

Reuband, K. H. 1990. Interviews, die keine sind: "Erfolge" und "Mißerfolge" beim Fälschen von Interviews. *Kölner Zeitschrift für Soziologie und Sozialpsychologie* 42:706–33.

Sarracino, F., and M. Mikucka. 2016. Estimation bias due to duplicated observations: A Monte Carlo simulation. *Munich personal RePEe archive*, 1–18. https://mpra.ub.uni-muenchen.de/69064/1/MPRA_paper_69064.pdf (accessed November 30, 2019).

Schäfer, C., J. P. Schräpler, and K. R. Müller. 2004. Identification, characteristics and impact of faked and fraudulent interviews in surveys. Discussion Paper.

Schnell, R. 1991. Der Einfluß gefälschter Interviews auf Survey-Ergebnisse. *Zeitschrift für Soziologie* 20:25–35.

Schnell, R. 2012. *Survey-interviews: Methoden standardisierter Befragungen*. Wiesbaden: VS Verlag für Sozialwissenschaften.

Schräpler, J. P., and G. G. Wagner. 2005. Characteristics and impact of faked interviews in surveys: An analysis of genuine fakes in the raw data of SOEP. *Allgemeines statistisches archiv* 89:7–20.

Schreiner, I., K. Pennie, and J. Newbrough. 1988. Interviewer falsification in Census Bureau surveys. *Proceedings of the survey research method section*, 491–96.

Simmons, K., A. Mercer, S. Schwarzer, and C. Kennedy. 2016. Evaluating a new proposal for detecting data falsification in surveys. *Statistical journal of the IAOS* 32(3):327–38.

Slomczynski, K. M., P. Powalko, and T. Krauze. 2017. Non-unique records in international survey projects. *Survey research methods* 11(1):1–16.

Stokes, S. L., and P. Jones. 1989. Evaluation of the interviewer quality control procedure for the post-enumeration survey. *Proceedings of the survey research method section*, 696–98.

Storfinger, N., and P. Winker. 2013. Assessing the performance of clustering methods in falsification using bootstrap. In *Interviewers' deviations in surveys*, ed. P. Winker, N. Menold, and R. Porst, 46–65. Frankfurt am Main: Internationaler Verlag der Wissenschaften.

Swanson, D., M. Cho, and J. L. Eltinge. 2003. Detecting possibly fraudulent or error-prone survey data using Benford's Law. *Proceedings of the survey research method section*, 4172–77.

Tourangeau, R., F. Kreuter, and S. Eckman. 2012. Motivated underreporting in screening interviews. *Public opinion quarterly* 76(3):453–69.

Turner, C. F., J. N. Gribble, A. A. Al-Tayyib, and J. R. Chromy. 2002. Falsification in epidemiologic surveys: Detection and remediation. *Technical papers on health and behavior measurement* 53:1–12.

Winker, P. 2016. Assuring the quality of survey data: Incentives, detection and documentation of deviant behavior. *Statistical journal of the IAOS* 32(3):295–303.

Winker, P., K. W. Kruse, N. Menold, and U. Landrock. 2015. Interviewer effects in real and falsified interviews. *Statistical journal of the IAOS* 31(3):423–34.

Ziegler, M., C. J. Kemper, and B. Rammstedt. 2013. The Vocabulary and Overclaiming Test (VOC-T). *Journal of individual differences* 34:32–40.

8

Examining the Utility of Interviewer Observations on the Survey Response Process

Brady T. West, Ting Yan, Frauke Kreuter, Michael Josten, and Heather Schroeder

CONTENTS

8.1 Introduction

Surveys of representative samples of persons selected from general populations represent a vital source of information for policy makers and program evaluators. However, the utility of survey data depends heavily on the quality of the data collected. Survey data of poor quality could produce misleading, if not erroneous, estimates and inferences related to the characteristics of larger populations. Unfortunately, evaluating survey data quality can sometimes be difficult. Perhaps the best way to assess response quality would be to use true values available from an external source to validate answers reported by survey respondents. Unfortunately, these types of external validation data are usually unavailable or difficult to access.

In these situations, survey researchers and practitioners often turn to other sources of data that provide indirect indicators of data quality. These data sources capture breakdowns in the survey response process, which arise due to the inability and unwillingness of respondents to answer survey questions (Tourangeau, Rips, and Rasinski 2000).

One such data source is *paradata*, or data describing the survey data collection process (Kreuter 2013). Various types of paradata have been used to evaluate the quality of survey data, including response latency data (e.g., Callegaro et al. 2009), linguistic expressions of doubt and uncertainty (e.g., Schaeffer and Dykema 2011), call record data (such as contact attempts and call histories; e.g., Kreuter and Kohler 2009), patterns of attrition (e.g., Olson and Parkhurst 2013), and doorstep concerns (e.g., Yan 2017). These studies have all suggested that the various paradata were in fact useful indicators of data quality.

This chapter focuses on a different type of paradata that could provide information about breakdowns of the survey response process: post-survey interviewer observations of respondents and their behaviors during the interviewing process. Survey organizations often ask an interviewer to answer several closed-ended questions about the respondent and his/her behaviors (based on their observations) once the interviewer has finished an interview. For example, in the Health and Retirement Study (HRS), interviewers evaluate how interested the respondent was in the survey topic using a three-point scale (not at all interested, somewhat interested, and very interested) and assess how attentive he/she has been during the interview process using a four-point scale (excellent, good, fair, and poor). This practice assumes that respondents observed by the interviewers as showing undesirable behaviors (e.g., "not attentive") or having negative attitudes (e.g., "not interested") will provide data of poor quality.

The collection of these types of interviewer observations and evaluations dates back to 1948, when interviewers in a study were first asked to rate the dependability of the interviews that they obtained after each interview (Bennett 1948). More than seven decades later, many prominent government-sponsored surveys in the United States routinely collect post-survey interviewer observations, including the National Survey of Family Growth (NSFG). Several international survey programs also collect these observations, including the European Social Survey (ESS).

Survey programs generally design post-survey interviewer observations to tap into data quality issues, such as the overall quality of the information provided (e.g., Maclin and Calder 2006), respondents' understanding of the questions, interest in the interview, and attention to the questions. Interviewers record the observations at the end of the survey, after having just spent the entire interview witnessing the respondent's behaviors and recording their responses. Theoretically, these types of observations should correlate well with the quality of the answers provided by respondents. This potential led Billiet, Carton, and Loosveldt (2004) to recommend that interviewer observations of the survey response process be one evaluation criterion in the Total Quality Management paradigm.

Given their potential, several methodological studies have evaluated the utility of these observations. Fisk (1950) provided the first evaluation of interviewer observations of the survey response process, focusing on the variation of interviewer ratings of interest between interviewers. Later studies considering the relationships of individual interviewer observations with various indirect indicators of data quality found that respondents who received positive or favorable ratings tended to provide data of better quality, in terms of less missing data (Antoun 2012; Tarnai and Paxson 2005; Yan 2017), less measurement error (Wang, West, and Liu 2014), higher validity (Andrews 1990), more consistent reports (Antoun 2012), less "heaping" (Holbrook et al. 2014; Sakshaug 2013), less "satisficing" (Josten 2014), higher response propensity in later waves of panel studies (Plewis, Calderwood, and Mostafa 2017), and more codable answers to open-ended questions (Tarnai and Paxson 2005). A recent study has shown that interviewers do in fact use their observations of respondent behaviors during the interview process to answer the post-survey observation questions (Kirchner, Olson, and Smyth 2018).

Even though several methodological studies have now demonstrated the utility of individual observations for indicating data quality, little evidence exists of secondary analysts of survey data using these observations for substantive analysis purposes. The observations are generally not included in public-use data files or only available via restricted-access data user agreements (see, for example, "Other Data Files" at the web site www.cdc.gov/nchs/nsfg/nsfg_2011_2015_puf.htm). This raises questions about the cost-benefit tradeoffs of collecting these observations. For example, in the NSFG, interviewers spend an average of 5 to 6 minutes per interview completing these post-survey observations. Given that 22,682 interviews were completed in the NSFG from 2006 to 2010, an estimated 1,887 hours of production time (almost one full year of 40-hour work weeks) were spent on this task alone. Recent personal communication with NSFG managers at the National Center for Health Statistics suggests that public data users rarely use these observations, despite their public availability and the cost that it takes to collect them.

In this chapter, we apply latent class analysis (LCA) to multiple post-survey interviewer observations recorded in each of two major surveys (the NSFG and the ESS) to derive respondent-level data quality classes that could benefit secondary analysts of these data. The individual post-survey judgments and estimates recorded by interviewers that have been the focus of prior methodological work may be prone to quality issues themselves (e.g., Kirchner, Olson, and Smyth 2018; O'Muircheartaigh and Campanelli 1998; West and Peytcheva 2014). LCA offers the benefit of accounting for potential measurement error in the individual observations (Kreuter, Yan, and Tourangeau 2008), and (assuming that a given latent class model fits the data well) produces smoothed respondent-level predictions of the probability of membership in one of a small number of classes defined by patterns in the observations. If variables containing predicted response quality classes for the respondents based on the LCA effectively distinguish between respondents in terms of data quality, survey organizations could include these variables in public-use data files, enabling analysts to adjust for overall response quality in their analyses (e.g., analyzing changes in estimates across the quality classes). We therefore also compare the derived classes in terms of indirect indicators of data quality from each survey.

8.2 Methods

8.2.1 Data Sources

This chapter analyzes interviewer observations from two surveys – the ESS and the NSFG. The ESS is a cross-national survey of attitudes, beliefs, and behavior patterns of persons aged 15 years and older living in private households in more than 30 European nations. For this analysis, we analyzed data from the fifth wave of the ESS. The subset of ESS countries analyzed here (response rates* in parentheses) included Belgium (53.4%, n = 1,704), Bulgaria (81.4%, n = 2,434), Estonia (56.2%, n = 1,793), Germany (30.5%, n = 3,031), Ireland (65.1%, n = 2,576), Slovakia (74.7%, n = 1,856), and the United Kingdom (56.3%, n = 2,422). All countries targeted the general population. Fieldwork for the ESS is centrally monitored,

---

* Response rate = valid interviews/[total sample – (sum H, I, K, L, M)]; H = institutions, I = unoccupied, K = ineligible, L = moved, M = deceased.

and fieldwork procedures are centrally reported and documented following strict guidelines (www.europeansocialsurvey.org/data/; see Chapter 22 for additional analyses using ESS data). Computer-assisted personal interviewing (CAPI) was used in all countries except for Bulgaria and Slovakia, which each used paper-and-pencil interviewing (PAPI). Across this subset of countries, ESS teams collected data from 15,816 respondents between 2010 and 2011. After conducting an interview, the interviewer recorded post-survey observations; we recoded five of these observations capturing respondent behaviors and interviewing environments.

The NSFG is a national survey measuring sexual and reproductive health in 15- to 49-year-olds living in the United States. The survey takes approximately 60 minutes to complete, and has two sections: an in-person CAPI section, where all interviewers are female, and an audio computer-assisted self-interview (ACASI) section. Data were collected from 15,820 interviews over 12 quarters of data collection from January 2016 through December 2018 (AAPOR RR3 = 65.3%; see Lepkowski et al. (2013) for design details). After conducting an interview, the interviewer is asked to record 30 observations about the conditions surrounding the data collection, including details about the environment, the respondent's response behaviors, and the respondent's mood. We chose 22 of the 30 observations that explicitly described respondent behaviors and interviewing environments for the LCA (details below). A small number of cases ($n = 52$) with missing values for the interviewer observation data were excluded, resulting in a final sample size of 15,768.

8.2.2 Post-Survey Interviewer Observations in the ESS

In the ESS, interviewers recorded the five post-survey interviewer observations on a five-point scale (ranging from "never" to "very often"). Of these five observations, one captured the interviewing environment (an *objective* observation of how often someone else was present and potentially interfering during the interview). The remaining four were *subjective* interviewer perceptions and observations of respondent behaviors during the interview:

- Whether the respondent asked for question clarification.
- Whether the respondent seemed to understand the questions.
- The respondent's reluctance to answer any questions.
- The respondent's effort to answer the questions to the best of his/her ability.

8.2.3 Post-Survey Interviewer Observations in the NSFG

We analyzed 22 post-survey interviewer observations collected in the NSFG. Of these observations, the 12 *objective* observations of the interviewing environment included:

- Four indicators (yes, no) of distractions: (1) whether the television was on, (2) whether the respondent received phone calls, (3) whether children were present and needed attention, and (4) other distractions.
- Observations of the interview: (1) location (on the respondent's property, in the interviewer's car, in another public place), (2) atmosphere (chaotic, noisy; some interruptions; ideal – quiet and calm), and (3) language (English, Spanish, both).

- Observations of (1) the seating arrangement (next to respondent, facing the same way; next to respondent, at a right angle; across from the respondent; other) and (2) the presence of others within hearing range (no one present; others present, not able to hear; others present, able to hear part of the interview; others present, able to hear the entire interview).
- Additional reports of the respondent's ability to see the computer screen during the face-to-face CAPI section (yes, all questions; most, not all questions; a few questions; none), the interviewer's ability to see the computer screen during the ACASI section (yes, no), and in general what the interviewer's self-reported mood during the interview was (happy, neutral, or sad or unhappy).

The 10 *subjective* interviewer perceptions and observations of respondent behaviors included the following:

- Observations of behaviors during ACASI, including (1) how much help the respondent needed from the interviewer (none, a little, a lot, interviewer-administered); (2) the respondent's use of headphones, which may help respondents with literacy problems better understand the questions being asked (at least some of the time, never); (3) the respondent's use of text and audio, both of which may help with comprehension (text only, text and audio, audio only, don't know); and (4) what support was used to hold the laptop, which could measure respondent comfort during the interview (table, lap, other).
- Observations of difficulty using the ACASI application (any, none) and the interviewer's opinion of respondent attentiveness (not at all, some, very).
- Three observations of the respondent's mood that may affect his/her ability to clearly think about responses to the questions: (1) whether the respondent was upset (yes, no) or (2) tired (yes, no), and (3) in general how the respondent acted during the interview (hostile or neutral, friendly).
- An overall observation of the quality of information provided by the respondent (excellent, good, fair, or poor).

8.2.4 Dependent Variables in the ESS

In the ESS, we computed six dependent variables, including indirect measures of data quality, measures of response styles, and proxies of the respondent's cognitive response process. The first indirect data quality indicator captured *item nonresponse*, measuring the percentage of survey variables where respondents indicated "refusal," "don't know," or "no answer." The second indirect data quality indicator captured a lack of differentiated answers. We chose from the ESS several battery questions (see the Online Appendix), each of which included several attitudinal items using the same response scale. We only used item batteries that were comprised of at least four items. We defined a response pattern to an item battery as *non-differentiating* if a respondent chose the same answer throughout the battery. The overall indicator was the percentage of batteries where non-differentiation/straight-lining occurred.

For the third dependent variable, we assessed the *number of extreme answers* as a measure of a sub-optimal response style that can bias scale scores and affect scale reliability, in addition to affecting relationships between variables (Alwin and Krosnick 1991;

Diamantopoulos, Reynolds, and Simintiras 2006; Leventhal 2018). We considered the same survey items used to measure non-differentiation and simply calculated the percentage of questions answered with a 1 or 5 (on the five-point scale), or with a 1 or 10 (for ten-point scale items). ESS respondents could therefore have high measures of non-differentiation and extreme answers simultaneously, but this was rare; the correlation of these two dependent variables was only 0.06. For 23 attitudinal items within the ESS for which affirmative answers were possible, we computed the percentage of items where the respondent agreed or supported the affirmative as a fourth indirect indicator of data quality (Van Vaerenbergh and Thomas 2013). This type of *acquiescent response style* is a source of measurement error that, being a systematic behavior on the part of the respondents, can lead to bias in estimation (Reynolds and Smith 2010; Roberts 2016), and has also been studied as a type of socially desirable responding (Holbrook 2008). We did not include items bearing any reference to factual events as it is less likely that respondents affirm a factual question falsely.

For our fifth dependent variable, we constructed a measure of *internal consistency* in the answers provided by the respondent based on the following two items: "All laws should be strictly obeyed" and "Doing the right thing sometimes means breaking the law." Both items had to be answered using a five-point agree–disagree scale. The idea behind these items is the same, except that the statements are reversely worded. Agreeing to both statements can therefore be regarded as a sign that the cognitive process necessary to provide an optimal answer is incompletely or superficially performed. The resulting variable is a dummy variable that equals 1 if the answers to the items were inconsistent. Finally, *interview length*, operationalized in the ESS as the number of seconds spent per question asked, was also included in the analyses as an indirect measure of data quality, where longer surveys tend to yield data of lower quality (Galesic and Bosnjak 2009; Herzog and Bachman 1981). To deal with the extreme skewness of some of the variables, we applied natural log transformations to the item nonresponse rate, the extreme answer rate, and interview length. Collectively, these six measures tended to capture different indirect aspects of data quality and the response process, having pairwise correlations that were no greater than 0.28 in absolute value (see Table Table A8A.2 in the Online Appendix).

8.2.5 Dependent Variables in the NSFG

In the NSFG, our dependent variables measuring indirect indicators of data quality included one paradata variable (total interview length in minutes; Galesic and Bosnjak 2009; Herzog and Bachman 1981) and four proxy indicators of measurement error. These proxy indicators were binary indicators of inconsistent responses (1 = different responses, 0 = identical responses) for four survey variables that were measured in both CAPI and ACASI: number of sexual partners in the past year (females), number of sexual partners in the past year (males), number of live births to date (females), and total number of pregnancies fathered (males).

8.2.6 Analytic Approach

We first applied latent class analysis (e.g., Kreuter, Yan, and Tourangeau 2008) to the interviewer observations. Briefly, this multivariable modeling approach takes as input multiple categorical variables (e.g., the post-survey interviewer observations from each survey listed above) that an analyst believes indicate some latent (unmeasured) categorical trait (e.g., overall response quality). The analyst specifies a hypothesized number of categories, or classes, for the latent categorical trait, and the estimation procedure results in predicted probabilities of membership in each class for each case in the data set (e.g., survey respondent). In addition, the estimation procedure generates conditional probabilities of each category on the input items, depending on the predicted class membership; this allows analysts to profile the latent classes in terms of distributions on the observed categorical measures. One can evaluate the fits of competing latent class models with different counts of hypothesized categories for the latent trait and select the model that provides the best fit to the observed data.

For the purposes of the present study, we identified the best-fitting latent class model in terms of the number of latent response quality classes of respondents (given the input interviewer observations from each survey); we considered models with between two and seven classes. Given the best-fitting model for one of the two data sets, we then computed predicted probabilities of belonging to each class for the respondents. We also computed predicted probabilities for each of the values of the input observations as a function of class membership based on the best-fitting model. We assigned each survey respondent to the class for which they had the highest predicted probability (as one possible "modal" method of predicting class membership), and then profiled the resulting classes (e.g., high quality, low quality, etc.) based on the predicted probabilities for the values of each interviewer observation for each class.

We used PROC LCA in SAS (Lanza et al. 2007) to fit all latent class models, including the CLUSTER option to account for potential correlations in the observed data due to the interviewers. We compared model fit statistics (including the log-likelihood, g-squared, AIC, BIC, and adjusted BIC) and entropy (where higher entropy values indicate better class separation) across the competing models. In the NSFG, we found that the model with seven latent classes had by far the best fit in terms of minimizing all of the model fit criteria (see Table A8A.1 in the Online Appendix) and had an entropy estimate of 0.89, indicating good separation of the classes. For the ESS, we found that the best-fitting model had three latent classes when using a similar approach (fitting models with 4+ classes was not possible with only five input variables).

Next, we compared the predicted marginal means of the dependent variables between the derived latent response quality classes using PROC SURVEYREG or PROC SURVEYLOGISTIC in SAS (again accounting for clustering by interviewer), depending on the type of dependent variable. In these models, we adjusted for age, race/ethnicity (only measured in the NSFG), gender (only relevant in the ESS, since the NSFG measures were gender-specific), and education, to allow for possible associations of both the interviewer observations and the dependent variables with these observable socio-demographic characteristics and isolate the relationships of the response quality classes with the dependent variables when adjusting for these characteristics. We then performed pairwise comparisons of the model-based marginal predictions of the means and proportions for each of the dependent variables across the derived classes.

We used the distal LCA macro in SAS (Lanza, Tan, and Bray 2013) to assess the robustness of our findings about differences between the classes to the assumption that the latent response quality classes for each respondent were known with certainty (i.e., using predicted probabilities of class membership, rather than fixed assignments based on the "modal" approach). This macro currently does not have the ability to account for interviewer clustering, so we focused on general patterns of differences in means and proportions between the classes in these supplemental analyses. In general, we found that our primary results did not change when allowing for uncertainty in the predicted class membership.

8.3 Results

8.3.1 ESS: Latent Class Analysis

Figure 8.1 illustrates the results of the LCA in ESS, where we found the model with three classes to have the best fit. We plot conditional probabilities of specific types of observations on the vertical axis for the three derived classes, where the response options on the horizontal axis are the observations indicating the highest response quality. The three-class LCA solution shows respondents differing strongly with respect to the interviewer observations of respondent behaviors. Respondents assigned to the "Low Quality" class (mean age = 53.2, percent male = 43.2, percent lower education = 48.4) frequently ask for clarification, show reluctance, show low effort, and do not understand the questions. These respondents were significantly older and less educated. Our expectation is that this class will provide data of lower quality. Of the 15,816 ESS respondents analyzed, 12% had the highest posterior probability of belonging to this class.

Respondents assigned to the "High Quality" class (mean age = 46.8, percent male = 46.3, percent lower education = 21.9) are more likely to be rated by the interviewers as having understood questions very often, using effort very often, never asking for clarification, and never showing reluctance. These respondents were significantly younger and more highly educated. Roughly 57% of the 15,816 ESS respondents were assigned to this class, which we expect to provide responses of higher quality. The remaining 31% of respondents assigned to the "Moderate Quality" class (mean age = 51.5, percent male = 43.6, percent lower education = 33.9) are expected to provide responses of "moderate" quality, given the profile of this class in Figure 8.1. We note that the three derived classes did not differ in terms of the one observation of the interviewing environment (others present and potentially interfering).

8.3.2 ESS: Class Comparisons on Dependent Variables

We compare the model-based marginal predictions of the means for the data quality indicators and interview length across the three derived classes in Table 8.1. As we expected,

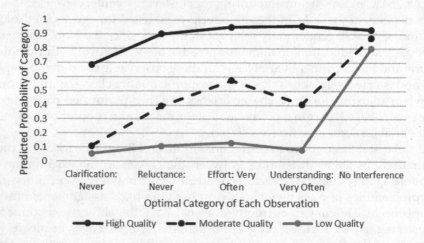

FIGURE 8.1
Conditional probabilities of receiving the rating category indicating the highest response quality from the interviewer (ESS).

TABLE 8.1

Comparisons across the Three Derived Quality Classes of Model-Based Marginal Predictions of Means and Probabilities for the Six Response Quality Indicators (ESS)

Derived Quality Class	Mean % of Batteries with Non-Differentiation	Mean % of Items with Extreme Answers	Probability of Inconsistency	Mean % of Items Agreeing/ Affirming	Mean % of Items Missing	Mean Interview Length (Sec./ Item)
High	6.87[a]	24.02[b]	0.028[a]	59.54[a]	2.99[a]	17.11[a]
Moderate	7.74[b]	23.21[a]	0.026[a]	57.16[b]	3.95[b]	17.42[b]
Low	10.69[c]	24.74[b]	0.041[b]	53.77[c]	6.72[c]	17.34[b]

Note: Different superscripts indicate significant differences of marginal means/probabilities at $p < 0.01$.

after adjusting for age (which had a significant positive relationship with each dependent variable, except for internal consistency), education (which had a significant *negative* relationship with each dependent variable, except for internal consistency), and sex (where males tended to be more inconsistent, more acquiescent, and had less missing data) in the multivariable models, respondents assigned to the "Low Quality" class had significantly higher rates of missing data, non-differentiation, extreme answers, and inconsistent answers than respondents in the other two classes. However, inconsistent with expectations based on the literature, respondents assigned to the "Low Quality" class exhibited *less* acquiescence than respondents assigned to the other two classes. Acquiescence may arise out of deference to the interviewer (Holbrook 2008), so individuals disinterested in the interview and struggling to understand the questions may not have cared about social desirability or pleasing the interviewer.

In terms of mean interview length, measured as seconds per question asked, respondents assigned to the "Moderate Quality" and "Low Quality" classes took more time on average to answer questions than respondents assigned to the "High Quality" class. This is consistent with the class profiles: respondents assigned to these classes needed more clarification, were more reluctant, and generally did not have clear understanding of the questions. Collectively, these ESS results demonstrate the ability of the classes derived based on the post-survey observations of respondent behaviors to distinguish between respondents based on their data quality.

8.3.3 NSFG: Latent Class Analysis

Table 8.2 profiles the seven latent classes derived in the NSFG based on the best-fitting model. We found that 11 of the 22 observations had distributions that varied substantially across the seven derived classes. Five were objective observations of the environment (location of the interview, seating arrangement, distractions due to kids, presence of others, interviewer not happy), and six were subjective observations of respondent behaviors (overall data quality rating, use of headphones in ACASI, respondent attentiveness, respondent tired, respondent not happy, and need for assistance during ACASI). These results suggest that data quality in the NSFG may be a function of both the interviewing environment and specific respondent behaviors.

Table 8.2 shows that nearly two-thirds of the NSFG respondents had higher posterior probabilities of belonging to either class 1 or class 2 than the posterior probabilities of belonging to any other class (like the ESS analysis). Given the characteristics of classes 1 and 2 as described in Table 8.2, we expect these respondents to provide data of relatively

TABLE 8.2

Latent Class Profiles in the NSFG Data

Latent Class (% of Sample)	Distinct Patterns on Interviewer Observations (Socio-Demographics)	Expected Data Quality
1 (28%)	Private interview, used headphones in ACASI, data quality rated as excellent (mean age = 30.6, % white = 68.8, mean years of education = 13.2)	High
2 (36%)	Private interview, no headphones in ACASI, data quality rated as excellent (mean age = 30.8, % white = 71.7, mean years of education = 13.5)	High
3 (5%)	Private interview conducted in respondent's car, interviewer and respondent seated next to each other, no respondent problems (mean age = 28.8, % white = 62.4, mean years of education = 12.3)	Moderate
4 (9%)	Distractions due to kids needing attention, no respondent problems, used headphones in ACASI (mean age = 30.7, % white = 67.5, mean years of education = 13.0)	Moderate
5 (9%)	Distractions due to kids needing attention, no respondent problems, used text in ACASI (mean age = 31.5, % white = 69.5, mean years of education = 13.6)	Moderate
6 (8%)	Frequent interruptions, others present during the survey, respondent inattentive, respondent tired, respondent not happy, interviewer unhappy, data quality rated as low, respondents needed assistance during ACASI and used headphones (mean age = 32.1, % white = 57.0, mean years of education = 12.0)	Low
7 (5%)	Frequent interruptions, others present during the survey, respondent inattentive, respondent tired, respondent not happy, interviewer unhappy, data quality rated as low, respondent sat next to the interviewer during ACASI and did not use headphones (mean age = 30.8, % white = 58.1, mean years of education = 13.0)	Low

high quality. Respondents assigned in the same way to classes 3 through 5 are expected to provide data of moderate quality, primarily due to child-related distractions and non-conventional settings for the interview (e.g., the respondent's car). Respondents in the last two classes (6 and 7) will likely provide data of questionable quality, for a variety of reasons captured in the interviewer observations and indicated in Table 8.2. Interestingly, while the derived classes are similar in terms of mean age and mean education, the last two classes also have significantly lower percentages of white respondents. Figures A8A.1 and A8A.2 in the online supplemental materials provide additional illustrations of the differences between these seven derived response quality classes.

8.3.4 NSFG: Class Comparisons on Dependent Variables

Table A8A.2 in the online supplemental materials presents comparisons of the model-based marginal predictions of the means and proportions for the NSFG dependent variables across the seven derived response quality classes. Included in Table A8A.2 are indications of which pairwise differences were found to be significant at the $0.05/21 = 0.002$ level (using a Bonferroni correction to account for the 21 pairwise comparisons), suggesting robust differences in the means and proportions across the classes after adjusting for age, race/ethnicity, and education. The lower-quality classes had interviews that took significantly longer on average after adjusting for the covariates, which is generally consistent with the ESS results. In addition, inconsistencies between the CAPI and ACASI responses on items measured in both modes became significantly more likely in the lower-quality classes, for all four indicators of inconsistent reporting. Figure 8.2 illustrates these trends in the

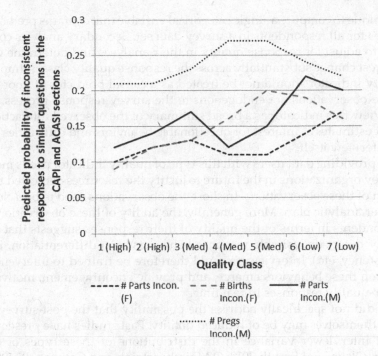

FIGURE 8.2
Differences across the derived quality classes in terms of the marginal predicted probabilities of inconsistent responses in CAPI and ACASI on four NSFG measures.

marginal predicted probabilities of inconsistency of responses across the derived quality classes. Finally, older, African-American, and lower-educated respondents tended to have longer interviews on average, while higher-educated and white respondents tended to provide more consistent reports in CAPI and ACASI.

8.4 Discussion

This chapter presented an application of LCA to multiple post-survey interviewer observations from two major surveys (the NSFG and the ESS). The results suggest that applying this type of approach to post-survey interviewer observations of the survey response process can produce meaningful classes of respondents that vary in terms of the quality of their reporting. This approach can also identify those observations (whether they are observations of the environment or respondent behaviors) that are the most important for defining the response quality classes, in that they have distributions that vary substantially across the derived classes. In the NSFG, both types of observations contributed to defining the quality classes, suggesting that both respondent behaviors and the interviewing environment can affect response quality; this makes sense given the sensitive subject matter about sexual health. In the ESS, only the observations of respondent behaviors were found to vary across the derived quality classes.

If a latent class model fitted to the observations in a given survey has a reasonably good fit, and meaningful distinctions between the response quality classes emerge, one could use this technique to compute a single categorical variable that contains predicted response quality classes for all respondents in a survey data set. Secondary analysts could then use this variable to adjust for response quality in their analyses and determine whether estimates of interest change substantially across the response quality classes. Importantly, this type of quality indicator should not be treated as a causal predictor of poor data quality, given that the observations are endogenous to the survey response process. Rather, analysts should view this indicator as a broad summary of the observed interaction and use it only to adjust estimates of interest for questionable behaviors and difficulties encountered during the interview itself.

The idea of providing data users with this type of quality indicator needs more consideration by survey organizations in the future to justify the resources dedicated to the continued collection of these observations. In short, the observations need to be a clearly defined part of a larger analytic plan. More generally, the ability of these observations to broadly classify respondents in terms of the quality of their responses suggests that interviewers can accurately detect undesirable response behaviors (non-differentiation, item-missing data, inconsistency, etc.). Interviewers could therefore be trained to intervene during the interview when these behaviors emerge, and provide encouragement, motivation, etc., to prevent lower-quality responses in real time.

Finally, we did not specifically address the possibility that the post-survey interviewer observations themselves may be of reduced quality. Past studies have presented evidence of significant interviewer variance in the distributions of these types of observations (e.g., Kirchner, Olson, and Smyth 2018; O'Muircheartaigh and Campanelli 1998; West and Peytcheva 2014). Future research should investigate this issue in more detail and focus on sources of any unexplained variance among interviewers in terms of distributions on these observations.

Acknowledgments

We acknowledge financial support for this work from a methodological research modification to the contract between the University of Michigan and the National Center for Health Statistics (NCHS) that supports the National Survey of Family Growth. The National Survey of Family Growth is conducted by the Centers for Disease Control and Prevention's (CDC's) National Center for Health Statistics (NCHS), under contract # 200-2010-33976 with University of Michigan's Institute for Social Research with funding from several agencies of the U.S. Department of Health and Human Services, including CDC/NCHS, the National Institute of Child Health and Human Development (NICHD), the Office of Population Affairs (OPA), and others listed on the NSFG webpage (see www.cdc.gov/nchs/nsfg/). The views expressed here do not represent those of NCHS nor the other funding agencies.

References

Alwin, D. F., and J. A. Krosnick. 1991. The reliability of survey attitude measurement: The influence of question and respondent attributes. *Sociological methods and research* 20(1):139–181.

Andrews, F. M. 1990. Some observations on meta-analysis of MTMM Studies. In *Evaluation of measurement instruments by meta-analysis of multi-trait multi-method studies: Proceedings of the international colloquium*, ed. W. E. Saris, and A. van Meurs, 172–184. New York: North-Holland.

Antoun, C. 2012. The utility of interviewer ratings about survey data quality. In *JSM proceedings*, Survey Research Methods Section. Alexandria, VA: Amer. Stat. Assoc.

Bennett, A. S. 1948. Toward a solution of the "cheater problem" among part time research investigators. *Journal of marketing* 12(4):470–474.

Billiet, J., A. Carton, and G. Loosveldt. 2004. Assessment of survey data quality: A pragmatic approach focused on interviewer tasks. *International journal of market research* 46(1):65–82.

Callegaro, M., Y. Yang, D. S. Bhola, D. A. Dillman, and T. Y. Chin. 2009. Response latency as an indicator of optimizing in online questionnaires. *Bulletin de methodologie sociologique* 103(1):5–25.

Diamantopoulos, A., N. L. Reynolds, and A. C. Simintiras. 2006. The impact of response styles on the stability of cross-national comparisons. *Journal of business research* 59(8):925–935.

Fisk, G. 1950. Interviewer ratings of respondent interests of sample surveys. *Journal of marketing* 14(5):725–730.

Galesic, M., and M. Bosnjak. 2009. Effects of questionnaire length on participation and indicators of response quality in a web survey. *Public opinion quarterly* 73(2):349–360.

Herzog, A. R., and J. G. Bachman. 1981. Effects of questionnaire length on response quality. *Public opinion quarterly* 45(4):549–559.

Holbrook, A. L. 2008. Acquiescence response bias. In *Encyclopedia of survey research methods*, ed. P. Lavrakas, 4–5. Thousand Oaks, CA: Sage Publications, Inc.

Holbrook, A. L., S. Anand, T. P. Johnson, Y. I. Cho, S. Shavitt, N. Chavez, and S. Weiner. 2014. Response heaping in interviewer-administered surveys: Is it really a form of satisficing? *Public opinion quarterly* 78(3):591–633.

Josten, M. 2014. Can interviewers tell? Using post-survey interviewer observations to assess survey data quality. Master's thesis completed at University of Mannheim, Germany.

Kirchner, A., K. Olson, and J. D. Smyth. 2018. Do interviewer postsurvey evaluations of respondents' engagement measure who respondents are or what they do? *Public opinion quarterly* 81(4):817–846.

Kreuter, F. 2013. *Improving surveys with paradata: Analytic uses of process information*. Hoboken, NJ: John Wiley and Sons.

Kreuter, F., and U. Kohler. 2009. Analyzing contact sequences in call record data: Potential and limitation of sequence indicators for nonresponse adjustment in the European Social Survey. *Journal of official statistics* 25(2):203–226.

Kreuter, F., T. Yan, and R. Tourangeau. 2008. Good item or bad – Can latent class analysis tell? The utility of latent class analysis for the evaluation of survey questions. *Journal of the royal statistical society, Series A (Statistics in Society)* 171(3):723–738.

Lanza, S. T., L. M. Collins, D. R. Lemmon, and J. L. Schafer. 2007. PROC LCA: A SAS procedure for latent class analysis. *Structural equation modeling: A multidisciplinary journal* 14(4):671–694.

Lanza, S. T., X. Tan, and B. C. Bray. 2013. Latent class analysis with distal outcomes: A flexible model-based approach. *Structural equation modeling: A multidisciplinary journal* 20(1):1–26.

Lepkowski, J. M., W. D. Mosher, R. M, Groves, B. T. West, J. Wagner, and H. Gu. 2013. Responsive design, weighting, and variance estimation in the 2006–2010 National Survey of Family Growth. *Vital and health statistics* 2(158). National Center for Health Statistics.

Leventhal, B. C. 2018. Extreme response style: A simulation study comparison of three multidimensional item response models. *Applied psychological measurement* 43(4):322–335.

MacLin, M. K., and J. C. Calder. 2006. Interviewers' ratings of data quality in household telephone surveys. *North American journal of psychology* 8(1):163–170.

Olson, K., and B. Parkhurst. 2013. Collecting paradata for measurement error evaluations. In *Improving surveys with paradata: Analytic use of process information*, ed. F. Kreuter, 43–72. New York, NY: Wiley and Sons.

O'Muircheartaigh, C. A., and P. Campanelli. 1998. The relative impact of interviewer effects and sample design effects on survey precision. *Journal of the royal statistical society, Series A* 161(1):63–77.

Plewis, I., L. Calderwood, and T. Mostafa. 2017. Can interviewer observations of the interview predict future response? *Methods, data, analyses* 11(1):29–44.

Reynolds, N., and A. Smith. 2010. Assessing the impact of response styles on cross-cultural service quality evaluation: A simplified approach to eliminating the problem. *Journal of service research* 13(2):230–243.

Roberts, C. 2016. Response styles in surveys: Understanding their causes and mitigating their impact on data quality. In *The SAGE handbook of survey methodology*, ed. C. Wolfe, D. Joye, T. W. Smith, and Y. Fu, 579–596. Los Angeles, CA: SAGE Publishing.

Sakshaug, J. W. 2013. Using paradata to study response to within-survey requests. In *Improving surveys with paradata: Analytic use of process information*, ed. F. Kreuter, 171–190. Hoboken, NJ: John Wiley and Sons.

Schaeffer, N. C., and J. Dykema. 2011. Response 1 to Fowler's chapter: Coding the behavior of interviewers and respondents to evaluate survey questions. In *Question evaluation methods: Contributing to the science of data quality*, ed. J. Madans, K. Miller, and G. Willis, 23–39. New York, NY: Wiley.

Tarnai, J., and M. C. Paxson. 2005. Interviewer judgments about the quality of telephone interviews. In *JSM proceedings*, Survey Research Methods Section, 3988–3994. Alexandria, VA: American Statistical Association.

Tourangeau, R., L. Rips, and K. Rasinski. 2000. *The psychology of survey response*. Cambridge, UK: Cambridge University Press.

Van Vaerenbergh, Y., and T. D. Thomas. 2013. Response styles in survey research: A literature review of antecedents, consequences, and remedies. *International journal of public opinion research* 25(2):195–217.

Wang, Y., B. T. West, and M. Liu. 2014. Interviewer perception of survey data quality in a long face-to-face survey. Paper presented at the annual conference of the American association for public opinion research, Anaheim, CA.

West, B. T., and E. Peytcheva. 2014. Can interviewer behaviors during ACASI affect data quality? *Survey practice* 5(7):1–13.

Yan, T. 2017. Using doorstep concerns data to evaluate and correct for nonresponse error in a longitudinal survey. In *Total survey error in practice*, ed. P. Biemer, E. de Leeuw, S. Eckman, B. Edwards, F. Kreuter, L. Lyberg, C. Tucker, and B. West, 413–430. New York, NY: John Wiley and Sons.

Section IV

Interviewer Effects and Interview Context and Mode

9

Why Do Interviewers Vary in Achieving Interview Privacy and Does Privacy Matter?

Zeina N. Mneimneh, Julie A. de Jong, and Yasmin A. Altwaijri

CONTENTS

9.1 Introduction and Background

Surveys that collect sensitive information often require the interview to be conducted in private; that is, only between the respondent and the interviewer. Having a third party present during the interview may affect the response process, leading to measurement error within and across samples (Aquilino 1997; Mneimneh, et al. 2015). Numerous investigations have looked at the effect of third-party presence on reporting sensitive information. Some studies found that third-party presence reduces the reporting of undesirable outcomes (Aquilino 1993; Aquilino, Wright, and Supple 2000; Moskowitz 2004), others observed increased reporting of such outcomes (Bulck 1999; Edwards, Slattery, and Ma 1998; Hoyt and Chaloupka 1994), and some showed no association between third-party presence and reports of sensitive information (Aquilino 1997; Pollner and Adams 1997). Recent work has revealed that the effect of third-party presence is not systematic; rather, it depends on the cultural background of the respondent and his or her need for social conformity (Mneimneh, et al. 2015). As Aquilino (1997) proposed in his framework on third-party effects, such variations in the effect are theorized to be driven by the nature of the information held by the third party. According to Aquilino's (1997) framework, the effect of third-party presence depends on whether the bystander is already knowledgeable about the information requested during the interview, and on the likelihood that the respondent might experience negative consequences by revealing new and unwelcome information to the bystander. Whether the respondent has disclosed such sensitive information to the bystander depends on the type of information (e.g. behavior, attitude, observable, not observable), the relationship between the third person and the respondent, and the respondent's need for social conformity.

There is also extensive literature examining the association between respondent and household characteristics and third-party presence. Male, older, married, and unemployed respondents who live in larger households are more likely to have a third person present during an interview (Aquilino 1993, 1997; Hartmann 1995; Mneimneh, et al. 2018a), and respondents with lower levels of education and income are more likely to have a spouse present (Aquilino 1993, 1997; Mneimneh, et al. 2018a). The level of income of the country where the interviews are conducted also seems to be related to the interview setting; non-private interviews are more common in low- and middle-income countries than in high-income countries (Mneimneh, et al. 2018a). However, very little research discusses the role of the interviewer in establishing a private setting. In this chapter, we focus on the interviewers and their contribution to achieving interview privacy.

Interviewers are key in enforcing the privacy requirement that is part of the protocol for many surveys. Yet achieving and maintaining privacy can be difficult. Interviewers are essentially guests, and household members do not normally expect them to dictate the interview setting. In spite of the difficulty of the task, many study protocols only require interviewers to conduct the interview in private without providing interviewers with specific instructions or details training material on how to request and achieve interview privacy (Aquilino 1993; Mneimneh, et al. 2018b; Smith 1997; Taietz 1962). Thus, interviewers have to apply their own judgment and skills in enforcing the privacy requirement for the interview.

With such minimal training, one would expect interviewers to vary in their attitudes toward the importance of establishing and maintaining privacy and their ability to do so. In fact, a recent publication (Mneimneh, et al. 2018a) found the estimated between-interviewer variance in interview privacy to be significant. Yet, very little is known about what interviewer characteristics predict the privacy setting of the interview. Only one known study has investigated the relationship between interviewer characteristics (age, gender, and years of experience) and interview privacy, showing that more experienced interviewers achieved higher rates of interview privacy than less experienced interviewers in three out of the five surveys analyzed from different countries (Lau, et al. 2017). However, the study did not assess between-interviewer variance in interview privacy or the contribution of interviewer characteristics to such variations. Moreover, we know of no studies that investigate whether an interviewer's own opinions and attitudes toward privacy contribute to the between-interviewer variation. Measuring between-interviewer variation in interview privacy and the contribution of interviewer characteristics to such variation is essential for identifying modifiable factors for interventions aimed to reduce the presence of a third party and for designing training materials or protocols to aid interviewers in establishing privacy.

This study uses data from a national face-to-face interviewer-administered study in the Kingdom of Saudi Arabia (KSA) to: (1) investigate what interviewer characteristics predict third-party presence during the interview; (2) quantify the between-interviewer variation in third-party presence; (3) estimate the relative percentage of between-interviewer variance (in third-party presence) that is explained by different types of interviewer characteristics; and (4) investigate the effect of third-party presence on reporting desirable and undesirable attitudes and behaviors.

Given the collectivist nature of the culture in KSA where the self is defined in terms of the relationship with others and where harmony is maintained by paying close attention to others in the social context (especially among family and friends) (Smith, Bond, and Kağıtçıbaşı 2006; Triandis 1995), investigating interview privacy predictors and the effect on reporting is highly relevant.

9.2 Methods

Sample. The data come from the Saudi National Mental Health Survey (SNMHS). The SNMHS is a multi-stage area probability, cross-sectional survey administered to Saudi citizens, aged 15–65, living in urban and rural areas. The survey was fielded between 2013 and 2016 with an overall response rate (RR2) of 61% (American Association for Public Opinion Research 2016).*

Questionnaire. The SNMHS used the Composite International Diagnostic Interview 3.0 (CIDI 3.0), a fully structured interview that assesses mental health disorders and is used in all World Mental Health Surveys (Kessler and Üstün 2004).[†] The source English-language CIDI was adapted and translated into Arabic using the TRAPD (Translation, Review, Adjudication, Pretesting, and Documentation) model (Harkness, et al. 2010). The instrument was administered by interviewers using computer-assisted personal interviewing (CAPI) with two audio computer-assisted self-interviewing (ACASI) sections that collected sensitive information. Interviewers were gender-matched to respondents. The average interview duration was about two hours. The instrument included two main parts. Part I, the core sections, was administered to all respondents. Part II, additional sections, was administered to a sub-sample of respondents in an effort to reduce the overall interview length. Part II respondents were weighted by the inverse of their probability of selection to adjust for differential sampling. Additional weights were used to adjust for differential probabilities of selection within households, for nonresponse, and to match the samples to population socio-demographic distributions.

In addition to the respondent instrument, a subset of interviewers was required to complete a self-administered questionnaire capturing their own sociodemographic characteristics and attitudes about the survey process itself.[‡]

Interviewers were trained over a two-week period on general interviewing techniques, locating housing units and gaining respondent cooperation, and administering the questionnaire. While the training emphasized the importance of conducting the interview in a private setting, it did not include detailed instructions on how to request and achieve a private interview.

9.3 Measures

We begin with models predicting third-party presence. At the end of each questionnaire section, interviewers were required to record whether a third person was present at any time during the section administration.[§] These reports were used to create the primary

---

* The field period extended from 2013 to 2016 because of several interruptions in the data collection phase due to both weather and funding cuts.

† See www.hcp.med.harvard.edu/wmh for more information.

‡ The decision to collect interviewer characteristics was made later in the data collection phase. Only interviewers who were working in the field during that time (76 out of 141) were asked to complete the interviewer questionnaire. Interviewers who completed the self-administered questionnaire were responsible for the majority of the completed interviews (70%).

§ In this questionnaire, a "section" is a set of questions about a common topic such as depression, marriage, childhood adversity, etc.

dependent variable, an indicator for whether (1) or not (0) a third party was ever present during the interview. Overall, 36% of the interviews had a third party present at any point during the interview.

The main independent variables for predicting third-party presence are interviewers' sociodemographic characteristics and attitudes toward privacy. Interviewer's sociodemographic characteristics included: citizenship (Saudi citizenship vs. others), marital status (currently married vs. not), education (completed a college degree vs. lower levels of education), age (30–44 years old vs. younger), income (categorized as high, middle, and low), prior survey experience (yes or no), whether the interviewer held another job in addition to interviewing (yes or no), and religiosity (whether the interviewer considers religion very important vs. not). Details on how these were measured in the self-administered questionnaire and coded for analysis, as well as their distribution, are included in Online Appendix 9A.

Interviewers' attitudes toward privacy included whether the interviewer reported that: (1) the culture in KSA respects one's privacy (disagree vs. agree), (2) it is difficult to ask for a private interview setting (disagree vs. agree), (3) s/he likes to ask for a private setting (like very much, like somewhat, vs. like a little or dislike), and (4) s/he prefers to complete an interview in the presence of a third person vs. be confronted with a refusal for a private interview. Measurement, coding details, and distribution of these variables as well as the rest of the control variables (described below) are included in Online Appendix 9A.

Respondent characteristics that were found to be associated with privacy in the interview setting as described in the Introduction were also added to the model as control variables. These include gender (male vs. female), marital status (currently married but never in a polygamous marriage, currently in a polygamous marriage, currently married and previously in a polygamous marriage vs. currently not married), age in years (15–17 vs. 18–29, 30–44, 45–54, 55–65), years of education (6 years or less vs. 7–9, 10–15, and 16 and over), having a diagnosis of social phobia in the past 12 months (yes vs. no), presence of a disability (physical disability, other disability, vs. no disability), number of household members (1–4 vs. 5–7, 8 or more), household urbanicity (rural vs. urban), and region of residence (eastern, northern, southern, western, vs. central).

Time of interview, defined as whether the interview took place during or outside of working hours in KSA (Sunday through Thursday, 7:30AM–5:30PM), was also included as a control variable.

To examine the effects of third-party presence on reporting desirable/undesirable outcomes, we consulted with local Saudi collaborators to identify survey items judged to be desirable or undesirable in the Saudi culture. For example, given that the Islamic law that allows men to be married to multiple women governs KSA, some men practice polygamy and citizens are expected to express positive views about this religious law. To assess these views, two types of attitudinal measures were constructed. The first is the average endorsement score for a set of 14 statements about polygamy (using a five-point Likert scale with higher numbers indicating stronger endorsement) (for more details, refer to Appendix 9A). The mean endorsement score was 2.9. The second polygamy measure is a binary indicator for whether respondents report it is always or sometimes (1) vs. never (0) acceptable for a man with financial means to marry a second wife. Eighty-one percent indicated any acceptance (81%). In addition to attitudes toward polygamy, currently married respondents were asked to rate their marriage(s) on a scale from "0" (worst possible marriage) to "10" (best possible marriage), with a resulting mean of 8.3.

Five undesirable behavioral outcomes were also identified: ever having smoked (23%), ever having considered suicide (5%), ever been so angry that the respondent threatened to hit or hurt, or actually hit or hurt someone (32%), ever been a victim of spousal physical abuse (15%), or ever physically abused a spouse (19%).*

Given that the specific effect of a third person present on any given outcome depends on the relationship between the person present and the respondent, the third-party presence variable for these analyses consists of a set of four indicators (for different types of people): a parent or a parent-in-law, an adult family member (including a spouse),† a child aged 5–17, or another unspecified person, with the reference category being no one over the age of 5.‡ In addition, social conformity was included as a main independent variable and its interaction with third-party presence was also tested to replicate past findings (Mneimneh, et al. 2015). Social conformity was measured using a shortened version of the Marlow Crowne scale (Crowne and Marlowe 1960), composed of five true/false statements. A summary score was calculated with a score of "1" given to each endorsed statement indicating social conformity. The score ranged from zero to five, with a mean of 1.80 (standard deviation = 1.47).

9.4 Analysis

A weighted two-level logistic regression model was used (respondents in level 1, interviewers in level 2) to estimate the between-interviewer variance and predict the presence of a third party during the interview using interviewer sociodemographic characteristics and attitudes toward privacy (with respondent characteristics and the time of the interview as control variables).§ To calculate the relative reduction in between-interviewer variance, a base model with the control variables only was compared to: (a) a model with the control variables and interviewer sociodemographics, (b) a model with the control variables and interviewer attitudes and views toward privacy, and (c) the full model. These models were based on Part I respondents who were interviewed by interviewers who completed the self-administered questionnaire and have full observed information on all the variables used in the model (n = 2475).¶ Weights were scaled within each interviewer to sum up to the workload of the interviewer.

To predict responses to each of the eight desirable/undesirable outcomes, design-adjusted regression models were used; specifically, logistic regression models for binary outcomes and linear regression models for scale outcomes or computed scores. Respondent

---

* Weighted percentages are calculated using only those respondents who were asked these questions. For example, questions about spousal physical abuse were only measured among ever married respondents.

† Spouses and other adult family members were combined in one group since only a small percentage of interviews had a spouse present given that all interviews were gender matched.

‡ A small percentage of interviews had more than one third party present. For those interviews, we prioritized coding in the following order: parent, other adult family member, child, unspecified other. For example, if both a parent and a child were present, the third-party status was recoded to indicate that a parent was present.

§ Primary stage units were not considered as a level because the size of clusters was small and because cluster-level weights were not available to rescale; only the composite respondent-level weight was provided in the data set.

¶ The total number of respondents interviewed by interviewers who completed the interviewer questionnaire is 2817; 100 cases are missing on the time of interview entered as a control variable, bringing the sample of cases to 2717. The reduction of sample from 2717 to 2475 is caused by missing data on any of the interviewer characteristics. The highest rate of missing data (6.6%) was on interviewer's income. Missing cases were dropped from the analysis.

sociodemographics were entered as control variables for each of the models. The main independent variables were the type of the third person present (parent or a parent-in-law, other adult family member, child or teenager, unspecified member) as well as the respondent's social conformity score and an interaction term between the type of third person present and the social conformity score. Only significant interaction terms (at a 0.05 significance level) were retained in the models. Respondent-level weights were used in all the models. Each model was based on the Part II sample that was administered each of the outcome measures: ever smoke (n = 1965), ever considered suicide (n = 1922), ever had an anger attack (n = 1960), ever physically abused a spouse (n = 1238), ever physically abused by a spouse (n = 1238), marital rating (n = 1104), overall polygamy endorsement (n = 874), and polygamy endorsed if finances allow (n = 870).* Predicted probabilities were generated for the models with significant interactions and displayed in Figure 9.1. All analyses were conducted using SAS 4.2.

9.5 Results

Table 9.1 presents estimates from the full model predicting third-party presence with interviewer characteristics and attitudes about privacy. None of the interviewer sociodemographic characteristics were significantly associated with third-party presence. One measure of interviewer attitudes about privacy did significantly predict third-party presence and another was marginally significant. Interviewers who disagreed (vs. agreed) with the statement that the Saudi culture respects one's privacy needs were more likely to have a third person present (regression coefficient = 0.734). On the other hand, interviewers who reported that they very much like to ask for a private interview were less likely to have a third party present than interviewers who reported that they like that part of their job only a little or not at all (regression coefficient = −0.495). As a set of variables, while interviewer sociodemographic characteristics were not found to be statistically associated with third-party presence (F-test = 0.76; p-value = 0.653), interviewer attitudes and views toward privacy were (F-test = 2.39; p-value = 0.036).

We turn next to quantifying the between-interviewer variation in third-party presence and estimating the relative percentage of between-interviewer variance that is explained by interviewer characteristics and attitudes. The base model included only respondent-level variables and resulted in an estimated between-interviewer variance of 0.552 (details for all models are available in Online Appendix 9B). When interviewer sociodemographics were added to the base model, the variance was reduced by about a quarter (to 0.407). When only interviewer attitudes and opinions toward privacy were added to the base model, the variance was also reduced by a quarter (to 0.410). When both sets of variables were added to the base model, the total reduction was 41% (to 0.321).

---

* The total number of Part II respondents is 1971. The drop in the total number of respondents for ever smoke, ever considered suicide, and ever had an anger attack outcome is because of missing data on any of the variables used in the models (rates are less than 2%). Ever abused a spouse and ever abused by a spouse were only measured among ever married respondents in Part II and had minimal missing data (less than 1%); marital rating was restricted to those who were currently married and had 1.3% missing data. Items measuring attitudes toward polygamy were administered to a random sub-sample of respondents, with no more than 1% missing data for either dependent variable. Given the low rate of missing data, analyses were conducted on completed cases only.

TABLE 9.1

Random Intercept Two-level Logistic Regression Model Predicting Third-Party Presence during the Interview (n = 2475) and Controlling for Respondent Sociodemographic Characteristics and Time of the Interview

	Coefficient	S.E.
Have a college degree vs. lower education level	0.266	0.260
Interviewer's age (30–44 years old) vs. younger	0.062	0.279
Saudi citizen vs. foreign	−0.144	0.277
Currently married vs. not	−0.191	0.346
Income (ref: low)	0.002	0.323
High income		
Middle income	−0.046	0.252
Prior interviewing experience vs. none	−0.088	0.308
Concurrent job vs. none	0.375	0.308
Religion very important vs. not	0.568	0.400
Disagree that the KSA culture respects privacy vs. agree	0.734**	0.275
Find it difficult to ask for privacy in KSA vs. not	−0.006	0.214
Like asking for a private setting	−0.495†	0.300
Very much like vs. a little/none		
Somewhat vs. a little/none	−0.367	0.266
Preference for non-private interviewer vs. accepting a refusal on an interview	−0.071	0.216

Note: Weighted estimates with weights scaled by interviewer workload; Level 1 = respondent, Level 2 = interviewer. Models controls for respondent gender, age, marital status, education, having social phobia in the past 12 months, having disability, household size, residence in rural or urban area, region of residence, and interview time; for full models, see Online Appendix 9B.
†p < 0.1; * p < 0.05; **p < 0.01.

Table 9.2 presents models predicting each of the five undesirable behavioral outcomes by type of third-party person present. Having a third person present during the interview was significantly (p < 0.05) associated with reporting three behaviors (smoking, ever physically abused a spouse, and ever being abused by a spouse). The direction of the association differed depending on the type of third person present and the social conformity level of the respondent. Controlling for respondent and household characteristics, respondents who were interviewed in the presence of an adult family member (other than a parent or a parent-in-law) were less likely to report smoking (coefficient = −0.837) than respondents interviewed in private. Having other unspecified people present also had a negative effect on reporting smoking, but the effect was moderated by the respondent's social conformity score (main coefficient = −1.523 and interaction coefficient = 0.888). Figure 9.1A displays the mean predicted probability of smoking by social conformity scores for respondents who have an unspecified other present and those without an unspecified other present. Among respondents without an unspecified other present, the mean predicted probability of smoking ranged from 0.24 (for those who score the lowest on social conformity) to 0.17 (for those who score the highest). However, among respondents who had an unspecified other person present, the mean predicted probability of smoking increased from 0.08 (for those who score the lowest on social conformity) to 0.68 (for those who score the highest). It is likely that respondents who reported a high score of social conformity and who had an unspecified other present had already confided in that other person (who might be their close peer and also engaged in the same behavior) leading them to respond more truthfully.

TABLE 9.2

Design-Adjusted Logistic Regression Model Predicting Each of the Behaviors

	Ever Smoke (n = 1965)		Ever Suicide (n = 1922)		Anger Attack: Hit or Threaten (n = 1960)		Ever Abuse (n = 1238)[a]		Ever Abused (n = 1238)[a]	
	Coefficient	S.E.	Coefficient	S.E.	Coefficient	S.E.	Coefficient	S.E.	Coefficient	S.E.
Social conformity score	−0.236**	0.069	−0.073	0.093	−0.243**	0.064	−0.103	0.081	−0.076	0.108
Parent/parent-in-law present	0.304	0.501	−0.339	0.511	−0.152	0.302	−1.128†	0.572	1.829†	0.950
Parent/parent-in-law *social conformity score	−0.837*	0.336							−0.958*	0.400
Non-parent adult family member present	0.054	0.412	1.171	0.831	0.046	0.619	0.637	0.703	0.355	0.659
Child/teenager present	−1.523**	0.454	−0.726	0.477	−0.037	0.353	−0.088	0.392	−0.048	0.304
Unspecified other present	0.888**	0.161	0.251	0.463	−0.086	0.439	−0.936	0.791	0.137	0.411
Unspecified other*social conformity score							0.824*	0.323		

Note: Model controls for respondent sociodemographics (gender, polygamy status, age, education, social phobia, disability, household size, and use of ACASI). Only significant interactions were entered in the model. The full models are available upon request.

[a] Model restricted to those ever married since the measure is related to marital physical violence.

†p < 0.1; *p < 0.05; **p < 0.01.

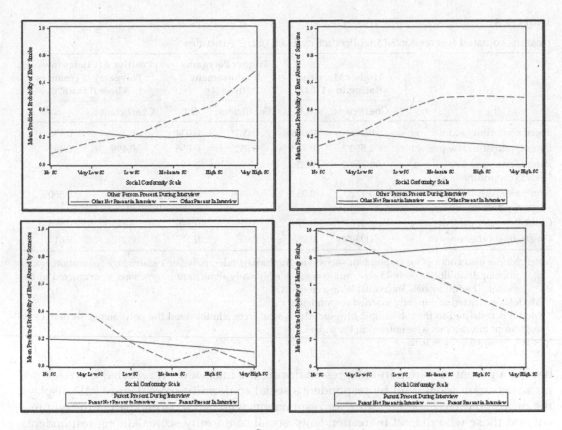

FIGURE 9.1
Predicted probabilities and scores from design-adjusted regression models controlling for respondent and household characteristics.

A similar interaction effect was observed for reporting perpetration of physical abuse against a spouse. The negative effect of having other unspecified people present during the interview on reporting physical abuse of a spouse was reduced as the respondent's social conformity score increased (main coefficient = −0.936 and interaction coefficient = 0.824). This interaction is graphed in Figure 9.1B. Moreover, having a parent or parent-in-law present was marginally related to reduced reporting of spousal physical abuse (main coefficient = −1.128).

There was also a significant interaction effect between respondent's social conformity score and having a parent or a parent-in-law present on reporting being a *victim* of spousal physical abuse, albeit in a different direction. As Figure 9.1C shows, the decline in the mean predicted probability for reporting being a victim of spousal abuse as the respondent score of social conformity increased was steeper for respondents who had a parent or a parent-in-law present during the interview than those who did not. Respondents who were high on social conformity might not have confided to their parents/parents-in-law about being a victim of spousal physical abuse, reducing the reporting of such sensitive information in the presence of a parent/parent-in-law.

Table 9.3 presents estimates for each of the three desirable marriage-related attitudes. Here, only rating of one's marriage was found to be significantly associated with third-person presence. Compared to having no third person present, having a child/teenager or an unspecified other present was positively associated with a more favorable rating of one's current marriage (coefficients = 0.876 and 1.059, respectively). However, the effect of

TABLE 9.3

Design-Adjusted Regression Model Predicting Each of the Attitudes

	Higher Marriage Rating (n = 1104)[a]		Higher Polygamy Endorsement (n = 874)[b]		Positive Attitudes toward Polygamy If Finances Allow It (n = 870)	
	Coefficient	S.E.	Coefficient	S.E.	Coefficient	S.E.
Social conformity score	0.201**	0.068	0.067**	0.024	−0.288*	0.138
Parent/parent-in-law present	2.550**	0.763	0.052	0.096	0.498	0.441
Parent/parent-in-law *social conformity	−1.754**	0.259				
Non-parent adult family member present	0.172	0.353	0.503†	0.290	0.094	0.560
Child/teenager present	0.876**	0.279	0.010	0.149	−1.44†	0.815
Unspecified other present	1.059**	0.255	−0.035	0.117	0.210	0.614

Note: Model also controls for respondent sociodemographics (gender, polygamy status, age, education, social phobia, disability, household size, and use of ACASI). Only significant interactions were entered in the model. The full models are available upon request.

[a] Model is restricted to currently married respondents.

[b] Model is restricted to the sub-sample of respondents who were administered the polygamous section. Only significant interactions were entered in the model.

†p < 0.1; * p < 0.05; ** p < 0.01.

having a parent or parent-in-law present during the interview on reported marriage ratings was again moderated by respondent's social conformity score. Figure 9.1D displays the mean predicted marital rating for respondents who had a parent/parent-in-law present and those who did not by respondents' social conformity score. Among respondents who did not have a parent or parent-in-law present, the mean predicted marital ratings slightly increased as the respondent's score of social conformity increased (from 8.15 to 9.24); however, a reverse effect was observed among respondents who had a parent or a parent-in-law present during the interview. Among this group, ratings of one's marriage became less favorable as social conformity increased. This effect is similar to the observations for two of the other behavioral outcomes described above, and could be attributed to respondents who are high on social conformity confiding this type of information (one's marriage rating) to only certain types of family members such as parents.

9.6 Discussion

This is the first study that investigates between-interviewer variation in interview privacy and the contribution of interviewer's characteristics to such variation. Interviewers varied significantly in the rate of interview privacy; while some interviewers had no private interviews, others conducted all their interviews in a private setting. This is consistent with our earlier finding of a significant between-interviewer variance in interview privacy (Mneimneh, et al. 2018a). While we speculated in our earlier work that such variation could be attributed to interviewer attitudes, traits, and skills, the data did not allow us to test this hypothesis. In this study, we could conduct such a test because a range of interviewer characteristics were collected. Our results revealed that none of the interviewer's sociodemographic characteristics were significantly associated with the private setting of the interview. This is partially consistent with Lau, et al.'s (2017) work that found that, for the most part, neither interviewer's age nor gender were consistently associated with interview privacy in five surveys.* However, unlike Lau, et al. (2017), who found that interviewers with more years of experience were more likely to report a private interview in three out of the five surveys, our study did not find an association between interviewer experience and interview privacy. This could be attributed to the specific pool of interviewers employed for this study; the majority did not have any experience (70%) and the rest had only minimal experience.

Our results, however, showed that among this pool of interviewers (who showed variation in attitudes toward privacy), those who disagreed that the Saudi culture respects one's privacy were less likely to conduct an interview in private. Such pre-existing beliefs might hold back interviewers from asking for or insisting on a private interview and lead to acceptance of the presence of others even if it is against the survey protocol. On the other hand, those who reported more positive attitudes toward asking for a private interview setting were marginally more likely to conduct an interview in private. Given the lack of specific training on requesting and achieving a private interview setting (Aquilino 1993; Mneimneh, et al. 2018b; Smith 1997; Taietz 1962), it is important to address interviewer opinions and attitudes toward privacy and empower them with culturally appropriate approaches to request and achieve a private interview setting. Considering the cultural background of the respondent is important, as the rate of interview privacy has been found to be higher in lower-income countries compared to high-income countries (Mneimneh, et al. 2018a for a review). Thus, if higher rates of interview privacy are desired, interviewer trainings need to: (1) emphasize interviewers' role in achieving a private interview setting, (2) identify and address interviewers' concerns about making such requests, and (3) practice role play on how and when to request privacy. Such trainings might equip interviewers with the needed skill set to enforce a study's privacy protocol.

The underlying argument for increasing interview privacy is that the lack of privacy might affect the reporting of desirable and undesirable information, thereby increasing measurement error. In our study, such effects were found for three out of the five behavioral outcomes and one out of the three attitudinal outcomes. The effects varied by the type of information requested, the relationship of the third person present to the respondent, and the respondent's level of social conformity. The presence of certain types of third people such as children and teenagers seems to have a direct (unmoderated) effect on reporting desirable information such as favorable ratings of one's current marriage. However, for other people, such as an unspecified other or a parent/parent-in-law, the effect is moderated by the respondent's reported social conformity score. Among respondents with higher scores of social conformity, those who were interviewed in the presence of an unspecified other person reported higher rates of smoking and spousal physical abuse than those interviewed without an unspecified other present. It is likely that respondents with a high social conformity score have shared information about these types of behaviors (e.g. smoking) only to non-family members (such as their peers who might also engage in this behavior) and feel compelled to be more truthful. However, such differences in reporting are not observed for respondents with very low social conformity scores; such respondents might not feel the need to differentially share sensitive information with different types of people.

A moderation effect was also found between parent or parent-in-law presence and respondent's social conformity score in relation to reports of marital rating and being

---

* Interviewer's age was significantly associated with interview privacy in one out of the five surveys. A significant interaction between interviewer gender and respondent gender was also found in one out of the five surveys.

a victim of spousal abuse, albeit in different directions. The findings show that among respondents who scored high on social conformity, those who had a parent/parent-in-law present reported less favorable views of their marriage (compared to those who did not have such a family member present), yet their reports of being a victim of spousal physical abuse were similar to those who did not have a parent/parent-in-law present. Thus, while such respondents might have confided to their parent/parent-in-law their true attitudes about their current marriage (making them feel compelled to report a more realistic marital rating in their parents/parent-in-law's presence), they might not have disclosed that they were a victim of spousal physical abuse.

In summary, when certain types of people (who already know the information in question) are present during the interview, respondents might feel compelled to report a more truthful answer. In fact, higher reports of undesirable outcomes in the presence of certain third parties during the interview have been reported in earlier studies (Aquilino 1997; Hoyt and Chaloupka 1994; Menimneh, et al. 2015). On the other hand, when interviewed in the presence of a family member who does not have knowledge of the sensitive information, respondents might decide not to report truthfully to avoid any potential repercussions from divulging this information.

Given that the effect of the presence of a third person might go in either direction depending on the type of information requested and the type of person present, one might argue that the presence of certain types of family members or close friends might enhance the quality of reporting of certain measures. We argue that given the confidentiality assurance provided to consented respondents, it is important to train our interviewers on ways to enhance an interview environment where respondents do not feel coerced to report certain information and where they are not worried about sharing information that might affect the dynamics of the relationship with anyone else who is present. Thus, we encourage the request for a private interview setting whenever sensitive information is collected.

A number of limitations are important to consider when interpreting these results. First, interview privacy measures were based on interviewer observations. If the study protocol explicitly calls for interview privacy, some interviewers might underreport the presence of a third person to avoid a negative evaluation. Such systematic misreporting in addition to the possibility of unintentional measurement error in recording the type of third person present might contribute to some of the observed variation. However, given that we have collected this information at the end of each questionnaire section rather than at the end of the entire interview, we believe that any potential unintentional recall issues were reduced. Second, the effect of interviewers' characteristics and estimated between-interviewer variance is conditional on the pool of recruited interviewers. While this specific pool of interviewers allowed us to explore the effect of interviewers' views on the interview privacy setting, the limited level of experience among our interviewer pool might have reduced our ability to detect any possible effect of interviewing experience on establishing privacy. Third, while we believe that interviewers' views and attitudes toward privacy affect their privacy-related behaviors, we cannot ascertain this as we did not have a precise measure of how privacy was requested since interviews were not recorded. Fourth, though the study collected rich information about interviewers, other unavailable information such as the interviewer's level and type of social skills and personality traits might also be associated with the interview privacy setting and could partially explain the residual between-interviewer variance. Finally, while we believe that the mechanism of some of the reporting effects is due to the fact that certain respondents might have already confided the requested information to certain types of bystanders and thus act as a prompt for truthfulness, without the ability to measure what information is actually held by these people, confirming such a mechanism is not possible.

In conclusion, given that the presence of a third person during the interview affects the reporting of certain types of sensitive information in different directions, we argue that more attention needs to be given to the privacy setting of the interview. First, we encourage researchers to collect such privacy measures and control for them in their models. Second, we urge practitioners to design appropriate training procedures and materials that are relevant to the culture where the interviews are being conducted and that give interviewers the necessary skills to request and maintain a private interview setting that is consistent with the confidentiality assurance provided to consented respondents.

References

American Association for Public Opinion Research. 2016. *Standard definitions: Final dispositions of case codes and outcome rates for surveys* (9th ed.), AAPOR.

Aquilino, W. 1993. Effects of spouse presence during the interview on survey responses concerning marriage. *Public opinion quarterly* 57(3):358–376.

Aquilino, W. A. 1997. Privacy effects on self-reported drug use: Interactions with survey mode and respondent characteristics. In *The validity of self-reported drug use: Improving the accuracy of survey estimates*, ed. L. Harrison, and A. Hughes, 383–415. Rockville, MD: National Institute on Drug Abuse.

Aquilino, W. S., D. L. Wright, and A. J. Supple. 2000. Response effects due to bystander presence in CASI and paper-and-pencil surveys of drug use and alcohol use. *Substance use and misuse* 35(6–8):845–867.

Bulck, J. 1999. Does the presence of a third person affect estimates of TV viewing and other media use? *Communications* 24(1):105–115.

Crowne, D. P., and D. Marlowe. 1960. A new scale of social desirability independent of psychopathology. *Journal of consulting psychology* 24(4):349–354.

Edwards, S. L., M. L. Slattery, and K. N. Ma. 1998. Measurement errors stemming from nonrespondents present at in-person interviews. *Annals of epidemiology* 8(4):272–277.

Harkness, J. A., A. Villar, and B. Edwards. 2010. Translation, adaptation, and design. In *Survey methods in multinational, multicultural and multiregional contexts*, ed. J. A. Harkness, M. Braun, B. Edwards, T. P. Johnson, L. E. Lyberg, P. Ph. Mohler, B-E Pennell, and T. W. Smith, 117–140. Hoboken, NJ: John Wiley & Sons.

Hartmann, P. 1995. Response behavior in interview settings of limited privacy. *International journal of public opinion research* 7(4):383–390.

Hoyt, G. M., and F. J. Chaloupka. 1994. Effect of survey conditions on self-reported substance use. *Contemporary economic policy* 12(3):109–121.

Kessler, R. C., and T. B. Üstün. 2004. The World Mental Health (WMH) Survey initiative version of the World Health Organization (WHO) Composite International Diagnostic Interview (CIDI). *International journal of methods in psychiatric research* 13(2):93–121.

Lau, C. Q., M. Baker, A. Fiore, D. Greene, M. Lieskovsky, K. Matu, and E. Peytcheva. 2017. Bystanders, noise, and distractions in face-to-face surveys in Africa and Latin America. *International journal of social research methodology* 20(5):469–483.

Mneimneh, Z. N., M. R. Elliott, R. Tourangeau, and S. G. Heeringa, 2018a. Cultural and interviewer effects on interview privacy: Individualism and national wealth. *Cross-cultural research* 52(5):496–523.

Mneimneh, Z. M., R. Tourangeau, B. Pennell, S. G. Heeringa, and M. R. Elliott. 2015. Cultural variations in the effect of interview privacy and the need for social conformity on reporting sensitive information. *Journal of official statistics* 31(4):673–697.

Mneimneh, Z, J. Wittrock, K. T. Le, and A. Elmaghrabi, 2018b. Investigating face-to-face interview privacy perceptions and preferences. Presentation at the Comparative Survey Design and Implementation workshop, Limerick, Ireland.

Moskowitz, J. M. 2004. Assessment of cigarette smoking and smoking susceptibility among youth: Telephone computer-assisted self-interviews versus computer-assisted telephone interviews. *Public opinion quarterly* 68(4):565–587.

Pollner, M., and R. E. Adams. 1997. The effect of spouse presence on appraisals of emotional support and household strain. *Public opinion quarterly* 61(4):615–626.

Smith, T. W. (1997). The impact of the presence of others on a respondent's answers to questions. *International journal of public opinion research* 9(1):33–47.

Smith, P. K., M. H. Bond, and Ç. Kağıtçıbaşı. (2006). *Understanding social psychology across cultures: Living and working in a changing world.* London, England: Sage.

Taietz, P. 1962. Conflicting group norms and the "third" person in the interview. *American journal of sociology* 68(1):97–104.

Triandis, H. C. (1995). *Individualism and collectivism.* Boulder, CO: Westview Press.

10

Unintended Interviewer Bias in a Community-Based Participatory Research Randomized Control Trial among American Indian Youth

Patrick Habecker and Jerreed Ivanich

CONTENTS

10.1 Introduction

Unlike traditional survey interviews where an interviewer is hired and trained by a survey organization, community-based participatory research (CBPR) often starts from the community (study area and social group of interest), in some cases including hiring interviewers who are members of the community of interest. CBPR aims to alter power distributions between researchers and those who participate in the research (Minkler and Wallerstein 2008). That is, historically, research has been driven entirely from academic, governmental, or commercial institutions downward to the communities of interest. CBPR aims to genuinely incorporate community partners with research institutions in designing and conducting research studies. To this end, CBPR assumes that: (a) genuine partnerships means co-learning, (b) research efforts include capacity building, (c) findings and knowledge should benefit all partners, and (d) CBPR involves long-term commitments (Wallerstein and Duran 2006). In general, CBPR values and aims to incorporate all parties into the research project for a holistic outcome that promotes community health and pushes rigorous advancements in science. However, the use of CBPR opens questions about the introduction of unintended negative consequences to the overall study design, implementation, recruitment, and data quality. This chapter tests if differences in the privacy of an interview – some of which are directly related to CBPR practices – are associated with several key outcomes of an ongoing family-based and culturally adapted evidence-based substance use prevention program for American Indian pre-adolescents aged 8–10.

The benefits of a CBPR approach in the social and behavioral sciences are considerable. A 2004 Agency for Healthcare Research and Quality report on CBPR evidence found that 78% of studies that took a CBPR approach to health outcomes reported increased community capacity after the studies were conducted (Viswanathan, et al. 2004). Successful CBPR interventions have reduced health disparities while building capacity among the partnering community (Barlow, et al. 2014) and developed culturally appropriate prevention programs (Allen, et al. 2018). CBPR is a key factor in the ability of these programs to develop tailored and localized programs.

Although there is not a strict "gold standard" method of conducting CBPR, scholars have provided many insights on how meaningful collaborations can be enacted. One recommendation is to "build on the strengths and resources within the community" (Israel, et al. 2008). A common implementation of this advice is to employ local community members to assist in conducting research with the academic institutions. This can involve hiring local program facilitators (Ivanich, et al. 2020), interviewers (Sittner, Greenfield, and Walls 2018), research councils (Fong, Braun, and Tsark 2003), or a combination to build local partnerships. Employing local community members builds sustainability, increases capacity, allows for transparency, and opens dialogues of meaningful feedback learning between parties (Minkler and Wallerstein 2008).

10.1.1 Social Desirability

A major concern for researchers who hire local interviewers is social desirability effects. Error from social desirability occurs when a participant alters their true response in order to present a more desirable view of themselves to those collecting data. An early example comes from W. E. B. Du Bois' study of the seventh ward of Philadelphia in the late 19th century (Du Bois 1899). The ability to read was a point of pride among residents and Du Bois was concerned that illiterate participants might say they are literate in order to appear more socially desirable.

Participants from Du Bois' study and others who engage in socially desirable responses are often thought to edit their answer in the latter phase of the cognitive response process (Sudman, Bradburn, and Schwarz 1996; Tourangeau, Rips, and Rasinski 2000). In this last phase, a participant translates the response they have in mind to fit the question and decides if they need to make any edits to their response before they finally answer. Here, participants think about how their answer will be perceived by the interviewer and whether they should change their answer to avoid a negative response from the interviewer (real or perceived). A major focus of social desirability research is therefore on circumstances where a participant decides to edit their answer to be more socially desirable.

There are two general views on social desirability. The first is that social desirability may be a characteristic of a person. This view proposes that there are some people who wish to be viewed by others in a favorable light in almost all of their actions with others (Johnson and Van de Vijver 2003; Tourangeau and Yan 2007). Assuming the participant has an idea (real or perceived) about what the other wishes to hear, they will shift their responses accordingly. A different perspective suggests that social desirability is specific to a given question (Johnson and Van de Vijver 2003). This view suggests that a socially desirable response is provoked by the property of the question itself, triggering a socially desirable response, and not a personality trait.

Socially desirable answers are more likely when questions focus upon sensitive issues and may result in under- or overreporting. Underreported responses have been detected when questions have been asked about illegal and deviant substance use (Davis, Thake, and Vilhena 2010; Johnson and Van de Vijver 2003; Krumpal 2013; van de Mortel 2008) and questions about victimization (Bell and Naugle 2007; Krumpal 2013; Sugarman and Hotaling 1997; van de Mortel 2008). Overreporting is more likely with desirable responses such as voter participation or seat belt use (Johnson and Van de Vijver 2003; Krumpal 2013). However, the sensitivity of a question may also vary between participants depending upon their history and current context (Kreuter, Presser, and Tourangeau 2008).

A major potential source of social desirability in an interview is a lack of privacy. Whenever a participant perceives that an interview is not private, there may be higher potential for an edited response, particularly with sensitive questions. A common threat to privacy is the presence of a third party during an interview. In these settings participants have often been found to edit their responses when asked about sensitive questions, but less so when asked about neutral questions (Aquilino 1993; Gfroerer, Wright, and Kopstein 1997; Hartmann 1995; Mneimneh, et al. 2015; Turner and Martin 1984). Where the interview takes place also affects privacy. Gfroerer, Wright, and Kopstein (1997) speculate that differences between national surveys on drug use among minors may in part be due to where the interviews take place, one in the home and one in the classroom. In this example, the classroom is presumed to be more private than the home as there may be parents present or nearby during an at-home interview.

A third area where privacy may affect social desirability in an interview may occur when the interviewer and the participant have a pre-existing relationship that is external to the interview context. This situation is exceptionally rare in standardized interviewing and most interview protocols assume that the interviewer and the participant are complete strangers (Fowler and Mangione 1990; Hyman 1954). In a CBPR context, with local interviewers hired from small communities, a participant knowing the interviewer becomes more likely. In such an interview, the participant has to consider what they are comfortable telling an interviewer who is also a member of their own community and with whom they will continue to have contact after the interview is complete. This goes beyond considering risk of divulgence to an outside entity and to the real consideration that a question response may affect their day to day life.

10.1.2 Research Questions

In this chapter, we test if the primary outcomes measured through interviewer-administered surveys are associated with three measurements of privacy in a family-based substance use prevention program for American Indian youths. We measure the presence of a third party during the interview, whether the interview took place at home or somewhere else, and if the interviewer knew the participant before the interview. We use data from the baseline wave of the program to test for associations between interview privacy and differences in mental health, substance use, cultural participation, and cultural discrimination among 8- to 10-year-olds. We expect that social desirability effects might decrease reports of externalizing and internalizing behaviors, reduce reports of substance use, and increase reports of cultural participation.

10.2 Methods

10.2.1 Sample

Data are from the baseline interview of the multi-site randomized control trial Bii-Zin-Da-De-Dah (bē-zen-dä-dē-dä; Listening to One Another; BZDDD) program. BZDDD is a family-based culturally adapted evidence-based substance use prevention program for American Indian pre-adolescents aged 8–10 delivered in four communities that share a common language, culture, and history (for detailed discussion of the adaptation process, see Ivanich, et al. 2020).

Youth and their families were recruited through a school-based and community outreach recruitment strategy. Working with community partners, we created a web-based interest form for local recruiters to use at events, school take-home flyers, and other community outreach activities, resulting in 497 online interest forms across four community partner locations. These recruiting activities are commonly used in CBPR and with research in American Indian communities (Whitesell, et al. 2019). Local recruiters attempted visits with 440 families for an official recruitment home visit and to screen for eligible youth living in the home. All families received wild rice for their time learning about the program and each youth received a Bii-Zin-Da-De-Dah drawstring bag. Of the 440 attempted visits, 363 families (85.5%) were successfully visited, eligible, and were initially interested in participating in the program.

A total of 679 interviews were conducted with 303 (68.9%) families that met eligibility criteria (i.e., youth living in the home of the target 8–10 age range, and a caregiver willing to attend the 14-week program) and did not refuse enrollment, move, or become unreachable after the initial visit. Baseline data collection began in May 2017 and was completed in April 2018. Data were collected using in-person interviews with a trained interviewer from the local community in which the respondent lived. Each caregiver and child who completed the baseline survey received a $20.00 Visa gift card. All materials, processes, and surveys were reviewed and approved by our local research advisory boards called Prevention Research Councils for community and culture appropriateness prior to university institutional review board approval (IRB #20140214158FB). All participants were informed of policies, risks, benefits, and compensation before signing consent/assent agreement forms for enrollment and data collection.

10.2.2 Measures

The BZDDD program focuses on substance use, mental health, and cultural/traditional engagement. Mental health was measured using the Achenbach System of Empirically Based Assessment (ASEBA) youth self-reports (Achenbach 1991). A standardized scale of 102 questions, ASEBA captures information about psychopathology of youth, specifically internalizing and externalizing behavior. Due to the young age of participants (8–10 years old), permission was granted by ASEBA to remove a total of six questions: two focused on suicidal thoughts and attempts (1 and 2), two focused on sex and sexual thoughts (3 and 4), and two related to hearing things that others do not hear and seeing things that others do not see (5 and 6). The latter two (5 and 6) were removed upon advice from community members serving as field interviewers and the local research advisory boards. In this American Indian community, there is a history of being shamed, and institutionalized for similar cultural practices, and such questions were therefore not appropriate for this study.

Substance use is a dichotomous measure if youth said yes (1) to any of the following questions. If they have ever "had a drink of any beverage that contains alcohol, such as beer, wine, hard liquor, or mixed drinks," "smoked a cigarette," "used marijuana (pot/weed)," "used any other type of drug or substance to get high," and "used prescription pills to get high." Youth that indicated that they had not tried any of these were assigned a value of (0).

Cultural participation is measured with a sum of dichotomous measures. Participants were asked if they had offered tobacco, participated in ceremonial songs, smudged or saged, participated in a ceremonial dance, gone to a traditional healer, sought advice from a spiritual advisor, or participated or sung in a drum group in the past 12 months. A yes is counted as a (1) and a no as a (0). The sum of these responses provides a measure of how much each youth participated in cultural practices.

Cultural discrimination is measured with a series of questions about being a tribal member. Individuals were asked how often you have been treated as if you were not smart because you are [redacted to maintain confidentiality of the communities], you have been treated as if you were not as good as others because you are [redacted], you have been treated as if you were not an honest person because you are [redacted], you have been treated with less respect than others because you are [redacted], you have been called names or insulted because you are [redacted], you have been treated negatively because you are [redacted], your family talks about incidents of discrimination, and you feel confident in discussing discrimination with your family. Youth could respond, never (0), rarely (1), sometimes (2), and often (3). We summed the numeric responses of all eight measures to create a single cultural discrimination measure. The Cronbach's alpha for this measure is 0.77.

The interviewer was asked a few questions once the interview was complete and they were alone. We use three questions about privacy in this post-interview questionnaire. We consider interviewers to be known (1) by the participant if they answered "very well" or "somewhat" to the question: "how well did you know the person you interviewed before you interviewed them?" and unknown (0) if they said, "not at all." The presence of an engaged third party was assessed by asking: "Was anyone else present during any portion of the interview? By present, we mean not just walking through the area where the interview was taking place, but listening to or taking part in the interview process," and coded yes (1) or no (0). Interviewers also noted if the interview took place in the participant's home (1) or somewhere else (0).

10.2.3 Analytic Strategy

We first present univariate descriptive statistics of all variables used in this study in Table 10.1. Then we present a series of multilevel linear random-effects models for four of the dependent variables and a logistic random-effects model for the substance use outcome. In these models, the participants of the study are treated as level 1 and the interviewers are level 2. The only information we have about interviewers is their unique ID, all other information used in the analysis is about the participant. Results for internalizing and externalizing measures are presented in Table 10.2. Results for the remaining dependent variables are presented in Table 10.3. All analytic models include covariates for the participant's gender and age. For every outcome, each privacy measure is tested on their

TABLE 10.1

Descriptive Statistics of Participants (N=365)

Statistic	Mean/%	Std. Dev.	Min	Max
Female	52%	0.50	0	1
Age	9.10	0.91	7	11
Know interviewer	26%	0.44	0	1
Present third party	31%	0.46	0	1
Interviewed in home	73%	0.45	0	1
Internalizing – total	16.77	8.99	0	53
Anxiety/depressed	5.53	3.96	0	20
Withdrawn	4.67	2.92	0	15
Somatic complaints	6.57	3.89	0	19
Externalizing – total	7.95	6.92	0	35
Rule breaking	2.15	2.50	0	15
Aggressive behavior	5.80	4.91	0	24
Substance use – yes	11%	0.31	0	1
Cultural participation	2.41	2.24	0	7
Cultural discrimination	13.59	5.44	0	31

own (with demographic covariates) in their own model, and then all three privacy measures are placed in a combined model. All analyses were conducted in R (R. Core Team 2014) using the glmmTMB package.

TABLE 10.2

Linear Mixed-Effects Models Predicting Internalizing and Externalizing Behaviors

	Internalizing				Externalizing			
	(1)	(2)	(3)	(4)	(5)	(6)	(7)	(8)
Female	0.874	0.808	0.848	0.952	−1.358[+]	−1.348[+]	−1.418[+]	−1.400[+]
	(0.918)	(0.933)	(0.934)	(0.923)	(0.724)	(0.724)	(0.725)	(0.728)
Age	−1.573**	−1.594**	−1.557**	−1.555**	−0.253	−0.260	−0.277	−0.279
	(0.506)	(0.514)	(0.514)	(0.508)	(0.399)	(0.399)	(0.399)	(0.400)
Know interviewer	−3.539***			−3.459**	−0.201			−0.200
	(1.042)			(1.049)	(0.822)			(0.827)
Present third party		0.684		0.530		0.443		0.289
		(1.045)		(1.052)		(0.811)		(0.829)
Interviewed in home			−0.942	−0.924			0.812	0.770
			(1.013)	(1.016)			(0.786)	(0.801)
Constant	31.556***	30.352***	30.780***	31.228***	11.011**	10.696**	10.956**	10.824**
	(4.669)	(4.777)	(4.732)	(4.723)	(3.682)	(3.709)	(3.672)	(3.723)
N	365	365	365	365	365	365	365	365
AIC	2,628.065	2,639.116	2,638.674	2,631.101	2,454.627	2,454.386	2,453.611	2,457.409

[+]$p < 0.1$; *$p < 0.05$; **$p < 0.01$; ***$p < 0.001$.

TABLE 10.3

Logistic and Linear Mixed-Effects Models Predicting Substance Use, Cultural Participation, and Cultural Discrimination

	Substance Use[a]				Cultural Participation				Cultural Discrimination			
	(1)	(2)	(3)	(4)	(5)	(6)	(7)	(8)	(9)	(10)	(11)	(12)
Female	1.19	1.17	1.18	1.17	0.162 (0.22)	0.156 (0.22)	0.158 (0.22)	0.153 (0.22)	0.274 (0.57)	0.267 (0.57)	0.290 (0.57)	0.251 (0.57)
Age	1.04	1.03	1.03	1.03	-0.034 (0.12)	-0.032 (0.12)	-0.037 (0.12)	-0.035 (0.12)	-0.804* (0.31)	-0.788* (0.31)	-0.800* (0.31)	-0.794* (0.31)
Know interviewer	0.57				0.112 (0.27)			0.111 (0.27)	0.597 (0.65)			0.538 (0.66)
Present third party		1.19		1.22		-0.156 (0.26)		-0.166 (0.26)		-0.684 (0.64)		-0.648 (0.65)
Interviewed in home			0.97	0.92			0.067 (0.26)	0.093 (0.26)			-0.017 (0.62)	0.075 (0.63)
Constant	0.10	0.09	0.09	0.10	2.524* (1.15)	2.655* (1.16)	2.555* (1.15)	2.625* (1.16)	20.605*** (2.88)	21.110*** (2.91)	20.718*** (2.88)	20.982*** (2.92)
N	365	365	365	365	365	365	365	365	365	365	365	365
AIC	195	197	197	199	1,599	1,599	1,599	1,603	2,275	2,275	2,276	2,278

[a] Odds ratios presented.

+ $p < 0.1$; * $p < 0.05$; ** $p < 0.01$; *** $p < 0.001$.

10.3 Results

Just over half of the youth who participated were female (52%), and the average age of the youth is 9.10 years old (Table 10.1). In 26% of the interviews, the interviewers knew the youth prior to the interview, 31% of the interviews took place while someone else was present, and 73% of the interviews took place in the youth's home. The average score for the total internalizing measure is 16.85 on a scale that ranged from 0 to 53. The first subscale of the internalizing measure, anxiety/depressed, showed an average score of 5.57 with a high score of 20. The second subscale, withdrawn, showed an average score of 4.68 out of 15. The last internalizing subscale, somatic complaints, has an average score of 6.61 and a high value of 19. The overall externalizing scale averaged 7.98 and ranged from 0 to 35. The first subscale of the externalizing measure, rule breaking, has an average of 2.17 and ranged from 0 to 15. The final subscale of externalizing, aggressive behavior, has an average of 5.81 with a range of 0–24. Only 11% of the youth indicated that they have ever tried any form of substance use. On a scale of 0–7, the average cultural participation score for youth was 2.41. Lastly, the average reported discrimination score was 13.65, which ranged from 0 to 31.

Table 10.2 presents the results of random-effects models predicting overall measures of internalizing (linear) and externalizing (linear) behavior, and Table 10.3 presents substance use (logistic), cultural participation (linear), and cultural discrimination (linear). All models include random effects for the 19 independent interviewers.

10.3.1 Internalizing and Externalizing Behaviors

Model 1 of Table 10.2 shows that youth who were known to the interviewer prior to the interview had −3.539 fewer reported internalizing behaviors (p < 0.001) after adjusting for gender and age. Models 2 and 3 show that the other measures of privacy (presence of a third party and interview location) are not significantly associated with internalizing behavior. Model 4 shows that knowing an interviewer prior to an interview still significantly reduced the total internalizing score by −3.459 (p < 0.001), after adding the other two measures of privacy to the model.

None of the measures of privacy are associated with externalizing behavior in any of the models (Table 10.2, Models 5, 6, 7, 8). Neither gender nor age significantly predicts the externalizing aggregate scale. In separate models we also explore the relationship between our privacy measures and the three internalizing subscales and the two externalizing subscales (see Online Appendix 10A). In brief, we find that knowing an interviewer was negatively associated with all three internalizing subscales and that none of the privacy measures were associated with the externalizing subscales.

10.3.2 Substance Use, Cultural Participation, and Cultural Discrimination

Table 10.3 presents the random-effects models predicting substance use (Models 1–4), cultural participation (Models 5–8), and cultural discrimination (Models 9–12). We see that none of the privacy measures are associated with reports of substance use, cultural participation, or cultural discrimination. There are also no associations between gender and the outcome variables in any of the models in Table 10.3. However, a one year increase in age is associated with an average decrease in reports of cultural discrimination of −0.794 (p < 0.05). This association appears in all four models in this set with a similar effect size (Models 9–12).

10.4 Discussion

We find that when an interviewer knew the participant in a CBPR intervention study among American Indian communities, the reports of internalizing behavior were on average 3.5 points lower (Table 10.2, Models 1 and 4) than instances where the participant was unknown to the interviewer. Knowing a participant is negatively associated with all of the internalizing subscales (Online Appendix Table A10A.1). This is critical as the ASEBA measurement of internalizing behavior includes thresholds for borderline and clinical ranges on all its scales. Underreporting of internalizing behavior means that borderline or clinical participants may be incorrectly classified. Programs that rely on these classifications for intervention eligibility would then be missing key participants when they are using interviewers who know the participants outside of the survey context who would otherwise be included.

Although knowing an interviewer was associated with lower internalizing scores, the same privacy concern was not associated with reports of externalizing behavior, substance use, cultural participation, or cultural discrimination. The presence of a third party during the interview and whether the interview took place in the home was also not associated with any of the core measures. The majority of the tested associations between privacy concerns and the core BZDDD outcomes among youth are not significant. These outcomes indicate that social desirability effects due to these three privacy concerns are largely non-existent among youth during the baseline period of this study.

A longstanding challenge of CBPR projects is that there is greater potential for breaches of confidentiality and privacy when the lines between researcher and participant blur. This is particularly true when we employ interviewers and other research members for a research project situated in their own community (Banks, et al. 2013; Holkup, et al. 2004). In such settings, participants may edit their responses to prevent their neighbor or someone else that they know who is working as an interviewer from discovering information they would not normally disclose outside of an interview. The findings of this text are therefore largely reassuring as they show few associations between reduced privacy and edited responses.

However, there is still work to be done to reduce the potential for social desirability concerns related to privacy. In this study, 26% of the youth interviews took place with a youth whom the interviewer already knew. For research projects with the potential for this situation to occur, we suggest that blocks of sensitive questions should be administered with a self-administered data collection mode when feasible. Using either a paper questionnaire or preferably a form of computer-assisted self-interviewing (CASI) would prevent the interviewer from hearing or seeing how a participant responds to a sensitive question. This would retain the advantage of employing interviewers from the local community and discourage edited answers.

Addressing the 31% of interviews that took place with a third party present is more difficult. Youth interviewed for the baseline data collection were between 8 and 10 years old, and 73% of the interviews took place in their home. Although we lack data on who the third person was and their relationship to the youth, it is reasonable to assume the presence of caretakers or other relatives. It can be exceptionally difficult to request privacy in many interview settings in the home, and even more difficult when interviewing younger children (Mauthner 1997; Mneimneh, et al. 2018). Fortunately, we see no evidence that the presence of a third party was associated with any of the key outcomes for this study, nor any associations with the location of the interview. Additionally, although 31% of interviews with a third party present is high, it is remarkably similar to the percent (30.2%) of third party presence reported for the World Mental Health Interviews conducted in the United States (Mneimneh, et al. 2018).

10.4.1 Limitations

The sample for this study is unique in that it represents a group of people who share a language, culture, and history. Further, it is a sample of youth aged 8–10. The extent to which the associations tested here can be generalized to other groups and age ranges is questionable. However, this is one of the few studies to look for social desirability effects among American Indians in a survey context. A review of social desirability in cross-cultural research identified no studies that examined social desirability among American Indians (Johnson and Van de Vijver 2003).

Another limitation is that although we find that many of the key outcomes are unassociated with our privacy measures, many other questions were asked in the survey. Our focus here is to assess the core outcomes only. Future studies using these data or data gathered with similar methods should assess item-specific privacy effects whenever possible.

Finally, there were multiple interviewers who worked in each of the study's primary locations. However, these interviewers were not randomly assigned to different participants and we do not have information about how interviewers decided to select a family to interview out of the pool that was available for a given community. This means that we cannot assess if some interviewers picked participants that they knew or if the outcomes we examined in this chapter were associated in some way with the interviewer's choice to interview a family.

10.5 Conclusions

This study demonstrates that with one key exception, there does not appear to be an association between privacy-motivated social desirability effects and the key outcomes of the BZDDD project. The exception is where internalizing reports are significantly lower when the interviewer knew the participant before the interview. This association is present among all three of the internalizing subscales as well. To address the source of error, we suggest employing self-report modules for sensitive questions in the future. Overall, the lack of significant associations is very good news. Each study has unique aims and protocols, and we certainly acknowledge that American Indians and Alaska Native peoples are not a monolithic ethnic group and have a great deal of geographic, historical, cultural, and political variation. However, these results provide early reassurance to those using community interviewers in rural reservation-based settings, particularly for researchers concerned about introducing sources of error related to privacy and the employment of community interviewers in association with the key outcomes tested here. The results discussed in this chapter are item- and study-specific. We strongly encourage future researchers to plan for similar tests in their own studies if they employ community interviewers and when privacy during an interview may not be absolute.

References

Achenbach, T. M. 1991. *Integrative Guide for the 1991 CBCL/4–18, YSR, and TRF Profiles*. Burlington, VT: Department of Psychiatry, University of Vermont. http://www.getcited.org/pub/102906443.

Allen, J., S. M. Rasmus, C. C. T. Fok, B. Charles, D. Henry, and Q. Team. 2018. Multi-level cultural intervention for the prevention of suicide and alcohol use risk with Alaska native youth: A nonrandomized comparison of treatment intensity. *Prevention Science* 19(2):174–85.

Aquilino, W. S. 1993. Effects of spouse presence during the interview on survey responses concerning marriage. *Public Opinion Quarterly* 57(3):358–76.

Banks, S., A. Armstrong, K. Carter, H. Graham, P. Hayward, A. Henry, T. Holland, C. Holmes, A. lee, A. McNulty, N. Moore, N. Nayling, A. Stokoe, and A. Strachan. 2013. Everyday ethics in community-based participatory research. *Contemporary Social Science* 8(3):263–77.

Barlow, A., B. Mullany, N. Neault, N. Goklish, T. Billy, R. Hastings, S. Lorenzo, C. Kee, K. Lake, C. Redmond, A. Carter, and J. T. Walkup. 2014. Paraprofessional-delivered home-visiting intervention for American Indian teen mothers and children: 3-year outcomes from a randomized controlled trial. *The American Journal of Psychiatry* 172(2):154–62.

Bell, K. M., and A. E. Naugle. 2007. Effects of social desirability on students' self-reporting of partner abuse perpetration and victimization. *Violence & Victims* 22(2):243.

Davis, C. G., J. Thake, and N. Vilhena. 2010. Social desirability biases in self-reported alcohol consumption and harms. *Addictive Behaviors* 35(4):302–11.

Du Bois, W. E. B. 1899. *The Philadelphia Negro; a Social Study*. Philadelphia, PA: University of Pennsylvania Press.

Fong, M., K. L. Braun, and J. Tsark. 2003. Improving native Hawaiian health through community-based participatory research. *Californian Journal of Health Promotion* 1(1):136–48.

Fowler, F. J., and T. W. Mangione. 1990. *Standardized Survey Interviewing*. Newbury Park, CA: SAGE Publications, Inc.

Gfroerer, J., D. Wright, and A. Kopstein. 1997. Prevalence of youth substance use: The impact of methodological differences between two national surveys. *Drug & Alcohol Dependence* 47(1):19–30.

Hartmann, P. 1995. Response behavior in interview settings of limited privacy. *International Journal of Public Opinion Research* 7(4):383–90.

Holkup, P. A., T. Tripp-Reimer, E. M. Salois, and C. Weinert. 2004. Community-based participatory research. *ANS. Advances in Nursing Science* 27(3):162–75.

Hyman, H. H. 1954. *Interviewing in Social Research*. Chicago, IL: University Of Chicago Press.

Israel, B. A., A. J. Schulz, E. A. Parker, and A. B. Becker. 2008. Critical issues in developing and following community-based participatory research principles. In: *Community-Based Participatory Research for Health*, ed. M. Minkler, and N. Wallerstein, 47–62. San Francisco, CA: Jossey-Bass.

Ivanich, J. D., A. C. Mousseau, M. Walls, L. Whitbeck, and N. R. Whitesell. 2020. Pathways of adaptation: Two case studies with one evidence-based substance use prevention program tailored for indigenous youth. *Prevention Science*, 21:43-53.

Johnson, T. P., and F. J. R. van de Vijver. 2003. Social desirability in cross-cultural research. *Cross-Cultural Survey Methods* 325:195–204.

Kreuter, F., S. Presser, and R. Tourangeau. 2008. Social desirability bias in CATI, IVR, and web surveys: The effects of mode and question sensitivity. *Public Opinion Quarterly* 72(5):847–65.

Krumpal, I. 2013. Determinants of social desirability bias in sensitive surveys: A literature review. *Quality & Quantity* 47(4):2025–47.

Mauthner, M. 1997. Methodological aspects of collecting data from children: Lessons from three research projects. *Children & Society* 11(1):16–28.

Minkler, M., and N. Wallerstein. 2008. *Community-Based Participatory Research for Health*. San Francisco, CA: Jossey-Bass.

Mneimneh, Z. M., R. Tourangeau, B. Pennell, S. G. Heeringa, and M. R. Elliott. 2015. Cultural variations in the effect of interview privacy and the need for social conformity on reporting sensitive information. *Journal of Official Statistics* 31(4):673–97.

Mneimneh, Z. N., M. R. Elliott, R. Tourangeau, and S. G. Heeringa. 2018. Cultural and interviewer effects on interview privacy: Individualism and national wealth. *Cross-Cultural Research* 52(5):496–523.

van de Mortel, T. F. 2008. Faking it: Social desirability response bias in self-report research. *The Australian Journal of Advanced Nursing* 25(4):40–8.

R. Core Team. 2014. *R: A Language and Environment for Statistical Computing*. Vienna, Austria: R Foundation for Statistical Computing. 2013.

Sittner, K. J., B. L. Greenfield, and M. L. Walls. 2018. Microaggressions, diabetes distress, and self-care behaviors in a sample of American Indian Adults with Type 2 Diabetes. *Journal of Behavioral Medicine* 41(1):122–9.

Sudman, S., N. A. Bradburn, and N. Schwarz. 1996. *Thinking about Answers: The Application of Cognitive Processes to Survey Methodology*. San Francisco, CA: Jossey-Bass Publishers.

Sugarman, D. B., and G. T. Hotaling. 1997. Intimate violence and social desirability: A meta-analytic review. *Journal of Interpersonal Violence* 12(2):275–90.

Tourangeau, R., L. J. Rips, and K. Rasinski. 2000. *The Psychology of Survey Response*. New York: Cambridge University Press.

Tourangeau, R., and T. Yan. 2007. Sensitive questions in surveys. *Psychological Bulletin* 133(5):859–83.

Turner, C. F., and E. Martin. 1984. *Surveying Subjective Phenomena*. Vol. 1. New York: Russell Sage Foundation.

Viswanathan, M., A. Ammerman, E. Eng, G. Garlehner, K. N. Lohr, D. Griffith, S. Rhodes, C. Samuel-Hodge, S. Maty, L. Lux, L. Webb, S. F. Sutton, T. Swinson, A. Jackman, and L. Whitener. 2004. *Community-Based Participatory Research: Assessing the Evidence*. Agency for Healthcare Research and Quality (US). https://www.ncbi.nlm.nih.gov/books/NBK11852/.

Wallerstein, N. B., and B. Duran. 2006. Using community-based participatory research to address health disparities. *Health Promotion Practice* 7(3):312–23.

Whitesell, N. R., A. C. Mousseau, E. M. Keane, N. L. Asdigian, N. Tuitt, B. Morse, T. Zacher, R. Dick, C. M. Mitchell, and C. E. Kaufman. 2019. Integrating community-engagement and a multiphase optimization strategy framework: Adapting substance use prevention for American Indian families. *Prevention Science* 20(7):1136–46.

11

Virtual Interviewers, Social Identities, and Survey Measurement Error

Frederick G. Conrad, Michael F. Schober, Daniel Nielsen, and Heidi Reichert

CONTENTS

11.1 Introduction

Many considerations can – and should – go into the critical decision of survey mode for empirical studies, including cost, data quality, efficiency, and the trade-offs between them. For example, interviewer-administration is more costly than self-administration (e.g., Baker 1998), yet survey interviewers remain indispensable for some types of surveys (e.g., when high response rates are a priority, when studying populations whose members cannot be reached via other modes, and when studying populations with limited literacy). Interviewers can increase response rates, motivate conscientious responding, explain how certain questions should be interpreted or how response tasks should be performed, and help assure respondents that answers will be maintained confidentially. However, interviewers can also introduce error into the measurement process, for example promoting socially desirable responding (e.g., Tourangeau and Yan 2007) about sensitive topics and introducing bias because of their perceived social identities, such as gender, age, race, and sub-group membership when the questions concern those identities (e.g., Davis, et al. 2009; Ehrlich and Reisman 1961; Groves, et al. 2009, 292–295; Kane and McCaulay 1993; Liu 2016; Schuman and Converse 1971; Wolford, et al. 1995).

The effects of interviewers' social identities on responses generally take the same form irrespective of the particular identity involved: more prevalent support for positions that seem more consistent with the views of an interviewer given their identity such as race or gender. For example, Schuman and Converse (1971) reported that 35% of Black respondents indicated they can trust most White people when a White interviewer read the question but only 7% reported that they could trust most White people when the interviewer was Black. Similarly, Kane and McCaulay (1993) reported 19.8% of respondents reported sharing childcare with their spouse when the interviewer was female compared to only 13.5% when the interviewer was male.

In this chapter, we explore an approach that may incorporate attractive features of both interviewer- and self-administration while reducing monetary costs and measurement error inherent in either approach. The proposal is that by implementing embodied, animated agents or "virtual interviewers" (*VI*s), which are considerably cheaper than live human interviewers, it may be possible to engage respondents more than in traditional text-based web surveys, while deliberately using *VI*s' social identities to improve the quality of answers. In fact, *VI*s may facilitate further improving data quality by (1) matching *VI*s' perceived social identities with respondents' self-reported social identities or even (2) allowing respondents to choose a *VI*.

Our research questions are: (1) Do *VI*s bias responses on the basis of their perceived social identities (in this case race and gender), to at least the same extent that human interviewers do? (2) Are respondents more likely to disclose undesirable information for questions concerning sensitive topics when they match the *VI* on a perceived identity compared to when there is no such match? (3) Do respondents disclose more undesirable information when they choose their *VI* than when they do not?

11.2 Virtual Interviewers

The three *VI*s pictured in Figure 11.1 vary in their capabilities and attributes, such as their facial realism (Victoria is more cartoon-like; Derek is more photo-realistic), amount of facial movement (there were two versions of Victoria and Derek, one high and the other low in facial movement, including movement of eyes, eyebrows, and mouth), autonomy (Ellie interprets the respondents' speech and determines what to say next; Derek seems similarly autonomous but is actually controlled by an experimenter who strategically plays particular video files to convey Derek's apparent autonomy), responsivity (Ellie reacts to the respondent's speech by nodding and raising her eyebrows at appropriate moments, while the comparable movements by Derek and Victoria are not conditional on respondent actions), and so on. Despite their human-like qualities no one would mistake these or other *VI*s for actual people. Yet in the three studies that deployed these *VI*s (Conrad, et al. 2015; Lind, et al. 2013; Lucas, et al. 2014), respondents reacted socially to them.

FIGURE 11.1
Three virtual interviewers. Left-to-right: "Victoria" (Lind, et al. 2013); "Derek" (Conrad, et al. 2015); "Ellie" (Lucas, et al. 2014).

In general, the attributes of *VI*s are largely under the control of researchers and designers. These include visual attributes such as eye color, facial features, clothing, and skin tone; vocal and linguistic features such as accent, tone, pitch, pitch variation, and prosodic variation; and their communicative sophistication such as comprehension ability and dialog capability. At least some of these attributes may convey social identities, including gender and race. How likely is it that respondents will be affected by *VI* design?

Computers are social actors. There is now considerable evidence that humans react socially to computers, i.e., as if the computers were human, leading to what Clifford Nass and his colleagues labeled the "Computers are Social Actors" (CASA) perspective. One of the most compelling demonstrations of the CASA phenomenon is that users are polite to computers despite recognizing that the computers cannot detect user politeness. Nass, Moon, and Carney (1999) required participants to complete a tutoring task with a desktop computer and then evaluate the success of the tutoring session on the computer they had just interacted with, a paper questionnaire, or a different but identical computer. When respondents evaluated the tutoring experience on the computer that also administered the tutoring, scores were higher and less variable than those provided in the other conditions, suggesting that users wished to avoid being rude or hurtful to the original computer. In another well-known CASA demonstration (Nass, Moon, and Green 1997), participants rated speaking computer tutors as more informative or effective when gender stereotypes conformed to the perceived gender of the voice output, e.g., participants rated female-voiced computer tutors to be more informative when the topic concerned love and relationships and male-voiced computers as more proficient in technical subjects. CASA effects seem particularly hard to suppress when a virtual agent is depicted in the interface (e.g., Antos, et al. 2011; Nass, Isbister, and Lee 2000).

Although evidence for the CASA phenomenon has been reported across a range of tasks, there is no guarantee that the phenomenon will extend to survey interviews, especially if attributing animacy to an interviewing agent might lead respondents to feel judged when their answers are socially undesirable. Tourangeau, Couper, and Steiger (2003) placed one of two photos in the interface of an online questionnaire, either a male or female researcher. Compared to a version of the questionnaire with no photo in the interface, the versions with a photo did not produce any more socially desirable responding, although the version with the female researcher's photo did produce a small but significant "gender of researcher" effect for a gender attitude scale borrowed from Kane and Macauley (1993).

Might a moving, speaking, embodied agent provide a more powerful trigger to respond socially than a still photograph, potentially making it hard to turn off the processes that are triggered by human–human interaction? The evidence suggests that this is indeed the case. In Lind, et al. (2013), respondents indicated they engaged in stigmatized behaviors equally often when Victoria or an in-person human interviewer asked the questions, but in both cases these behaviors were reported less often than when a disembodied voice asked the questions, suggesting that the *VI* discouraged candid responding to the same extent that a human interviewer did. In Conrad, et al. (2015), respondents produced more backchannels (acknowledging Derek's speech and consenting to his continued speaking through utterances such as "uh huh" or "okay" and nodding) as well as smiling marginally more when Derek was implemented with more facial movement. And in Lucas, et al. (2014), Ellie – who was designed to establish rapport with respondents – elicited more disclosure and more intense expressions of emotion when respondents (correctly) thought the agent was autonomous than when told a human controlled it.

11.3 Virtual Interviewer Identities and Measurement Error

It may be the case that respondents cannot "turn off" the CASA impulse, in which case race- and gender-of-*VI* effects will occur as frequently as race- and gender-of-human-interviewer effects do. It is also possible that respondents' impulse to treat *VI*s as social actors and their recognition that *VI*s are inanimate may be in conflict, moderating the frequency of social responses. How common are such identity-related effects with human interviewers?

Both race- and gender-of-interviewer effects tend to be observed for items that concern issues related to race or gender, respectively, but not for items on other topics. For example, Schuman and Converse (1971) observed that for questions that "employ racial terms bluntly" 39% showed a race-of-interviewer effect, but only 17% of other questions showed a race-of-interviewer effect. In 11 studies (see Online Appendix Table A11A.1) that reported race-of-interviewer effects, the percent of race-related items for which an effect was observed ranged between 36% and 100%.* Race-of-interviewer effects have primarily been observed in face-to-face interviews, but such effects have also been reported in telephone interviews. Gender-of-interviewer effects are more evenly divided between face-to-face and telephone interviews, presumably because voice can be a predictive – though not perfect – indicator of gender (see Davis, et al. 2009).

While such interviewer effects can undoubtedly harm data quality, interviewers' social identities may also be exploited to improve data quality. For example, respondents report more stigmatized behavior such as drug use and sensitive sexual behaviors – which is presumably more truthful – when they match the interviewer's social identities (e.g., age, gender, race/ethnicity, education) (Catania, et al. 1996; Johnson, et al. 2000). Moreover, the effect is stronger the more identities they share, presumably because there is increasingly less social distance between the respondent and the interviewer, increasing their comfort (Johnson, et al. 2000). While not practical with live interviewers because of the limited size and diversity of the interviewing corps and possible legal and ethical constraints, it is relatively straightforward to assign *VI*s to respondents so that their attributes (and perceived identities) match the respondent's self-reported social identities.

Catania, et al. (1996) demonstrated that enabling respondents to *choose* their interviewer on the basis of a social identity, in particular gender, increased interview completion rates and disclosure of sensitive sexual behaviors. While female respondents overwhelmingly chose female interviewers, male respondents were more evenly split, indicating that the data quality advantages observed for those in the choice condition had to do with choice, not the gender of the interviewer who was chosen. Catania, et al. (1996) suggest this is due to an increased sense of control, which in turn reduces the threat of the questions. While enabling respondents to choose a human interviewer is logistically and ethically complicated, enabling respondents to choose a *VI* seems far more practical. It is relatively straightforward at least conceptually to create *VI*s that vary systematically on many attributes, enabling respondents to choose a *VI* from a diverse set without explicitly considering these attributes or their implications for social identity (see Eyssel and Hegel 2012; Eyssel and Kuchenbrandt 2012).

---

* The studies all asked questions that explicitly used racial or ethnic terms; nine were reviewed by Davis, et al. (2009) – the only studies in that review whose items used explicit race/ethnicity-related terms – as well as the two subsequently published studies that we are aware of.

11.4 Research Design

To explore our three research questions, it was necessary to design *VI*s with at least two racial and gender identities and to recruit respondents with the same identities. The racial identities we implemented in the *VI*s were African American and White and the genders were binary male and female – recognizing that these race and gender categories are two of many possible identities but the two most often studied in the interviewer effects literature.

We built the 16 *VI*s by first creating two 3D facial models from photos of two interviewers, one an African-American female and the other a White male, and then transforming each into three other models using FaceGen Modeler, an off-the-shelf tool for creating facial models that can be animated. The initial African-American female face was transformed into a White female, an African-American male, and a White male face. The initial White male face was transformed into an African-American male, a White female, and an African-American female face. We then paired each of these eight models with one of two gender-congruent voices, from either one of two male (one White and one African American) or one of two female (one White and one African American) professional interviewers employed or recently employed at the University of Michigan's Survey Services (telephone) Laboratory. Creating the face–voice combinations in this way allowed us to reduce the chances that perceived *VI* gender or race might result from unknown attributes of one particular facial configuration or voice, or idiosyncratic facial movement. Finally, we animated the models by applying captured facial motion using Famous 3D Proface Video. The shape and movement of their lips were edited to correspond to the words being spoken, and their eyebrow and mouth movements were coordinated with their speech. See Online Appendix Table A11A.2 for video examples of all 16 *VI*s and respondent ratings of their eeriness and humanness using –5 to 5 scales from MacDorman, et al. (2009). Note that ratings vary mostly between –1 and 1. See Online Appendix Table A11A.3 for the rating questionnaire.

The data were collected in December 2009 by Market Strategies International (now Escalent) from 1,735 volunteer respondents from an online panel administered by Survey Sampling International.* The sample consisted of equal numbers of respondents from four (self-reported) race-by-gender groups. Respondents' mean age was approximately 50 years (range 18–90 years). Forty-three percent reported an education level of Associate Degree or higher. Most (93%) reported using a computer 5–7 days per week. A summary of all measured respondents' characteristics appears in Online Appendix Table A11A.4. Each respondent was randomly assigned to one of the 16 *VI*s, resulting in equal proportions of respondents of each gender and race assigned to *VI*s of each gender and race (see Online Appendix Table A11A.5); each confirmed they had audio capability before starting the interview.

The questions that the *VI*s asked (see Online Appendix Table A11A.6) had all been previously administered in published studies. These included ten race-related opinion questions (Krysan and Couper 2003), although any one respondent answered only eight because of branching. It contained six gender-related opinion questions (Kane and Macaulay 1993) and five questions about sensitive topics (Lind, et al. 2013), for three of which larger numerical responses were generally considered more desirable (e.g., news consumption) and for two of which less frequent/smaller numerical responses were assumed to be more desirable (e.g., weight).

---

* We contracted for roughly this number of completed cases, evenly assigned to the four gender × race combinations. We do not know how many panel members were invited in order to yield this number of completed cases so cannot calculate a response rate, but for a convenience sample such as this, response rates are not meaningful.

When each page was displayed, a video recording of the *VI*s automatically started to play in the upper half of the page. Once the video was finished playing, the response options and answer fields appeared below the video region of the page. The *VI* remained visible and static in the start position until the respondent answered the question or replayed the video. A next button became active once the respondent had selected or entered an answer. After answering the substantive questions, respondents rated their *VI* on several attributes, including its eeriness and humanness.

Respondents were then asked to choose a *VI* for a hypothetical future survey. We presented respondents one of two arrays of eight different *VI*s. The two arrays consisted of the same face models but complementary voices (see Figure 11.2). Respondents were asked to play a video clip (about 30 seconds) of each of the eight *VI*s introducing themselves, using identical wording. Once they had played all eight videos, respondents were asked, "If you could choose one of these interviewing agents to ask you questions for a future interview, which one would you choose? You will not actually take part in an interview; we just want to know which agent you would choose if you were going to participate in an interview. Please check the box below the interviewing agent you choose to conduct an interview." Respondents who did not play all eight videos were excluded from analyses of choice.

Because the choice was hypothetical and followed the main questionnaire, we are not able to examine the causal relationship between choice of *VI* and responses. However, we can see if the pattern of previously provided answers differed for respondents who chose VIs with certain social identities and not others, e.g., whether respondents who later chose a *VI* whose perceived race matched their own exhibited more extreme racial opinions than did other respondents. Finally, respondents were asked to provide data on their personal characteristics, including their age, gender, race, level of education, and frequency of computer use.

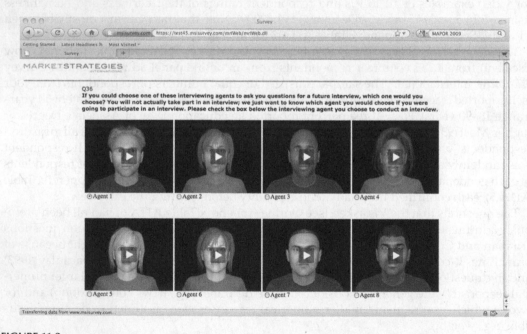

FIGURE 11.2
Eight *VI*s presented after respondents completed the main questionnaire. Respondents were asked to play the video of each *VI* and choose one of the eight to conduct a hypothetical future interview. The video consisted of each *VI* speaking "Hello, I conduct interviews for the University of Michigan's Institute for Social Research and today I am going to ask you a series of questions on different topics. When you are ready to begin, click 'Start.'"

11.5 Analytic Methods

Data were summarized using counts and percentages for categorical measures, and using means, medians, standard deviations, and ranges for continuous measures. Hierarchical models were used to derive intraclass correlations (ICC) and test for lack of independence. To evaluate whether *VI*s bias responses on the basis of their social identities, we modeled answers to the race-related questions on the basis of main effects of *VI* race, respondent race, and respondent gender. Additional control variables for each model included respondent age, education, frequency of computer use, and rating of interviewer eeriness; final models included those that survived backward selection using p > 0.40 as a criterion for exclusion. Answers to gender-related questions were modeled following the same approach, substituting *VI* gender in place of *VI* race. For analysis of socially desirable responding, we focused on the questions hypothesized or shown in prior literature to be sensitive, using the same inclusion criteria for control variables.

To examine how matching respondent and *VI* race and gender affects disclosure, we created a predictor variable for each of the four types of respondent–*VI* match as a measure of social distance: (1) respondent and *VI* match on race only; (2) respondent and *VI* match on gender only; (3) respondent and *VI* match on race and gender; (4) respondent and *VI* match on neither gender nor race. For each response the model included the new predictor of social distance and, as in the race- and gender-related question models, control variables for respondent age, education, frequency of computer use, and rating of interviewer eeriness, which were backward selected using the same criteria previously described.

To explore whether respondents who chose a *VI* with a racial identity that matched their own exhibited more extreme racial opinions than did other respondents, we employed race choice match (yes or no) as a predictor in our models. The same backward selection process was followed, with respondent race, respondent gender, and the interaction of respondent race and race choice match terms forced into the models. Linear regression was used for continuous responses, logistic regression for binary responses, and multinomial logistic regression for categorical responses with more than two categories.

Estimates of ICC for the questions with continuous and binary responses options were all approximately zero, indicating that all variation in the response was due to the respondent experience, as opposed to between-*VI* differences. This makes sense given that the 16 *VI*s were variants of the same two starting faces crossed with the same two male and two female voices. Likelihood ratio tests for the hierarchical models accounting for possible clustering by *VI* were insignificant, suggesting that the simpler fixed effects models were an equally good fit. Hence, fixed effect models were used to address our research questions.

11.6 Results

For brevity, we report only models where the primary findings are significant. Responses to the race-related questions are summarized in Table 11.1. Our first research question is whether the kinds of gender- and race-of-interviewer effects observed with human interviewers are reproduced with *VI*s, and if so whether they are as prevalent with *VI*s as with human interviewers. We observed no gender-of-*VI* effects in our data (not shown) so

TABLE 11.1

Models for Race-Related Questions

	Continuous Outcome Models		Binary Outcome Models							
	Feeling Therm, Whites	Feeling Therm, Blacks	For Prefs in Hiring (vs Against)	For Prefs in Hiring Strongly (vs Not Strongly)	Oppose Prefs in Hiring Strongly (vs Not Strongly)	Feel Particularly Close to Blacks	Feel Particularly Close to Whites	For Quotas (vs Against)	For Quotas Strongly (vs Not Strongly)	Oppose Quotas Strongly (vs Not Strongly)
	Coef. (SE)	Coef. (SE)	Odds Ratio (SE)	Odds Ratio (SE)	Odds Ratio (SE)	Odds Ratio (SE)	Odds Ratio (SE)	Odds Ratio (SE)	Odds Ratio (SE)	Odds Ratio (SE)
VI race (ref = White)										
African American	−0.081	0.376	1.019	0.990	0.761*	1.041	1.000	1.082	0.827	0.860
	(1.199)	(1.215)	(0.124)	(0.182)	(0.098)	(0.119)	(0.107)	(0.129)	(0.116)	(0.125)
Respondent gender (ref = Male)										
Female	0.042	2.089	0.989	0.938	0.965	0.811	0.904	1.171	1.017	0.814
	(1.192)	(1.211)	(0.119)	(0.170)	(0.124)	(0.092)	(0.096)	(0.138)	(0.143)	(0.119)
Respondent race (ref = White)										
African American	−10.800**	10.604**	15.162**	3.778**	0.260**	9.828**	0.239**	14.114**	3.862**	0.415**
	(1.265)	(1.286)	(2.159)	(1.006)	(0.035)	(1.239)	(0.027)	(1.671)	(0.705)	(0.078)
Respondent age	0.141**	0.175**				1.006	1.012**		1.008	1.008
	(0.043)	(0.044)				(0.004)	(0.004)		(0.005)	(0.005)
Respondent education (ref = Some high school or less)										

(Continued)

TABLE 11.1 (CONTINUED)

Models for Race-Related Questions

	Continuous Outcome Models		Binary Outcome Models							
	Feeling Therm, Whites Coef. (SE)	Feeling Therm, Blacks Coef. (SE)	For Prefs in Hiring (vs Against) Odds Ratio (SE)	For Prefs in Hiring Strongly (vs Not Strongly) Odds Ratio (SE)	Oppose Prefs in Hiring Strongly (vs Not Strongly) Odds Ratio (SE)	Feel Particularly Close to Blacks Odds Ratio (SE)	Feel Particularly Close to Whites Odds Ratio (SE)	For Quotas (vs Against) Odds Ratio (SE)	For Quotas Strongly (vs Not Strongly) Odds Ratio (SE)	Oppose Quotas Strongly (vs Not Strongly) Odds Ratio (SE)
High school graduate		-4.293 (4.647)			0.759 (0.408)	1.617 (0.691)	1.173 (0.476)		1.760 (0.935)	0.330 (0.223)
Some college (no degree)		-1.010 (4.558)			0.870 (0.461)	1.986 (0.833)	1.188 (0.472)		1.530 (0.792)	0.332 (0.222)
Associate degree		-4.627 (4.810)			0.738 (0.407)	1.443 (0.638)	0.638 (0.267)		1.269 (0.697)	0.379 (0.263)
Bachelor's degree		0.208 (4.638)			0.477 (0.255)	1.578 (0.673)	0.768 (0.310)		0.963 (0.507)	0.301 (0.204)
Graduate/professional degree		-0.292 (4.816)			0.603 (0.334)	1.249 (0.553)	0.842 (0.354)		1.327 (0.730)	0.273 (0.189)
VI eeriness	-0.686** (0.197)	-0.744** (0.200)	1.022 (0.020)	1.026 (0.029)	1.049* (0.023)	0.956* (0.018)	0.969 (0.017)	0.961* (0.019)		1.042 (0.026)
Intercept	65.986** (2.700)	53.834** (5.062)	0.087** (0.013)	0.619 (0.171)	4.381** (2.319)	0.262** (0.123)	1.895 (0.836)	0.259** (0.031)	0.232* (0.135)	4.253* (3.044)
N observations	1735	1735	1735	565	1170	1735	1735	1735	895	840
R²/pseudo R²	0.071	0.057	0.237	0.038	0.082	0.195	0.109	0.259	0.057	0.038

*p < 0.05, **p < 0.01. SE: standard error.

clearly this kind of effect is less prevalent than it is with human interviewers. There was one race-of-*VI* effect (Table 11.1, *VI* Race row). Specifically, when respondents indicated they were opposed to giving African-American people preferences in hiring and promotion, they were asked in a follow-up question: "Do you oppose preferences in hiring and promotion for blacks strongly or not strongly?" When the *VI* which asked the question was White, more respondents reported strongly opposing preferences (65.2%) than when the *VI* was African American (58.9%), $p < 0.036$, controlling for respondent race, respondent education, and *VI* eeriness.

The relatively infrequent effect of *VI*s' race on answers to race-related questions was not due to inattention: respondents' self-reported race significantly affected answers to all ten race-related questions in exactly the directions found in studies with human interviewers (see Table 11.1, "Respondent Race" row). It's possible that respondents' recognition that *VI*s are inanimate may have moderated the effect of the agents' social identities on answers. In any case, the frequency with which race-of-*VI* effects were observed, one in ten, is substantially lower than what has been observed in the race-of-interviewer effects literature (Online Appendix Table A11A.1). This suggests that while *VI*s' social identities *can* introduce measurement error, this is less of a concern than are the effects of human interviewers' social identities.

Our second research question asks whether the reduction in social desirability observed when respondent and interviewer social identities match (Catania et al. 1996; Johnson, et al. 2000) might be paralleled with *VI* social identities. We do in fact see this kind of effect for the question: "How would you describe your weight? Very underweight, slightly underweight, about the right weight, slightly overweight, or very overweight?" When respondents matched the *VI* on race, they were more likely to report being "slightly overweight" – a socially undesirable response – than the more socially desirable "about the right weight" ($p = 0.011$, Table 11.2). When respondents did not match the *VI* on either race or gender, they were no more likely to respond "slightly overweight" than "about the right weight." This is a social distance effect in the spirit of what Johnson, et al. (2000) reported with respect to drug use. Although we observed this type of effect for only one question, the implication is that intentionally matching respondents and *VI*s on one or more identities/attributes – much as Catania, et al. (1996) observed for gender – has the potential to lead to more candid reporting of sensitive behaviors and warrants further study.

Our third research question also concerns possible benefits of controlling how *VI* attributes are assigned to respondents. Based on Catania, et al.'s (1996) finding that allowing respondents to choose an interviewer on the basis of gender increased disclosure of sensitive information, we asked respondents to select one of eight *VI*s to conduct a hypothetical future interview so that we might explore, in a preliminary way, the potential of *VI* choice for increasing respondent comfort. When asked to report why they chose the *VI* they did, respondents made it plain that they attributed human qualities to the *VI*s, for example (preserving the comments' original spelling and punctuation), "because she is a black woman like myself and she looks young and hip but at the same time very mature," "Looked & sounded the friendliest," "The agent was comforting," "She is less eerie," "She is closer in age and racial background," "her voice is clear," "She's a cutie and she sounds like she's smart," "the others are scary looking," "laid back and i can relate to him," "has a more understanding expressional face also good one you can look at it is pleasant voice also good," "He seems to be very forward and not too impersonal like the rest," "He's expressions seemed more natural and the eye color wasn't as errie as some of the others."

TABLE 11.2

Multinomial Regression Model Results for "How Would You Describe Your Weight?"

	Response – Very Underweight	Response – Slightly Underweight	Response –Slightly Overweight	Response – Very Overweight
	Relative Risk Ratio (SE)	Relative Risk Ratio (SE)	Relative Risk Ratio (SE)	Relative Risk Ratio (SE)
Respondent/interviewer match condition (ref = Mismatch on race and gender)				
Match on race only	1.089	1.551	1.513*	1.009
	(0.519)	(0.466)	(0.248)	(0.206)
Match on gender only	0.724	0.905	1.102	1.099
	(0.367)	(0.291)	(0.179)	(0.213)
Match on race and gender	0.308	1.036	1.249	0.797
	(0.206)	(0.323)	(0.200)	(0.162)
Respondent age	1.018	1.004	1.025**	1.031**
	(0.014)	(0.007)	(0.004)	(0.005)
Interviewer eeriness	0.975	1.016	1.041	0.990
	(0.061)	(0.036)	(0.019)	(0.023)
Intercept	0.029**	0.137**	0.339**	0.126**
	(0.022)	(0.056)	(0.075)	(0.037)
Number of observations	1735			
Pseudo R^2	0.027			

Note: All coefficients describe the effect for the indicated response compared to the reference response, "About the right weight."

*p < 0.05, ** p < 0.01.

The full set of over 1700 open responses was coded into 16 categories (a single response could be assigned to more than one category) by three research assistants: two coders classified each open response. Interrater agreement was high, with an average kappa across all categories of 0.872. Somewhat to our surprise, respondents mentioned the *VI*'s voice as a reason for their choice more than any other category (in over 50% of the responses). Race and gender were mentioned in fewer than 5% of the choice explanations. Yet African-American respondents chose African-American *VI*s almost 80% of the time (Online Appendix Table A11A.7). This suggests that at least some respondents may have considered *VI* attributes without being willing to state so explicitly or perhaps even without necessarily being aware that these identities were at play. When asked to "rate Blacks" on a (101 point) feeling thermometer, White respondents who later chose a White *VI* assigned the lowest (coolest) ratings to the group, and African-American respondents who later chose an African-American *VI* assigned Blacks the highest (warmest) rating (Figure 11.3); this is consistent with an interpretation that these respondents' choice of *VI* was driven in part by the extremity of their racial opinions. The mirror image of this pattern was observed when respondents were asked to rate Whites on the feeling thermometer: White respondents who later chose a White *VI* rated "Whites" more warmly than other respondents and African-American respondents who later chose an African-American *VI* gave "Whites" the coolest rating (Figure 11.4). A similar interaction of respondent race and race of the chosen *VI* was observed when respondents were instructed to "Click next to the groups you feel particularly close to – people who are most like you in their ideas, interest

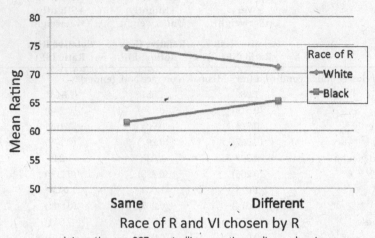

FIGURE 11.3
Ratings of Whites by White and Black respondents who chose a *VI* whose race was the same or different than theirs.

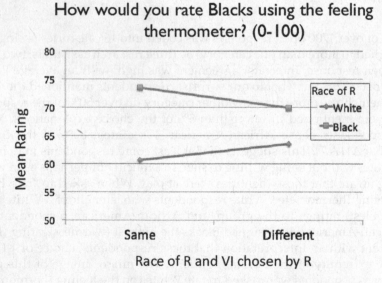

FIGURE 11.4
Ratings of Blacks by Black and White respondents who chose a *VI* whose race was the same or different than theirs.

and feelings about things," with African-American respondents who chose an African-American *VI* clicking next to "Blacks" most often, and when asked "Are you for or against preferential hiring and promotion of Blacks?" the same groups responded "For" more than the other groups.

11.7 Conclusions

The findings reported here show that respondents' answers *can* be affected by perceived social identities of *VIs* as with human interviewers, but for identity-related questions these effects seem to be less common than with human interviewers. We also found that when respondents' and *VIs'* social identities matched, respondents seemed to answer a question about their weight more candidly than when there was not a match. Finally, when offered a choice of *VI* for a hypothetical future interview, Black respondents primarily chose Black *VIs*, perhaps increasing their comfort. As we see it, the ease of giving respondents a choice of *VI* may allow researchers to help standardize respondents' experience, introducing a degree of comparability not typically available with human interviewers.

We cannot rule out the possibility that the effects of *VI's* social identities that we observed are artifactual; respondents may have inferred what was being manipulated and provided answers they believed to be consistent with the purpose of the study. But we see this as unlikely. Because respondents only saw one *VI* when responding, they had little basis on which to guess that the *VIs'* attributes were systematically different for other respondents. And the overall pattern of findings on racial attitudes (Table 11.1) was consistent with attitudes reported in studies using more conventional modes, suggesting that despite the novelty of *VIs*, the responses are plausible.

Finally, from one perspective our results might be considered a wash: one question produced a race of *VI* effect, increasing measurement error, and one produced a social distance effect, reducing measurement error by promoting the report of socially undesirable behavior when respondents and *VIs* matched on race. Our view is that the current results are a starting point and not the final word on how *VIs* and their attributes can be designed to improve data quality. The natural follow-up from the current study will be to explore ways of using *VIs* to increase data quality more broadly without also introducing harm. We are particularly intrigued by the prospect of allowing respondents to choose a *VI* and can see several ways to extend our preliminary findings. For example, voice was frequently reported as a criterion for choosing a *VI* even though we did not systemically vary voice characteristics. Offering respondents a more vocally diverse set of *VIs*, varied intentionally on dimensions such as speech rate, vocal pitch, pause rate, disfluency rate, and dialect may enable them to more finely calibrate their choice, presumably further increasing their comfort. Similarly, the choice instructions warrant investigation. In the current study, respondents were instructed to choose a *VI* with whom they would feel comfortable. But it could be that instructions to choose a *VI* with whom they would feel most secure when the interview concerns sensitive topics or best able to focus when many of the questions are cognitively demanding may strengthen the benefits of choice. At the very least, the current results provide sufficient promise for *VI* technology's potential to improve the data collected online to justify further investigation.

Acknowledgments

We are grateful to the National Science Foundation, Grant # SES 0551300, for supporting the research reported here. We thank Reg Baker, Rachel Davis, Jason Deska, Jon Gratch, Matt Jans, Maria Krysan, Roxy Shooshani, and Dave Vannette for their advice and assistance.

References

Antos, D., C. de Melo, J. Gratch, and B. J. Grosz. 2011. The influence of emotion expression on perceptions of trustworthiness in negotiation. Paper presented at the 25th annual AAAI Conference on Artificial Intelligence.

Baker, R. 1998. The CASIC future. In: *Computer Assisted Information Collection*, ed. M. P. Couper, R. P. Baker, J. Bethlehem, C. Z. F. Clark, J. Martin, W. Nicholls II, and J. O'Reily, 583–604. New York: John Wiley & Sons, Inc.

Catania, J. A., D. Binson, J. Canchola, L. M. Pollack, W. Hauck, and T. J. Coates. 1996. Effects of interviewer gender, interviewer choice, and item wording on responses to questions concerning sexual behavior. *Public Opinion Quarterly* 60(3):345–375.

Conrad, F. G., M. F. Schober, M. Jans, R. A. Orlowski, D. Nielsen, and R. Levenstein. 2015. Comprehension and engagement in survey interviews with virtual agents. *Frontiers in Psychology* 6:1–20.

Davis, R. E., M. P. Couper, N. K. Janz, C. H. Caldwell, and K. Resnicow. 2009. Interviewer effects in public health surveys. *Health Education Research* 25(1):14–26.

Ehrlich, J. S., and D. Riesman. 1961. Age and authority in the interview. *Public Opinion Quarterly* 24(1):99–114.

Eyssel, F., and F. Hegel. 2012. (S)he's got the look: Gender stereotyping of robots. *Journal of Applied Social Psychology* 42(9):2213–2230.

Eyssel, F., and D. Kuchenbrandt. 2012. Social categorization of social robots: Anthropomorphism as a function of robot group membership. *The British Journal of Social Psychology* 51(4):724–731.

Groves, R. M., F. J. Fowler Jr, M. P. Couper, J. M. Lepkowski, E. Singer, and R. Tourangeau. 2009. *Survey Methodology* (2nd ed.), Hoboken, NJ: John Wiley & Sons.

Johnson, T. P., M. Fendrich, C. Shaligram, A. Garcy, and S. Gillespie. 2000. An evaluation of the effects of interviewer characteristics in an RDD telephone survey of drug use. *Journal of Drug Issues* 30(1):77–101.

Kane, E. W., and L. J. Macaulay. 1993. Interviewer gender and gender attitudes. *Public Opinion Quarterly* 57(1):1–28.

Krysan, M., and M. P. Couper. 2003. Race in the live and the virtual interview: Racial deference, social desirability, and activation effects in attitude surveys. *Social Psychology Quarterly* 66(4):364.

Lind, L. H., M. F. Schober, F. G. Conrad, and H. Reichert. 2013. Why do survey respondents disclose more when computers ask the questions? *Public Opinion Quarterly* 77(4):888–935.

Liu, M. 2016. The effect of the perception of an interviewer's race on survey responses in a sample of Asian Americans. *Asian American Journal of Psychology* 7(3):167.

Lucas, G. M., J. Gratch, A. King, and L. P. Morency. 2014. It's only a computer: Virtual humans increase willingness to disclose. *Computers in Human Behavior* 37:94–100.

MacDorman, K. F., R. D. Green, C. C. Ho, and C. T. Koch. 2009. Too real for comfort? Uncanny responses to computer generated faces. *Computers in Human Behavior* 25(3):695–710.

Nass, C., K. Isbister, and E. J. Lee. 2000. Truth is beauty: Researching embodied conversational agents. In: *Embodied Conversational Agents*, ed. J. Cassell, J. Sullivan, E. Churchill, and S. Prevost, 374–402. Cambridge, MA: MIT press.

Nass, C., Y. Moon, and P. Carney. 1999. Are people polite to computers? Responses to computer-based interviewing systems. *Journal of Applied Social Psychology* 29(5):1093–1109.

Nass, C., Y. Moon, and N. Green. 1997. Are machines gender neutral? Gender-stereotypic responses to computers with voices. *Journal of Applied Social Psychology* 27(10):864–876.

Schuman, H., and J. M. Converse. 1971. The effects of black and white interviewers on black responses in 1968. *Public Opinion Quarterly* 35(1):44–68.

Tourangeau, R., M. P. Couper, and D. M. Steiger. 2003. Humanizing self-administered surveys: Experiments on social presence in Web and IVR surveys. *Computers in Human Behavior* 19(1):1–24.

Tourangeau, R., and T. Yan. 2007. Sensitive questions in surveys. *Psychological Bulletin* 133(5):859–883.

Wolford, M. L., R. E. Brown, A. Marsden, J. Jackson, and C. Harrison. 1995. Bias in telephone surveys of African Americans: The impact of perceived race of interviewer on responses. Paper presented at the Annual Meeting of the American Statistical Association, Orlando, FL.

12

Differences in Interaction Quantity and Conversational Flow in CAPI and CATI Interviews

Yfke Ongena and Marieke Haan

CONTENTS

12.1 Introduction

Extensive literature exists comparing computer-assisted personal interviewing (CAPI) and computer-assisted telephone interviewing (CATI) (for an overview, see Cernat 2015; Holbrook, Green, and Krosnick 2003; Scherpenzeel 2001). From the Total Survey Error (TSE) framework (Groves and Lyberg 2010), these studies have examined differences between CAPI and CATI for outcomes related to selection (i.e., coverage and nonresponse) and measurement.

Overall, the mode comparisons have yielded inconsistent effects, which Scherpenzeel (2001) argues may be due to four different causes. First, some studies are true experimental comparisons that only vary the medium for administration, while others involve a "real life" comparison in which each mode is optimized within its own common field-work procedures (e.g., sampling). Second, the experimental designs of studies differ, with the split-ballot design being much more common, but less informative than test–retest designs (see Martin, O'Muircheartaigh, and Curtice 1993) or Multitrait–Multimethod designs (Andrews 1984; Scherpenzeel 2001; Scherpenzeel and Saris 1995, 1997). Third, early mode comparisons of personal versus telephone interviewing did not always use computer-assisted techniques for the CAPI mode, which made it difficult to monitor the CAPI interviewers. Nowadays, computer audio-recorded interviewing (CARI) for CAPI allows for a rather unobtrusive means to monitor interaction (Pascale 2016). Fourth, studies differ in their criteria for comparison. While most studies evaluate costs, speed in terms of finishing field work, response rates, and similarity of the response distributions, only a few studies evaluate the quality of the data in terms of measurement error, such as by comparing data collected in interviews with official records (Körmendi 1988). Lacking validating information, studies evaluate data quality using other criteria such as comparing how response effects differ between modes (for an overview, see Cernat 2015; Holbrook, et al. 2003; Jäckle, Roberts, and Lynn 2010; Scherpenzeel 2001). Regardless of the criteria for comparison, inference of likely causes of mode effects are often limited, though satisficing and social desirability response bias have been suggested as underlying mechanisms. More specifically, the hypotheses tested in earlier works are based on the argument that differences are a result of greater trust and rapport and more effective nonverbal communication in face-to-face interviews as compared to telephone interviews (Holbrook, et al. 2003). An exhaustive review of 48 studies (all published before 2002) that compared face-to-face interviews with telephone interviews revealed confounding factors for 32 studies (Holbrook, et al. 2003), but a more recent systematic review that specifically takes these four factors into account is currently not available, and beyond the scope of our chapter.

In the current study, we use interaction quantity as a criterion for comparison to better understand mode effects. Interaction quantity is a rough indicator of actual behaviors of interviewers and respondents in interviewer–respondent interactions. In studies of interviewer–respondent interactions, generally referred to as "interaction analysis" or "behavior coding" depending on the level of detail employed (see Ongena and Dijkstra 2006), interactions are systematically evaluated for deviations from the so-called paradigmatic sequence (Schaeffer and Maynard 1996). A paradigmatic sequence, including one, two, or three turns, is the interaction as intended by the researcher: the interviewer reads the question exactly as worded, the respondent provides an answer that exactly matches one of the response options, and optionally, the interviewer acknowledges the response (e.g., "thank you"). Any deviation from this sequence may inform researchers about problematic aspects of the questionnaire or the interviewing procedure (Schaeffer and Maynard 1996).

Studies previously investigating interviewer–respondent interaction in CAPI and CATI have shown that the rate of standardization in CATI is usually higher than in CAPI (Pascale 2016; Pascale, Goerman, and Drom 2013; Snijkers 2002). To our knowledge, CAPI and CATI interactions have never been systematically compared in terms of the number of turns, events, and words spoken in the actual interaction, and therefore in this study we aim to fill this gap.

We aim to answer the following research questions:

(1) To what extent do CAPI and CATI interviewers differ in interaction quantity, hesitations, and uncertainty markers?

(2) How much variability in interaction quantity, hesitations, and uncertainty markers is due to mode versus respondents, or questions?

(3) What question and respondent characteristics are associated with interaction quantity, hesitations, and uncertainty markers in both modes?

12.1.1 Number of Turns, Events, and Words

A question–answer (QA) sequence consists of all the interactions between an interviewer and a respondent for a given question. QA sequences are made up of turns, separate and distinct utterances spoken by the interviewer versus the respondent. Within a given turn, the interviewer and the respondent can produce multiple actions or events (see Houtkoop-Streensta 2002). Events are discrete, identifiable, meaningful actions (such as questions, answers, and instructions, etc., see Ongena and Dijkstra 2006). Among CAPI and CATI interviews, we compare the QA sequences in terms of three different variables. First, we compare the number of turns taken by the interviewer and the respondent. Second, since both the interviewer and the respondent may, within one turn, produce multiple events, the number of events per question by mode is analyzed. Third, at an even more detailed level, we compare the number of words uttered (integrated in one variable for both the interviewer and the respondent) across modes and across different questions of the European Social Survey (ESS) questionnaire. This comparison may reveal differences in communication efficiency. It is expected that, due to more possibilities of nonverbal communication (Clark and Schober 1992), while discussing the questions, fewer words will be uttered by interviewers and respondents in CAPI than in CATI. However, the pace of question asking is likely to be slower in CAPI than in CATI, since silences are less problematic and risk of break-offs are lower (Holbrook, et al. 2003). Several studies have indeed found that face-to-face interviews take longer than telephone interviews (Groves and Kahn 1979; Holbrook, Green, and Krosnick 2003; Rogers 1976), but no data are available for differences across modes in length of individual questions.

12.1.2 Uncertainty Markers and Hesitations

In addition to examining outcomes that capture variation in the quantity of talk, we examine differences between CAPI and CATI in terms of the kinds of talk displayed in the question–answer sequences. Differences in the production of uncertainty markers (such as "I guess," "maybe," "might") by respondents are examined, as well as the hesitation of interviewers and respondents as captured by the filled pause "eh." These markers may be subtle displays of discomfort and are expected to be more likely to occur in CATI than in CAPI.

12.1.3 Question Characteristics

Based on the finding that question characteristics accounted for around 80 percent of response time in a CATI interview (Garbarski, et al., Chapter 18, this volume; Olson and Smyth 2015), we expect that question characteristics will also affect the number of turns, events, and words in the interaction. We use Olson and Smyth's (2015) distinction between necessary question features (i.e., features that are "largely determined by the survey topic and analytic goals," Olson and Smyth 2015) and features expected to affect respondent task complexity, interviewer task complexity, and respondent processing efficiency. Within the category of necessary question features, Olson and Smyth found that several characteristics were significantly associated with response time. The number of words in the question and the number of response options available were positively related to response time. Response time was longer for attitudinal and demographic questions than behavioral questions, presumably because the information asked for in behavioral questions requires more time for comprehension and retrieval (Olson and Smyth 2015). With respect to the format of the response options, open-ended numeric, closed-ended nominal, and yes/no questions took less time than open-ended text questions.

Within features expected to affect respondent task complexity, response time was positively affected by the questions' reading level, the number of response options, and question sensitivity, whereas within features expected to affect interviewer task complexity, only the presence of a backup (i.e., the interviewer returning to an earlier question to change the answer) at the question appeared to be significantly associated with response time. We expect to find similar relationships between these question characteristics and the interaction quantity. Further, due to greater possibility of nonverbal communication in CAPI, we expect effects to be smaller for CAPI than for CATI.

In CAPI, show cards are often used for batteries of items with the same scale or for questions with long lists of response options (Miller 1984; Sykes and Collins 1988). Through the visual presentation of response options, show cards are assumed to reduce the burden on the respondent to remember response categories, and they may facilitate comprehension (Roberts, Jäckle, and Lynn 2006). Nevertheless, show cards may also increase the cognitive burden on respondents, who may find it distracting to read the card while simultaneously listening to the interviewer (Dijkstra and Ongena 2006).

From studies showing that questions with show cards in face-to-face interviews elicit different answers when administered by telephone (Groves 1990; Groves and Kahn 1979), mode effects often cannot be disentangled from show card effects (Roberts, Jäckle, and Lynn 2006). To address this shortcoming, Roberts, Jäckle, and Lynn (2006) tested a face-to-face condition with show cards and one without show cards in an experimental evaluation of mode effects for CAPI and CATI. Results indicated that show cards had no effect on the extent of satisficing. Since show card questions are often longer and more complex than questions asked without them (Roberts, Jäckle, and Lynn 2006; Smith 1987), we also examine the effect of using a show card in CAPI.

12.2 Methods

Data were collected from March to June 2012 in the Netherlands by GfK Panel Services Benelux. Multistage cluster sampling was used (for more details see Haan, Ongena, and Aarts 2014). Of the 3,496 households selected, 824 participated in the survey. Data collection was part of a mixed-mode experiment using a slightly modified version of the ESS Round 5 questionnaire. Three modes were compared: CAPI, CATI, and a web version not included in the current analysis. One-third of the sample members were contacted by phone, and subsequently randomly assigned to CAPI, CATI, or web; one-third of the sample members were contacted by phone and allowed to choose between CATI and web; and one-third of the respondents were contacted face-to-face at their homes, and were allowed to choose between CAPI and web administration. Thus, while one-third of the sample members were randomly assigned to their CAPI, CATI, or non-interview administered condition, two-thirds of the sample members were offered a choice between an interviewer-administered mode and a non-interviewer-administered mode. The overall response rate was 37.5 percent (AAPOR 2016, RR1). Details about the response rates per experimental group can be found in Table A12A.1 in the supplementary online materials. More specific information about the sample can be found in an earlier study by Haan,

Ongena, and Aarts (2014). Similar to Olson and Smyth (2015), we include a control variable for the respondent's sex to control for inadequate representation of female respondents. In general, age and education are considered important factors in determining a respondent's working memory capacity (Salthouse 1991; Yan and Tourangeau 2008), and age and education have been shown to affect response time in CAPI (Couper and Kreuter 2013), CATI (Olson and Smyth 2015), and web surveys (Yan and Tourangeau 2008).

All interviews were audio-recorded; the 54 CAPI and 60 CATI interviews (with 27 interviewers and 81 different questions) comprise about 57 hours of interaction. The first half of the interview was transcribed using the Sequence Viewer Program (Dijkstra 2018). This program allows for division of interviews into separate QA sequences (i.e., all utterances involved in asking and answering one survey item), and subsequently dividing all utterances in a QA sequence into turns (i.e., utterances by the interviewer and those by the respondent), and events within turns (i.e., meaningful actions by the interviewer or respondents such as questions, answers, requests, comments, considerations, etc.). Thus, in the transcripts both for the interviewer and the respondent, the number of events and the number of words per QA sequence could easily be counted.

For missing data, list-wise deletion is used in some of our analyses, and therefore 6629 QA sequences were available for analysis (53 CAPI and 55 CATI interviews). The sequences included 24,195 turns containing 33,741 events, with an average of 3.65 turns (SD = 3.38) and 5.09 events (SD = 4.75) per sequence. The number of events for a sequence ranged from a minimum of one event (i.e., the interviewer briefly stating the assumed answer and continuing with the next question) to a maximum of 88 events (this occurred during the administration of a question about the respondent's satisfaction with life). In order to allow for computation of Pearson correlations, in subsequent analyses we used the trimmed number of turns and events (see Table A12A.2 in the supplementary online materials).

We tallied the number of respondents' uncertainty markers using the text search option of the Sequence Viewer program and counting the phrases and words indicating uncertainty in Dutch.* Transcribers were explicitly instructed to transcribe any noticeable hesitance of speech, i.e., filled pauses (Swerts 1998), that normally appear in Dutch as "uh" [ɔh] or "um" [əm] with "eh," and we also counted occurrences of these pauses using text search commands. All analyses were conducted using R, version 3.3.1 (R Core Development Team 2018). Statistical analyses have not been adjusted for the design effect of sampling.

For the analyses of interaction quantity, we use methods that include the complex structure of the data. In analyzing survey interview interactions, it is necessary to take cross-classification by respondents and questions, and nesting within interviewers into account (Olson and Smyth 2015; Yan and Tourangeau 2008), but in this case, interviewers, respondents, and questions are also nested within mode (i.e., interviewers interviewed only in one mode), as displayed in Figure 12.1. Since we expect that mode, interviewers, respondents, and questions each will have a unique effect on the interaction quantity, we take this nesting into account by estimating cross-classified random effects models with number of words uttered cross-classified by respondents and by questions and with questions and respondents nested within mode (see Online Appendix 12A for a full description of the model). Due to the nature of our data (108 respondents, 27 interviewers), including interviewers into the same model as respondents yielded zero variance at the interviewer level in all cases; therefore, in all models we did not include random effects at the interviewer level.

---

* In Dutch these comprised: "Ik denk" ("I guess"/ "I think"), "Ik geloof" ("I believe") "Ik weet + niet"("I know+negation"), " Misschien" ("maybe"), and "Zou kunnen" ("might")

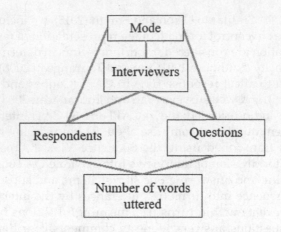

FIGURE 12.1
Data structure of the number of words uttered nested in mode and interviewers and cross-classified by respondents and questions.

12.3 Results

In this section, we first describe the differences between CAPI and CATI interviews regarding the number of turns, the number of events, and the number of words uttered by interviewers and respondents. Then we describe the exploratory analyses of differences between CAPI and CATI interviews with respect to disfluencies of interviewers and respondents and respondent uncertainty markers.

12.3.1 Number of Turns and Events

Overall, the average number of trimmed turns per QA sequence was higher for CAPI than for CATI (CAPI M = 3.42, SD = 2.09, CATI M = 3.06, SD = 1.89), and a similar difference was found for events (CAPI M = 4.31, SD = 2.93, CATI, M = 3.36, SD = 3.01).

The number of events was not normally distributed (as computed by the R package e1071) in CAPI (Shapiro W = 0.59, p < 0.01, skewness 4.45) or CATI (Shapiro W = 0.65, p < 0.01, skewness 3.56). Log transformations of the number of events did not yield distributions close enough to normal distributions (Shapiro W = 0.89, p < 0.01, skewness overall 0.94). Therefore, in testing differences between the modes, non-parametric tests (Mann Whitney U) and correlations were used (see Table A12A.3 in the supplementary online materials). The difference in the trimmed number of turns and events in CAPI and CATI was the same across almost all question characteristics, except for question characteristics that applied to only few questions (i.e., open-ended numeric questions, questions with a mismatch between question and response options). Except for yes/no questions and mismatch questions, all effects for questions were significant between the modes (see Table A12A.4 in the supplementary online materials).

12.3.2 Number of Words

The number of words per QA sequence was slightly lower for CAPI (35.57, SD = 35.55) than CATI (35.81, SD = 28.11). Although the number of words was not normally distributed

TABLE 12.1

Model Variance Components Predicting Log #Words Uttered, and the Occurrence of Interviewer and Respondent Filled Pauses and Respondents' Uncertainty Markers

	Log #Words Uttered		Interviewer Filled Pause		Respondent Filled Pause		Respondent Uncertainty	
	Var	ICC	Var	SD	Var	SD	Var	SD
Null model								
Mode	0.13***	0.07	0.63**	0.79	<0.01	<0.01	0.00	0.00
Question	0.63***	0.64	1.82*	1.35	1.29*	1.14	4.96*	2.23
Respondent	0.18***	0.05	1.62*	1.27	0.88	0.94	0.92	0.96
Model fit statistics								
Log-likelihood	−3,939.90		−2,160.2		−3,127.5		−1,176.1	
AIC	7,889.90		4,328.3		6,263.0		2,360.2	

Note: n – 108 respondents, 81 questions, 2 modes, total n = 6,629.
* p < 0.05, ** p < 0.01, *** p < 0.001.

(Shapiro W = 0.69, p < 0.01, skewness = 3.3), log transformation of the variable yielded nearly normally distributed data (Shapiro W = 0.98, p < 0.01, skewness 0.01). The log transformation enabled us to use parametric tests, which allowed for analysis of effects of several independent variables.

Base model. The base model (containing no covariates) predicting the logged number of words is shown in the first column of Table 12.1. Variance terms for the question, respondent, and mode were significant, indicating significant variability due to all three sources. The intraclass correlation coefficients show how much percent of the variance is due to the source; i.e., 6.7 percent was due to the mode of interviewing, 64.2 percent was due to questions, and 5.2 percent was due to respondents.

Question and respondent characteristics. Table 12.2 (first column) shows a model that includes all of the question and respondent characteristics. This model, compared to the base model, explained about 57 percent of the initial variability in number of words uttered due to questions, whereas 87.4 percent of the variability due to mode was explained. Several interactions between mode and question characteristics were significant. Mismatch and show card questions yielded longer interactions (coef = 0.96, p < 0.01, and 0.55, p < 0.01, respectively), but in CAPI these questions yielded shorter interactions (coef = −0.47, p < 0.01, coef = −0.35, p < 0.01, respectively). Similarly, closed nominal questions (as compared to open-ended questions) yielded longer interactions (coef = 0.86, p < 0.05), but shorter for CAPI interviews (coef = −0.96, p < 0.01). In CAPI, closed numeric, closed ordinal, and yes/no questions also yielded shorter interactions (coef = −1.27, p < 0.01, coef = −0.96, p < 0.01, and coef = −0.54, p < 0.01, respectively), whereas main effects for these question types were not significant. Questions with longer lists of response options yielded shorter interactions (coef = −0.09, p < 0.05), but in CAPI these interactions were longer (coef = 0.07, p < 0.01). In addition, sensitive questions yielded more words, regardless of mode (coef = 0.38, p < 0.01). Evaluation of respondent characteristics showed that age and gender were not associated with the number of words uttered, but level of education showed that respondents with an educational level of high school or lower uttered more words (coef = 0.10, p < 0.01) than respondents with higher levels of education. Inclusion of respondent characteristics reduced variance due to respondents by 4.24 percent.

TABLE 12.2

Model Coefficients and Standard Error (in Parentheses) of Question Characteristics and Respondent Characteristics Predicting Log (#Words Uttered) and the Occurrence of Interviewer and Respondent Filled Pauses and Respondents' Uncertainty Markers

	Log #Words Uttered Coef. (SE)	Interviewer Filled Pause Coef. (SE)	Respondent Filled Pause Coef. (SE)	Respondent Uncertainty Markers Coef. (SE)
Constant/Intercept	2.64 (0.30)***	−4.24 (0.98)***	−0.04 (0.74)	−5.57 (1.77)**
Mode (CAPI)	0.15 (0.10)	1.77 (0.76)*	−1.17 (0.06)	−0.20 (1.63)
Necessary question features				
Question length	−0.00 (0.00)	0.02 (0.00)***	−0.2 (0.06)	−0.01 (0.02)
# Response options	−0.09 (0.04)*	−0.10 (0.13)	0.01 (0.01)	0.28 (0.22)
Type of question				
Behavior (ref)	−	−	−	−
Attitude	−0.03 (0.21)	−0.84 (0.55)	−0.42 (3.75)	0.80 (1.02)
Demographic	−0.30 (0.17)	0.47 (0.51)	−1.70 (0.37)***	1.02 (0.94)
Format of response options				
Open-ended text (ref)	−	−	−	−
Open-ended numeric	−0.13 (0.45)	−0.34 (1.27)	−1.18 (0.26)	−1.46 (4.63)
Closed numeric	0.75 (0.45)	−0.34 (1.46)	−1.17 (1.07)	−4.26 (2.41)
Closed nominal	0.86 (0.36)*	−0.15 (1.31)	−0.91 (0.98)	−2.65 (2.09)
Closed ordinal	0.29 (0.31)	−0.74 (0.95)	−1.73 (0.69)*	−3.35 (1.69)*
Yes/no	−0.08 (0.29)	−1.01 (0.87)	−1.87 (0.68)**	−0.31 (1.65)
Respondent task complexity				
Question reading level	0.03 (0.02)	0.05 (0.05)	0.03 (0.03)	0.04 (0.08)
Mismatch Q and response options	0.96 (0.47)***	−0.54 (1.21)	2.31 (0.78)**	−3.12 (2.36)
Sensitive question	0.38 (0.13)***	0.02 (0.34)	−0.19 (0.22)	1.80 (0.64)**
Unknown terms	−0.06 (0.18)	−0.04 (0.45)	0.13 (0.29)	−0.34 (0.86)
Showcard question	0.55 (0.20)**	1.19 (0.61)	0.69 (0.40)	1.25 (1.04)
Interviewer instructions	0.41 (0.13)**	0.44 (0.34)	0.28 (0.21)	0.81 (0.63)
Female = 1	0.02 (0.04)	0.04 (0.00)	2.01 (0.19)	0.17 (2.42)
Age (centered)	0.00 (0.00)	−0.05 (0.03)*	−0.01 (0.02)	−0.05 (0.02)
HS degree or less = 1	0.10 (0.04)**	0.34 (0.29)	0.09 (0.22)	0.02 (0.26)
Choice of mode = 1	0.01 (0.05)	0.59 (0.37)	−0.43 (0.29)	0.18 (0.53)
Mode* question length	0.01 (0.00)***	0.04 (0.00)***	0.02 (0.00)**	0.02 (0.55)
Mode* # response options	0.07 (0.02)***	0.11 (0.11)	0.01 (0.09)	−0.26 (0.19)
Mode* attitude	0.09 (0.05)	0.67 (0.38)	−0.15 (0.30)	−0.18 (0.55)
Mode* demographic	−0.09 (0.04)*	−0.57 (0.41)	0.62 (0.40)	−0.26 (0.82)
Mode* open-ended numeric	0.00 (0.37)	−1.62 (1.99)	1.28 (2.59)	1.70 (4.63)
Mode* closed numeric	−1.27 (0.17)***	−1.90 (1.21)	−0.04 (1.07)	3.37 (1.67)*
Mode* closed nominal	−0.96 (0.17)***	−0.47 (1.21)	−0.07 (1.03)	1.70 (1.69)
Mode* closed ordinal	−0.96 (0.10)***	−1.52 (0.75)*	−0.12 (0.69)	1.42 (1.24)
Mode* yes/no	−0.54 (0.08)***	−0.55 (0.67)	−0.32 (0.62)	−0.08 (1.41)
Mode* question reading level	0.01 (0.00)*	0.00 (0.03)	0.01 (0.02)	1.58 (0.05)
Mode* mismatch Q and response options	−0.47 (0.10)***	−0.29 (0.83)	−1.36 (0.70)	−0.14 (2.15)

(Continued)

TABLE 12.2 (CONTINUED)

Model Coefficients and Standard Error (in Parentheses) of Question Characteristics and Respondent Characteristics Predicting Log (#Words Uttered) and the Occurrence of Interviewer and Respondent Filled Pauses and Respondents' Uncertainty Markers

	Log #Words Uttered Coef. (SE)	Interviewer Filled Pause Coef. (SE)	Respondent Filled Pause Coef. (SE)	Respondent Uncertainty Markers Coef. (SE)
Mode* sensitive question	0.01 (0.03)	0.11 (0.23)	0.24 (0.18)	−0.42 (0.37)
Mode* unknown terms	0.01 (0.04)	0.04 (0.30)	−0.15 (0.22)	−0.14 (0.48)
Mode* show card question	−0.35 (0.06)***	−0.83 (0.44)	0.37 (0.34)	0.14 (0.68)
Mode* interviewer instructions	0.02 (0.03)	0.21 (0.23)	0.21 (0.17)	−0.34 (0.35)
Random effects				
SD – mode	0.04***	0.01	0.02	0.00
SD – question	0.43***	0.91**	0.59	1.87
SD – respondent	0.17***	1.22**	0.93	0.97
SD – residual	0.40***	–	–	–
Log-likelihood	−3,896.2	−2,104.5	−3,067.0	−1,147.5
AIC	7,836.4	4,287.1	6,212.0	2,371.0
Wald chi-square	6.98**			

n = 108 respondents, 81 questions, 2 modes, total n = 6,629.

* p < 0.05, ** p < 0.01, *** p < 0.001.

12.3.3 Interviewer and Respondent Filled Pauses and Respondent Uncertainty Markers

In addition to the number of words uttered, we explored the occurrence of interviewers' and respondents' filled pauses and whether respondents uttered uncertainty markers in mixed effects logistic regression models.

Base model. The base models (containing no covariates) are shown in Table 12.1 (columns 2, 3, and 4). For interviewers' filled pauses, variance terms for the question, respondent, and mode were significant, indicating significant variability due to all three sources. For respondents' filled pauses and uncertainty markers, only the variance term for the question turned out to be significant.

Question and respondent characteristics. Table 12.2 (columns 2, 3, and 4) shows models for question and respondent characteristics for filled pauses and uncertainty markers. Only for interviewer filled pauses, mode showed a significant effect, indicating that odds of interviewer filled pauses were higher in CAPI than in CATI. A significant interaction effect was found between mode and question length for interviewer filled pauses. As expected, longer questions yielded more interviewer filled pauses (coef = 0.02, p < 0.001, odds ratio 1.02) and this effect was stronger for CAPI than for CATI (coef = 0.04, p < 0.01, odds ratio = 1.04). In addition, as compared to open-ended text questions, only in CAPI, closed ordinal questions decreased the odds of interviewer filled pauses (coef = −1.52, p < 0.05, odds ratio = 0.22). Respondent filled pauses increased due to question length only in CAPI (coef = 0.02, p < 0.01, odds ratio = 1.02), and demographic questions, as compared to behavior questions, decreased the odds of respondents' filled pauses (coef = −1.70, p < 0.01, odds ratio = 0.18). For respondents' uncertainty markers, a significant effect was found for sensitive questions, indicating that sensitive questions increased the odds of an uncertainty marker (coef = 1.80, p < 0.01, odds ratio = 6.05). Evaluation of respondent characteristics (Table 12.2, columns 2, 3, and 4) shows that for both interviewers' and respondents' filled pauses, gender and education of the respondent were not associated with the occurrence of filled pauses, but for age, a significant effect was found; with increasing respondents' age, the odds of an interviewers' filled pause were lower (coef = −0.05, p < 0.05, odds ratio = 0.95).

12.4 Conclusion

In this chapter, we investigated differences in interviewer–respondent interaction in CAPI and CATI interviews during the administration of the ESS Round 5, using slightly modified version of the ESS Round 5 questionnaire. In answering our first research question ("To what extent do CAPI and CATI interviewers differ in interaction quantity, hesitations and uncertainty markers?"), when not taking into account effects of question characteristics, we found a higher number of turns and events and more interviewers' filled pauses in CAPI than in CATI. However, again when not taking into account effects of question characteristics, the number of words uttered was slightly lower for CAPI than for CATI, and no differences between modes were found for respondents' filled pauses and uncertainty markers.

With regard to our second research question ("How much variability in interaction quantity, hesitations and uncertainty markers is due to mode, respondents, and questions?"), we found that more variance is due to questions rather than respondents or mode. This is very similar to findings in earlier studies investigating response time (Couper and Kreuter 2013; Olson and Smyth 2015). For interviewer filled pauses, we also found variability due to all three sources, but for respondent's filled pauses and uncertainty markers, only the variance term for the question turned out to be significant.

Finally, in answer to our third research question ("What question and respondent characteristics are associated with interaction quantity, hesitations and uncertainty markers in both modes?"), we found effects for length of question (in CAPI, positive effects for the number of words uttered, increased odds of interviewer filled pauses) and number of response options (negative effects for number of words uttered, but positive effects for CAPI). As compared to open-ended text questions, most closed-ended questions increased the number of words uttered but decreased the number of words uttered in CAPI, and decreased the odds of respondent filled pauses. This contrast in CAPI between open-ended text questions and closed-ended ones matches the finding in qualitative interviews that face-to-face interviews take longer than telephone interviews, due to participants being more forthcoming, and providing relevant information spontaneously (Irvine 2011). As opposed to previous research in which sensitive questions were answered more quickly compared to non-sensitive questions (Olson and Smyth 2015), in our study, sensitive questions lengthened the interaction in terms of number of words uttered, and increased the odds of respondents' filled pauses, regardless of mode. Although fast answers and more words might also mean faster speaking, this is not confirmed by Holbrook, et al. (Chapter 17, this volume). In addition, this finding is not a surprising one; sensitivity of a question topic may require respondents to think about their answers more thoroughly and respondents may also feel the need to elaborate to explain their answers. Further research into actual actions (e.g., elaborations of answers) could tell us what was specifically increasing the interaction quantity in sensitive questions.

Show card questions also increased the number of words in interaction, but not for CAPI. This is in correspondence with the finding that these questions are longer and more complex than questions without show cards (Smith 1987; Roberts, Jäckle, and Lynn 2006). Similarly, questions with mismatched response options yield fewer words in CAPI but more words in CATI. This question type has been shown to lower data quality, especially in CATI as compared to self-administered mail surveys (Smyth and Olson 2018), but effects might be different in CAPI. Our finding that demographic questions, as compared to behavior questions, decrease the interaction quantity in CAPI and the odds of respondents' filled pauses regardless of mode, is similar to Olson and Smyth's (2015) finding that these questions show decreased response time, presumably due to the likelihood that for demographics respondents often have an answer readily available, while behavioral questions require more effort in processing.

Including question characteristics accounted for 57 percent of the variability due to questions, and over 87 percent of the variability due to mode. For respondent characteristics, effects were found only for educational level (increased interaction quantity when respondents have a lower educational level) and age (decreased odds of interviewer's filled pauses with increasing respondents' age).

Our study is limited in the sense that only one-third of the sample comprised actual experimental manipulation of mode, and response rates between modes were different. However, including a control variable accounting for randomized versus non-randomized mode assignment did not show significant effects. In addition, question characteristics were not experimentally manipulated, and question topic may be confounded with other characteristics. Moreover, due to a small number of respondent–interviewer pairs, taking the interviewer as a separate level into account was not possible, and mode was taken as a separate level instead. Similar to Olson and Smyth (2015), our study did not include a direct measure of data quality. Evaluating response effects would have not have been likely to shed more light on effects on data quality, as an earlier study on the same data (Haan, et al. 2017) did not reveal substantial differences in satisficing indicators or socially desirable responding. Moreover, next to interaction quantity, by using some exploratory variables (filled pauses and uncertainty markers), we did provide more details in the interviewer–respondent interaction. Systematically coded data would indeed offer more information.

Looking for problems in interaction by listening to audio-recordings of interactions may be like looking for a needle in a haystack. In behavior coding research, therefore, analysis is done on coded data; coded data are created by having coders listen to recorded interactions and enter codes summarizing a list of potentially problematic aspects in a database. However, this requires quite labor-intensive coding. Another disadvantage with this approach is that the list of behaviors to be coded needs to be determined in advance. To tackle this latter disadvantage, in interaction analysis studies, therefore, an even more labor-intensive approach is chosen; first all interactions are fully transcribed, and next a detailed coding scheme is used to code all utterances of interviewers and respondents. Our approach in this chapter lies somewhere in between; we fully transcribed all interviews, but did not apply manual coding. Instead, we used text counts and automated word searches to look for patterns in behaviors of interest. This approach still yielded a wealth of information. Replicating earlier research (Olson and Smyth 2015), we find the question characteristics account for more than ten times the variance in interaction quantity in terms of number of words uttered than respondent characteristics. Our findings suggest that open-ended text questions increased interaction quantity in CAPI but not in CATI, while in CATI, absence of show cards or questions with mismatched response options increases interaction quantity. In addition, question wording can be improved by using fewer words, and reducing sensitive questions or finding ways to make those questions appear less sensitive.

Acknowledgments

This research is part of a project that was funded by the Netherlands Organization for Scientific Research (NWO), Grant #471-09-002.

References

American Association for Public Opinion Research (AAPOR). 2016. *Standard Definitions: Final Dispositions of Case Codes and Outcome Rates for Surveys*. 9th edition. Washington D.C.: AAPOR.

Andrews, F. M. 1984. Construct validity and error components of survey measures: A structural modeling approach. *Public Opinion Quarterly* 48(2):409–442.

Cernat, A. 2015. Evaluating mode differences in longitudinal data. Moving to a mixed mode paradigm of survey methodology. Dissertation. Institute for Social and Economic Research: University of Essex.

Clark, H. H., and M. F. Schober. 1992. Asking questions and influencing answers. In: *Questions about Questions: Inquiries into the Cognitive Bases of Surveys*, ed. J. M. Tanur, 15–48. New York: Russel Sage Foundation.

Couper, M. P., and F. Kreuter. 2013. Using paradata to explore item-level response times in surveys. *Journal of the Royal Statistical Society: Series A: Statistics in Society* 176(1):271–286.

Dijkstra, W. 2018. Sequence viewer. Amsterdam, Netherlands. http://www.sequenceviewer.nl/.

Dijkstra, W., and Y. Ongena. 2006. Question-answer sequences in survey-interviews. *Quality and Quantity* 40(6):983–1011.

Groves, R. M. 1990. Theories and methods of telephone surveys. *Annual Review of Sociology* 16(1):221–240.

Groves, R. M., and R. L. Kahn. 1979. *Surveys by Telephone: A National Comparison with Personal Interviews*. New York: Academic.

Groves, R. M., and L. Lyberg. 2010. Total survey error: Past, present, and future. *Public Opinion Quarterly* 74(5):849–879.

Haan, M., Y. P. Ongena, and K. Aarts. 2014. Reaching hard-to-survey populations: Mode choice and mode preference. *Journal of Official Statistics* 30(2):355–379.

Haan, M., Y. P. Ongena, J. T. A. Vannieuwenhuyze, and K. de Glopper. 2017. Response behavior in a video-web survey: A mode comparison study. *Journal of Survey Statistics and Methodology* 5(1):48–69.

Holbrook, A. L., M. C. Green, and J. A. Krosnick. 2003. Telephone vs. face-to-face interviewing of national probability samples with long questionnaires: Comparisons of respondent satisficing and social desirability response bias. *Public Opinion Quarterly* 67(1):79–125.

Houtkoop-Steenstra, H. 2002. Questioning turn format and turn-taking problems in standardized interviews. In: *Standardization and Tacit Knowledge: Interaction and Practice in the Survey Interview*, ed. D. W. Maynard, H. Houtkoop-Steenstra, N. C. Schaeffer, and J. Van der Zouwen, 243–261. New York: Wiley.

Irvine, A. 2011. Duration, dominance and depth in telephone and face-to-face interviews: A comparative exploration. *International Journal of Qualitative Methods* 10(3):202–220.

Jäckle, A., C. Roberts, and P. Lynn. 2010. Assessing the effect of data collection mode on measurement. *International Statistical Review* 78(1):3–20.

Körmendi, E. 1988. The quality of income information in telephone and face to face surveys. In: *Telephone Survey Methodology*, ed. R. M. Groves, P. P. Biemer, L. E. Lyberg, J. T. Massey, W. L. Nicholls, and J. Waksberg, 341–357. New York: Wiley and Sons.

Martin, J., C. O'Muircheartaigh, and J. Curtice 1993. The use of CAPI for attitude surveys: An experimental comparison with traditional methods. *Journal of Official Statistics* 9(3):641–661.

Miller, P. V. 1984. Alternative question forms for attitude scale questions in telephone interviews. *Public Opinion Quarterly* 48(4):766–778.

Olson, K., and J. D. Smyth. 2015. The effect of CATI questions, respondents, and interviewers on response time. *Journal of Survey Statistics and Methodology* 3(3):1–36.

Ongena, Y. P., and W. Dijkstra. 2006. Methods of behavior coding of survey interviews. *Journal of Official Statistics* 22:419–451.

Pascale, J. 2016. Behavior coding using computer assisted audio recording: Findings from a pilot test. *Survey Practice* 9(2):1–13.

Pascale, J., P. Goerman, and K. Drom. 2013. 2010 ACS content test evaluation: Behavior coding results. Studie Series *Survey Methodology* 7. Washington, D.C: Center for Survey Measurement Research and Methodology Directorate U.S. Census Bureau.

R Core Development Team. 2018. *R: A Language and Environment for Statistical Computing*. Vienna, Austria: R Foundation for Statistical Computing. http://www.R-project.org/.

Roberts, C., A. Jäckle, and P. Lynn. 2006. Causes of mode effects: Separating out interviewer and stimulus effects in comparisons of face-to-face and telephone surveys. In: *Proceedings of the Survey Research Methods Section of the American Statistical Association*, 4221–4228.

Rogers, T. F. 1976. Interviews by telephone and in person: Quality of responses and field performance. *Public Opinion Quarterly* 40(1):51–65.

Salthouse, T. A. 1991. *Theoretical Perspectives on Cognitive Aging*. Hillsdale, NJ: Lawrence Erlbaum Associates.

Schaeffer, N. C., and D. W. Maynard. 1996. From paradigm to prototype and back again: Interactive aspects of cognitive processing in standardized survey interviews. In: *Answering Questions: Methodology for Determining Cognitive and Communicative Processes in Survey Research*, ed. N. Schwarz and S. Sudman, 65–88. San Francisco: Jossey Bass.

Scherpenzeel, A. 2001. *Mode Effects in Panel Surveys: A Comparison of CAPI and CATI*. Bases statistiques et vues d'ensemble. Neuchâtel: Bundesamt, für Statistik, Office fédéral de la statistique.

Scherpenzeel, A. C., and W. E. Saris. 1995. Effects of data collection technique on the quality of survey data: An evaluation of interviewer- and self-administered computer assisted data collection techniques. In: *A Question of Quality. Evaluating Survey Questions by Multitrait-Multimethod Studies*, ed. A. C. Scherpenzeel. Dissertation. Rotterdam, Netherlands: Royal PTT Nederland NV, KPN Research.

Scherpenzeel, A. C., and W. E. Saris. 1997. The validity and reliability of survey questions: A meta-analysis of MTMM studies. *Sociological Methods and Research* 25(3):341–383.

Smith, T. W. 1987. The art of asking questions, 1936–1985. *Public Opinion Quarterly* 51(4 PART 2):S95–S108.

Smyth, J. D., and K. Olson. 2018. The effects of mismatches between survey question stems and response options on data quality and responses. *Journal of Survey Statistics and Methodology* 7(1):34–65.

Snijkers, G. J. M. E. 2002. *Cognitive Laboratory Experiences. On Pre-Testing Computerized Questionnaires and Data Quality*. Utrecht: Dissertation Universiteit Utrecht.

Swerts, M. 1998. Filled pauses as markers of discourse structure. *Journal of Pragmatics* 30(4):485–496.

Sykes, W., and M. Collins. 1988. Effects of mode of interview: Experiments in the UK. In: *Telephone Survey Methodology*, ed. R. M. Groves, P. P. Biemer, L. E. Lyberg, J. T. Massey, W. L. Nicholls, and J. Waksberg, 301–320. New York: Wiley and Sons.

Yan, T., and R. Tourangeau. 2008. Fast times and easy questions: The effects of age, experience, and question complexity on web survey response times. *Applied Cognitive Psychology* 22(1):51–68.

13

Interacting with Interviewers in Text and Voice Interviews on Smartphones

Michael F. Schober, Frederick G. Conrad, Christopher Antoun,
Alison W. Bowers, Andrew L. Hupp, and H. Yanna Yan

CONTENTS

13.1 Introduction

In the parts of the world with access to recent advances in mobile technologies, the daily communication habits of potential survey respondents have changed massively. An increasing majority of sample members are augmenting their face-to-face (FTF) and land-line telephone interactions (if they even have a landline any more) by communicating in multiple modes – talking, emailing, text messaging, video chatting, posting to social media – on a single mobile device, and switching between those modes on the same device or across devices (smartphone, tablet, desktop computer, etc.). People are growing accustomed to choosing and switching between communication modes appropriate to their current setting (a noisy environment? an unstable network connection?), communication goals (a professional communication with co-workers? a private conversation with a family member?), chronic or temporary needs (wanting a lasting record of what was communicated, or no record? not wanting to be seen right now? needing to care for a child while communicating?), and interlocutor (a partner who never responds to voice calls or emails?). People with access to advanced technologies also are more frequently engaging in human–machine interactions, whether with bank automated teller machines, ticket kiosks, and self-check-out at grocery stores, or in conversations with automated phone agents for travel reservations and tech support, or in online help "chat."

Many potential survey respondents have thus gotten used to modes of interacting that have quite different dynamics than the FTF "doorstep" interviews and landline telephone interviews that have formed the backbone of survey measurement – and even than the online "self-administered" surveys that are a growing part of the landscape. How will these transformations affect survey responding and the effects of interviewers? How should we think about the role of interviewers when potential respondents may want to – or even assume they can – participate in surveys in the modes of interaction they now use daily? Will interviewers in new modes enhance participation and respondent motivation, as they often do in current modes (e.g., Villar and Fitzgerald 2017)? What signs of their "humanness" do interviewers produce in modes with less social presence than FTF, and how do these affect respondents' participation and the quality of the data they provide?

13.2 Study

The study described here begins to address these questions by exploring the dynamics of interviewer–respondent interaction in a corpus of 634 US-based interviews on smartphones from Schober, et al. (2015). The interviews were carried out in four survey modes that work through two native iPhone apps with which all participants were likely to have experience (Phone and Messages), as opposed to through a study-specific survey app or as a web survey in the phone's browser; the study purposely limited itself to using a uniform interface for all respondents (iOS) rather than mixing mobile platforms to include Android, Windows, etc. The study varied the medium of interaction (voice vs. text messaging) and the interviewing agent (human interviewer vs. automated system), leading to four survey modes: Human Voice, Human Text, Automated Voice, and Automated Text. In each mode, respondents (who had screened into participating on their iPhone for a $20 (gift code) incentive and were randomly assigned to one of the modes) answered 32 questions selected from ongoing US social scientific surveys. (See Antoun, et al. 2016 for details about sample recruitment from online sources.)

The Human Voice and Human Text interviews were carried out by the same eight professional interviewers from the University of Michigan Survey Research Center, using a custom CATI interface for this study that supported voice and text interviews (see Schober, et al. 2015, supplementary materials, for more details). In the Human Voice interviews, interviewers read the questions and entered answers as they usually do. In the Human Text interviews, interviewers used the interface to select and send stored questions and prompts as well as to edit pre-existing prompts or type individualized messages. Respondents could text their answers with single characters rather than typing out full answers: "Y" or "N" (for "yes" or "no"), single letters ("a," "b," "c") corresponding to response options, or numbers for numerical questions. They could also text requests for clarification or help if they desired.

The Automated Voice mode was a custom-built Speech-IVR (Interactive Voice Response) dialogue system, implemented using AT&T's Watson speech recognizer and the Asterisk telephony gateway. Questions and prompts were presented by an audio-recorded human interviewer; respondents' spoken responses that were recognized were either accepted ("Got it," "Thanks") or presented for verification if the system's recognition confidence was lower than a minimum threshold ("I think you said 'nine' – is that right? Yes or no"). Responses that were not recognized led the system to re-present the question or response options (see Johnston, et al. 2013 for more details about the system and interface).

The Automated Text interviews were carried out via a custom text dialogue system, and the interface for respondents was the same as that for Human Text interviews. The system texted questions, accepted recognizable answers, prompted for an acceptable answer if what was provided was uninterpretable, and presented standard definitions of key survey terms if respondents requested help. The automated text interview introduction stated: "This is an automated interview from the University of Michigan," rather than the human text interview introduction: "My name is < first name, last name > from the University of Michigan."

As reported in greater detail in Schober, et al. (2015; see also Conrad, et al. 2017a, 2017b), text interviews differed from voice interviews in a number of ways, as did automated and interviewer-administered interviews, with a range of independent and additive effects. Text interviews (compared to voice interviews) led to higher interview completion rates whether automated or interviewer-administered, but also (in interviewer-administered interviews) to a higher breakoff rate. Respondents in text interviews – both automated and interviewer-administered – produced higher-quality data on several fronts, giving more precise (fewer rounded) numerical answers and more differentiated answers to battery questions (less straight-lining). They also disclosed more sensitive behaviors, consistent with West and Axinn's (2015) finding of greater disclosure in text than voice interviews in a quite different population and survey in Nepal. Respondents in automated (compared to interviewer-administered) interviewers completed interviews at a lower rate, and they had a higher rate of breakoff in both text and voice interviews. Respondents in automated interviews also reported more sensitive behaviors than those in interviewer-administered interviews (replicating the oft-reported finding of greater disclosure in self-adminis- tered interviews, e.g., Tourangeau and Smith 1996) – as a main effect independent of the increased disclosure due to texting. So the greatest level of disclosure was in Automated Text interviews, and the least in Human Voice.

What accounts for this pattern of differences in participation and data quality in these modes? Any or all of the many differences in timing and behavior between text and voice, and automated and interviewer-administered, interviews – alone or in combination – are plausible contributing factors. One could hypothesize, for example, that respondents report with greater precision (less rounding) in text because texting (vs. voice) reduces the immediate time pressure to respond, so the respondent has more time to think or to look up answers. Or one could hypothesize that respondents disclose more in text because text messaging reduces the interviewer's "social presence" – there is no immediate evidence of any interviewer reaction, and thus reduced salience of the interviewer's ability to evaluate or be judgmental.

The experimental design of the study helps rule in or rule out at least some such alterna- tive plausible accounts, because we can explore whether differences in interview timing or in particular interviewer or respondent behaviors across the modes correlate with par- ticular data quality or participation outcomes. We focus on whether interviewer behaviors that provide concrete evidence of interviewers' humanness – speech disfluencies (evidence that speech is spontaneous and that the interviewer might be fallible), laughter, chat that goes beyond the survey script – and that vary in the different modes in the study corre- late with or predict data quality. As potential correlates of interviewer–respondent rap- port (e.g., Garbarski, Schaeffer, and Dykema 2016), these interviewer behaviors may affect respondents' engagement and motivation, and thus the quality of respondents' answers. Theorizing about how interviewers might affect data quality (see, e.g., Kreuter, Presser, and Tourangeau 2008; Lind, et al. 2013; Tourangeau and Smith 1996, among many others) has proposed a range of possible effects, both positive and negative: interviewers may indeed increase respondents' motivation, but they may also cause respondents to feel embarrassed and reduce their candor, or distract their attentional focus from the content of survey questions. We see empirically examining effects of concrete interviewer behaviors associated with the "human touch" on data quality as a useful step.

We used two main analytic strategies. First, if interviewer behavior differs between two modes but our measures of data quality *don't* differ, interviewer behavior is unlikely to be causal. For example, one might hypothesize that respondents provided more precise numerical answers (rounded less) in text than voice because text interviewers (human or automated) may have been less likely to laugh (text "LOL" or "haha"), and laughter might suggest that less precise answers are sufficient. But this can't be the simple explanation if respondents round just as much in Human and Automated Voice interviews, but the Automated Voice system never laughs. Our second analytic strategy is to test whether interviewer behavior is associated with data quality. In combination, these two strategies help open a window onto what might account for observed mode differences.

13.3 Analyses of Interview Dynamics

Comparable transcripts of all interviewer and respondent speech and text during the course of each interview were assembled for analysis in Sequence Viewer (Dijkstra 2018), on the subset of 619 interviews for which we had full (no missing) audio or text recordings, and the onset of each "turn" was time stamped. What was immediately apparent was how differently text and voice interviews unfolded. Compare these two straightforward question–answer sequences for the same question asked in a Human Text vs. a Human Voice interview:

The two turns in the text interchange involve notably fewer total words, but they unfold over a much slower time frame (1 minute 21 seconds) than the 12 seconds it takes for the three turns in the voice interchange. See Online Appendix 13A for a more extended example with a request for clarification on what the question is meant to include; it exemplifies the same pattern of many more turns in voice unfolding over a shorter time period (1 minute 36 seconds in voice vs. 2 minutes in text). As Figure 13.1 (adapted from Schober, et al. 2015) plots, this pattern holds more generally, with far fewer turns after each question (a median of 1) in text than voice, but a much greater median duration of question–answer sequences.

To examine interview dynamics more closely, we developed a coding scheme for interviewer (or automated system) and respondent conversational "moves" and for interactional paradata that could apply across all four modes (see Online Appendix 13B for the complete coding manual). The scheme included 25 interviewer moves (e.g., asking the question exactly as worded, asking the question with minor or major deviation, presenting the

TABLE 13.1

Example Human Text and Voice Interchange

	Human Text		Human Voice
I:	During the last month, how many movies did you watch in any medium?	I:	During the last month, how many movies did you watch in ANY medium?
R:	Three	R:	OH, GOD. U:h man. That's a lot. How many movies I seen? Like thirty.
		I:	Thirty.

Note: Capital letters represent spoken emphasis; a colon (":") in a word represents a lengthened sound.

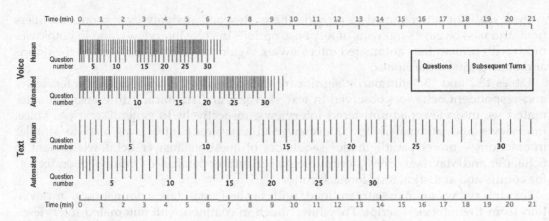

FIGURE 13.1
Interview duration and median number of turns per survey question.

response alternatives, providing clarification, and mode-appropriate versions of no-input ("I didn't hear that") and no-match ("I didn't understand that") and 30 respondent moves (e.g., providing an acceptable answer, answering the question not using the exact specified response alternatives, reporting a behavior instead of answering, and asking for clarification). Additional paralinguistic behaviors were also coded: speech disfluencies (fillers like *um* and *uh*, repairs) and typos, laughter, and hedges in responses (e.g., "maybe," "sort of").

Three authors triple-coded a subset of 400 question–answer sequences (all speech or text between an interviewer's presenting one question and the next) randomly selected from the 619 interviews, achieving very high interrater reliability (Cohen's kappa = 0.91–0.99), and resolving any discrepancies through discussion. Codes for all question–answer sequences in all interviews were entered and counted using Sequence Viewer. Summary counts for each code were compared in two different ways: (1) as the percentage of question–answer sequences per interview in each mode that included at least one instance of a particular code, and (2) as the percentage of interviews in each mode that included at least one instance of a particular code. For (1), comparisons of relative frequency across modes were carried out using two-way ANOVAs to examine effects of medium (text vs. voice), agent (human vs. automated interviewer), and interaction of medium x agent. For (2), the same comparisons and interaction analyses were carried out using logistic regression, with reference categories "text" and "automated."* To correct for the possibility of spurious findings given the multiple comparisons, a Bonferroni correction was applied such that effects were judged significant with an alpha level <0.00019, rather than 0.05.

Based on these comparisons, some behaviors are clearly mode-specific, only occurring in one of the four interview modes. For example, interviewers only ever repaired or restarted an utterance in Human Voice interviews – which makes sense, as Human Text interviewers can review and edit every utterance before they hit "send." Because Automated Voice interviews were programmed so that the "interviewer" acknowledged responses with "got it" or "okay" on a fixed schedule, it makes sense that it is in this mode that essentially every interview included such acknowledgments, with fewer in Human Voice interviews, far fewer in Human Text interviews, and none in Automated Text interviews, where the

---

* Because there was only one "interviewer" in the automated modes and eight interviewers in the human-administered modes, it was not possible to fit hierarchical models with respondents nested within interviewers for these comparisons, even though this was a feature of the human-administered modes.

system was not programmed to provide them. Respondents only ever produced fillers (*um*s and *uh*s) or gave synonyms of response options in voice interviews, and in both cases more with human than automated interviewers. Again, this makes sense given the norms and constraints of the modes.

Tables 13.2 and 13.3 summarize significant differences in the prevalence of interviewer and respondent behaviors observed in text vs. voice, and significant differences in automated vs. interviewer-administered interviews, respectively. In general, text (vs. voice) interviews and automated (vs. interviewer-administered) interviews lead to simpler interactions – more "paradigmatic" sequences of question–answer–acknowledgment in Schaeffer and Maynard's (1996) terminology. For further detail, see Online Appendix 13C for counts and statistical tests for each code.

Text interviews are more streamlined, with far less small talk, commentary, or deviation from the interview script. They have much in common with automated interviews, although speech recognition trouble in the Automated Voice interviews leads to more clarification sequences. It is striking that text respondents were sensitive to whether the questions were sent by a human interviewer or an automated system considering how similar

TABLE 13.2

Significant Differences in Text (vs. Voice) Interaction

Respondent Behaviors	Interviewer Behaviors
Fewer unacceptable answers	No misstatements of questions
Fewer descriptions of behaviors (reports) as opposed to answers that fit response options	Almost no repeats of questions or response alternatives
Fewer backchannels ("Uh-huh")	Fewer neutral probes
Almost no requests for repeat of survey question	Less commentary
Fewer "don't know" answers	Almost no laughter (e.g., "LOL")
Fewer requests for time to find answer	No speech disfluencies (*um*s and *uh*s, repairs, long pauses), few typos
Less commentary	
Less laughter	
Fewer hedges	
No speech disfluencies, few typos	

TABLE 13.3

Significant Differences in Automated (vs. Interviewer-Administered) Interaction

Respondent Behaviors	"Interviewer" Behaviors
Fewer unacceptable answers	No misstatements of questions
No reporting of behavior	Almost no repeats of questions or response alternatives
More changed answers (Automated Voice)	No neutral probes
Fewer backchannels ("Uh-huh")	No commentary
Fewer requests for repeat of survey question	No laughter (e.g., "LOL")
Fewer "don't know" answers	No speech disfluencies (*um*s and *uh*s, repairs, long pauses), few typos
Less commentary	
Less laughter	
Fewer hedges	
Fewer disfluencies	

the two kinds of text interviews often were – in some cases, almost the only difference was that human interviewers provided their names.

13.3.1 Do Interviewer Behaviors Predict Response Quality?

To what extent are these differences in coded behaviors connected with the improved data quality (less rounding, less straight-lining, more disclosure) in texting observed in this data set? The modes in which respondents disclosed more (text vs. voice, and automated vs. interviewer-administered interviews) and rounded less (text vs. voice) differ systematically in the various interviewer behaviors listed in Tables 13.2 and 13.3, and it is plausible that these differences are part of the bundle of differing features that led to the data quality differences. But texting and voice, of course, differ systematically on multiple fronts: in synchrony, in the medium (written vs. spoken), in the persistence and reviewability of the messages exchanged, in the potential impact of nearby others, and in how easy it is to multitask during the interaction (see Schober, et al. 2015, Table 1). Any or all of those features in combination might cause the observed patterns of precise responding and disclosure.

Using our first analytic strategy, among the behaviors we see as particularly demonstrating interviewers' humanness, fillers, repairs, and laughter only ever occurred in Human Voice interviews. Our mode differences in disclosure showed two independent effects: increased disclosure in text (vs. voice) interviews, and in automated (vs. human) interviews. Fillers, repairs, and laughter are, then, plausible (or at least not ruled out as potential) interviewer behaviors that correlate with reduced disclosure. For rounding, however, we only observed reduced rounding in text (vs. voice), and no effect of automation. Given this pattern, interviewer fillers, repairs, and laughter are unlikely to be connected with the greater precision in text vs. voice.

Following this same logic, the interviewer behaviors that were especially more frequent in Human Voice interviews relative to all the other modes (see Online Appendix 13C) also are plausible candidates for correlating with reduced disclosure – e.g., asking a question with a wording change or paraphrase, restating the response alternatives, presenting a neutral probe, continuing a previous move after a change of speakers, narrowing the response options, and providing nonstandard commentary – but not with the rounding effects. Interviewer behaviors that were particularly frequent only in voice interviews but not in text interviews (e.g., acknowledgments like "got it") can potentially contribute to the findings on rounding, which only found a voice vs. text difference. The findings in Online Appendix 13C can therefore be seen as providing a profile of which interviewer behaviors are potential contributors to the data quality findings.

Using our second analytic strategy, which focuses on the interviewer-administered modes where these "human touch" behaviors occurred, we carried out regression analyses to examine whether laughter, commentary, and disfluencies (fillers, repairs, pauses judged as notable by our coders) were linked with data quality (disclosure and rounding), as well as with respondent satisfaction with the interview. Focusing first on disclosure, the first two columns of data in Table 13.4 show that when voice interviewers produced more fillers for the 13 questions we used to measure disclosure,* respondents were significantly more likely to disclose more sensitive information.†

---

* Focusing the analyses only on interviewer behaviors in these 13 question–answer sequences runs the risk of missing potential effects of these behaviors that accumulate from prior or intervening question–answer sequences, but provides what we see as the cleanest test.

† Note that four interviewer behaviors – laughter, repairs, fillers, and pauses – did not occur in human text interviews, and so can't be predictors in any of the models of Human Text interviews.

TABLE 13.4

Regression Outputs for Models Predicting Respondents' Levels of Disclosure, Rounding, and Satisfaction with the Interview in Interviewer-Administered Interviews, Based on Interviewers' "Humanizing" Behaviors

	Disclosure		Rounding		Satisfaction	
	Voice	Text	Voice	Text	Voice	Text
Predictor	Est. (SE) p-value	Est. (SE) p-value	Est. (SE) p-value	Est. (SE) p-value	Est. (SE) p-value	Est. (SE) p-value
Intercept	2.778	3.234	3.956	3.203	2.717	2.499
	(0.348)	(0.181)	(0.212)	(0.138)	(0.101)	(0.056)
	<.001	<.001	<.001	<.001	<.0001	<.0001
Commentary	−0.098	0.045	0.332	−0.144	−0.062	0.027
	(0.147)	(0.185)	(0.202)	(0.163)	(0.042)	(0.057)
	.507	.808	.103	.378	.150	.634
Laughter	−0.151		−0.437		0.029	
	(0.307)		(0.189)		(0.089)	
	.623		.022		.748	
Repairs	−0.072		0.190		−0.070	
	(0.146)		(0.143)		(0.042)	
	.620		.186		.098	
Fillers	0.296		−0.066		0.013	
	(0.127)		(0.113)		(0.037)	
	.022		.560		.723	
Pauses	0.011		0.046		−0.004	
	(0.074)		(0.068)		(0.021)	
	.886		.501		.857	
Estimated variance of random interviewer effect	0.191	NA[a]	NA[a]	NA[a]	0.016	NA[a]
N interviewers	8	8	8	8	8	8
N respondents	148	156	148	156	148	156

NA[a]: random intercept was omitted because its estimated variance was less than zero when included in the model. Interviewer behaviors are included in the models as continuous variables. (When the behaviors are included in the model as binary predictors corresponding to whether a behavior ever occurred, the pattern of results is similar, with two exceptions as noted in the text.)

We see this finding as suggestive rather than definitive, in that further analyses show a less clear story: (a) the estimated effect of fillers on disclosure in voice interviews is no longer statistically significant if we treat the predictor as binary rather than continuous (coefficient = 0.304, p = .364); (b) when all interviewer disfluencies (repairs, fillers, pauses) are combined into one variable (not shown in Table 13.4), this combined variable shows no significant effect on disclosure (coefficient = 0.065, p = .207); and (c) if we log-transform our measure of disclosure to address the right-skew in the data, the estimated effect of fillers on disclosure in voice interviews is no longer statistically significant (p = .165). On the other hand, when fillers are added to the model as a binary variable (low: 0–2 fillers vs. high: 3 or more fillers), the estimated effect of three or more fillers is positive and significant (1.389, p = .016). So we (only very cautiously) interpret these findings as consistent with the possibility that one kind of evidence of interviewer fallibility – producing more *ums* and *uhs* – may increase willingness to disclose embarrassing information.

What about precise responding? As the middle data columns in Table 13.4 show, based on respondents' answers and interviewer behaviors in the eight question–answer sequences in voice interviews that we used to measure rounding, respondents provided more answers that were precise (fewer rounded answers) when interviewers laughed more. We see no evidence that interviewers' disfluencies affected precise responding, either individually or as a combined measure of repairs, fillers, and pauses (0.045, p = .367), nor that commentary affected precise responding. While one might think that an interviewer's laughter, as an indicator of informality, might license respondents to answer more casually and less thoughtfully, that is clearly not the case here; the evidence is instead consistent with an account that an interviewer's laughter may correlate with a feeling of reduced time pressure, perhaps giving respondents more time to retrieve instances and formulate their answers.

And what about respondent satisfaction? Perhaps linguistic evidence of interviewers' humanness makes respondents feel more comfortable, whether or not they disclose more or try harder to produce precise answers. As the final two data columns in Table 13.4 show, after voice interviews where interviewers produced more repairs (across all 32 question–answer sequences in the entire interview), respondents reported marginally less satisfaction with the interview in the post-interview online debriefing questionnaire (1 = very dissatisfied/somewhat dissatisfied; 2 = somewhat satisfied; 3 = very satisfied). This marginal effect becomes statistically significant if we treat the predictor as binary (presence or absence of repairs) (estimate −0.262, p = .010). Although we don't know why repairs (rather than fillers or a combined disfluency rate) should be particularly predictive, this finding suggests that interviewer disfluency either leads to respondent dissatisfaction – perhaps disfluent interviewers seem unprepared or unprofessional – or (conversely) that speech goes less smoothly in interviews that respondents are enjoying less. While analyses of telephone interview invitations (Conrad, et al. 2013) show that interviewers who are too perfectly fluent can sound "robotic" and off-putting, too much disfluency also can reduce agreement to participate. The evidence on satisfaction here is consistent with that finding.

While text interviews didn't provide the same robust opportunities for us to observe effects of interviewer humanizing behaviors on respondent data quality, we did have access to one aspect of the interactional dynamic in text interviews: the speed of back-and-forth. Some text interviews were close to synchronous, with more rapid-fire turn exchanges, while others featured longer delays between turns. (In most cases we attribute these speed differences to respondent speed; the automated text interviewing system always produced the next turn immediately, and interviewers were generally speedy in responding.) Although we didn't observe any effects of text interview speed on respondent disclosure or their satisfaction with the interview, we do see a consistent effect on rounding (reported in Schober, et al. 2015): respondents gave significantly fewer precise (more rounded) answers in faster interviews (interviews with shorter inter-turn intervals than the median of 15.75 seconds), and more precise answers in slower interviews. The direction of causality is of course unclear, but the clear link between interview speed and precise responding is that respondents who take longer to respond (perhaps because they are thinking harder about accurate answers or looking up records) give more precise answers.

Taken together, these additional analyses of the effects of interviewer behavior and texting speed lead us to a preliminary account. Part of what promotes precise responding in text (vs. voice) is the reduced time pressure to respond (evidenced by less rounding in slower text interviews), and a more relaxed less-time-pressured mode of interviewing in voice interviews (evidenced by greater precision in voice interviews when interviewers laughed more). Disclosure increases significantly in the modes that reduce interviewer social presence (text vs. voice, and automated vs. interviewer-administered), but within the interviewer-administered telephone interviews disclosure may be promoted by increased evidence of interviewer fallibility or humanness (more fillers). In contrast, increases in another kind of evidence of interviewer fallibility – repaired speech – corresponded with decreased respondent satisfaction with voice interviews.

13.4 Discussion

The evidence in this corpus is that text interviews have quite different dynamics than voice interviews on the same device: they take longer overall but with fewer turns of interaction – allowing respondents to answer when convenient for them and while multitasking, which a number of respondents report finding preferable (see Conrad, et al. 2017a) – and they are more "to the point" with less small talk. The interaction analyses reported here suggest (or are at least consistent with an explanation that) decreased social presence of the interviewer and the asynchrony of interaction in text may have important effects on data quality (in general, benefits) – but that in voice interviews interviewer behaviors that display human fallibility (laughter and repairs) may be associated with improved data quality (whether they cause or result from it). For precision of answers (rounding), interviews with *no* laughter (text interviews, automated interviews) were associated with more precise responding (better data quality), but *more* interviewer laughter in voice interviews was associated with more precise responding. For disclosure, respondents reported more socially undesirable behavior (and data quality was presumably better) in interviews with no or fewer disfluencies (text and automated interviews), but in voice interviews interviewers' speech disfluencies (in particular, fillers) seem to have been associated with greater disclosure.

From a Total Survey Error perspective, text interviewing (vs. voice interviewing) in this data set clearly improved both participation and measurement (Schober, et al. 2015). Although our corpus doesn't allow systematic calculation of interviewer effects (respondents were not assigned to interviewers randomly), the interaction analyses reported here suggest that text interviewing has the potential to reduce interviewer effects. To the extent that interviewer variance is related to interviewer behavior, texting simply has *less* interviewer behavior; in largely streamlining the interview to its essential question-asking and question-answering elements, text interviewing should lead to more standardized interviews than when interviewing is conducted via voice.

13.4.1 Questions and Implications

The analyses reported here raise at least as many questions as they answer. It remains to be seen whether the patterns of findings extend to other implementations of these modes (different variants on text messaging interviews or automated voice interviews) or to different survey questions, e.g., more or less sensitive behavioral questions or attitudinal or opinion questions. Will the findings generalize to non-convenience (probability) samples, to differently incentivized participants, or to subpopulations of respondents with different levels of experience in particular modes? One might suspect that respondents who are unfamiliar or uncomfortable with texting, or who only feel comfortable talking with a live human, might show different patterns of behavior and interaction than those observed in the current sample of respondents, who opted into participating in a study on their iPhone from advertisements in online sources.

A major challenge in interpreting studies of this sort is how much of a moving target they are attempting to hit. Communication devices themselves continue to evolve, with new versions of mobile devices and operating systems changing features frequently, which means that the features of the modes potentially available for use in survey interviews can be in flux even as a survey is deployed. Different groups or subpopulations may have different time courses for adopting a new device or a particular app, and so researchers deploying any particular mode at any moment may be tapping into unknown levels of experience among subgroups. And even among experienced users of a device or mode, norms for everyday communication evolve over time; people who once would have answered a voice call from an unknown caller may no longer be willing to pick up the phone, which may substantially change their willingness to participate in a particular mode, or their motivation to provide high-quality responses.

All the unknowns make this an exciting – if complex! – time to be exploring communication and interviewer effects in survey interviews. Despite how much more needs to be understood, the current findings do suggest a few main takeaway messages:

1. Interviewer effects (of the sort measured by the intraclass correlation) may take unexpected forms in different modes and as people's communication patterns and norms – not only with other people but with automated systems, in both personal and professional life – evolve. Modes that reduce interviewers' social presence and streamline the interaction (reducing time pressure to respond) have the potential to reduce interviewer effects – though the effects may vary for different measures of data quality in different modes.

2. Systematic methodological study over time and in multiple interviewing modes with a range of respondent populations will be needed to understand how interviewers will best be deployed moving forward. If interviews via text (or in any other new mode) prove to be popular, new interviewer training – and possibly even selecting interviewers with particular experience in or affinity for texting (or another mode) – may be needed.

3. Long-standing (if not always explicitly articulated) assumptions about FTF interviewing as the gold standard may need to be rethought (Schober 2018). The "human touch" in interviewing no doubt will continue to have important benefits for respondent motivation, rapport, and (as evidenced here) satisfaction with interviews. But the social presence of an interviewer can also have serious drawbacks as norms and practices of communication evolve in newer technologies. (The very fact that it is now routine in FTF interviews to switch to self-administration when the topic is judged to be sensitive demonstrates researchers' recognition that social presence matters.) At least some respondents may well end up finding interviewers' physical or virtual presence intrusive or burdensome in the way that many now already find interacting with human (vs. automated) bank tellers. How survey researchers should think about developing new gold standards will be an upcoming challenge.

Acknowledgments

NSF grants SES-1026225 and SES-1025645 (Methodology, Measurement, and Statistics program) to Frederick Conrad and Michael Schober. Many thanks to Patrick Ehlen, Stefanie

Fail, Michael Johnston, Courtney Kellner, Monique Kelly, Mingnan Liu, Kelly Nichols, Leif Percifield, Lucas Vickers, and Chan Zhang for advice and assistance, and to the editors for their helpful questions and comments.

References

Antoun, C., C. Zhang, F. G. Conrad, and M. F. Schober. 2016. Comparisons of online recruitment strategies for convenience samples: Craigslist, Google AdWords, Facebook, and Amazon Mechanical Turk. *Field Methods* 28(3):231–246.

Conrad, F., J. Broome, J. Benkí, F. Kreuter, R. Groves, D. Vannette, and C. McClain. 2013. Interviewer speech and the success of survey invitations. *Journal of the Royal Statistical Society: Series A* 176(1):191–210.

Conrad, F. G., M. F. Schober, C. Antoun, H. Y. Yan, A. L. Hupp, M. Johnston, P. Ehlen, L. Vickers, and C. Zhang. 2017a. Respondent mode choice in a smartphone survey. *Public Opinion Quarterly* 81(S1):307–337.

Conrad, F. G., M. F. Schober, A. L. Hupp, C. Antoun, and H. Y. Yan. 2017b. Text interviews on mobile devices. In: *Total Survey Error in Practice*, ed. P. P. Biemer, E. de Leeuw, S. Eckman, B. Edwards, F. Kreuter, L. E. Lyberg, C. Tucker, and B. T. West, 299–318. Hoboken, NJ: John Willey and Sons.

Dijkstra, W. 2018. Sequence viewer. Amsterdam, Netherlands. http://www.sequenceviewer.nl/.

Garbarski, D., N. C. Schaeffer, and J. Dykema. 2016. Interviewing practices, conversational practices, and rapport: Responsiveness and engagement in the standardized survey interview. *Sociological Methodology* 46(1):1–38.

Johnston, M., P. Ehlen, F. G. Conrad, M. F. Schober, C. Antoun, S. Fail, A. Hupp, L. Vickers, H. Yan, and C. Zhang. 2013. Spoken dialog systems for automated survey interviewing. In: *Proceedings of the 14th Annual SIGDIAL Meeting on Discourse and Dialogue (SIGDIAL 2013)*, 329–333.

Kreuter, F., S. Presser, and R. Tourangeau. 2008. Social desirability bias in CATI, IVR, and Web Surveys: The effects of mode and question sensitivity. *Public Opinion Quarterly* 72(5):847–865.

Lind, L. H., M. F. Schober, F. G. Conrad, and H. Reichert. 2013. Why do survey respondents disclose more when computers ask the questions? *Public Opinion Quarterly* 77(4):888–935.

Schaeffer, N. C., and D. W. Maynard. 1996. From paradigm to prototype and back again: Interactive aspects of cognitive processing in survey interviews. In: *Answering Questions: Methodology for Determining Cognitive and Communicative Processes in Survey Interviews*, ed. N. Schwarz, and S. Sudman, 65–88. San Francisco, CA: Jossey-Bass.

Schober, M. F. 2018. The future of face-to-face interviewing. *Quality Assurance in Education* 26(2):293–302.

Schober, M. F., F. G. Conrad, C. Antoun, P. Ehlen, S. Fail, A. L. Hupp, M. Johnston, L. Vickers, H. Y. Yan, and C. Zhang. 2015. Precision and disclosure in text and voice interviews on smartphones. *PLoS One* 10(6):e0128337.

Tourangeau, R., and T. W. Smith. 1996. Asking sensitive questions: The impact of data collection mode, question format, and question context. *Public Opinion Quarterly* 60(2):275–304.

Villar, A., and R. Fitzgerald. 2017. Using mixed modes in survey data research: Results from six experiments. In: *Values and Identities in Europe: Evidence from the European Social Survey*, ed. M. J. Breen, 273–310. New York: Routledge.

West, B. T., and W. G. Axinn. 2015. Evaluating a modular design approach to collecting survey data using text messages. *Survey Research Methods* 9(2):111–123.

Section V

Interviewers and Nonresponse

14

Explaining Interviewer Effects on Survey Unit Nonresponse: A Cross-Survey Analysis

Daniela Ackermann-Piek, Julie M. Korbmacher, and Ulrich Krieger

CONTENTS

14.1 Introduction

Interviewer effects on survey unit nonresponse have been detected in interviewer-mediated surveys for many decades (e.g., Durbin and Stuart 1951). Today, we know that interviewers affect the data collection process in various ways, both positively and negatively. On the positive side and in terms of response rates, interviewer-mediated surveys tend to achieve higher response rates than surveys using self-completion questionnaires; and face-to-face surveys, where the interviewers have the strongest involvement in the recruitment process, tend to achieve higher response rates than telephone surveys (Groves, et al. 2009). On the negative side, however, interviewers may have a differential impact on the representativeness of the achieved sample (e.g., Blom et al. 2011, Jäckle et al. 2013, Durrant et al. 2010). In addition, there is a growing body of studies attempting to *explain* the interviewer effects found and to develop methods for reducing them or adjusting for them in analyses (see West and Blom 2017 for an overview).

In this section, we provide an overview of the literature with regard to four types of interviewer characteristics that previous research has identified as predictors of survey unit nonresponse: socio-demographic characteristics, experience, attitudes and personalities, and skills and behaviors.

Many studies that investigate whether socio-demographic interviewer characteristics explain interviewer effects on unit nonresponse focus on age and gender, as this information is typically available from interviewer employment records. However, most of these studies find weak or nonsignificant relationships of these characteristics with unit nonresponse (West and Blom 2017). Some studies detect liking effects (Durrant, et al. 2010; Lord, Friday, and Brennan 2005; Moorman, et al. 1999; Webster 1996). For example, Durrant, et al.

(2010) find that a match between interviewer and respondent gender or education (see also West, et al. 2019) is associated with higher cooperation rates.

Research focusing on interviewer experience finds that interviewers with more experience achieve higher response rates (Couper and Groves 1992; Campanelli and O'Muircheartaigh 1999; Durbin and Stuart 1951; Durrant, et al. 2010; Jäckle, et al. 2013). Groves and Couper (1998) suggest that more experienced interviewers have learned the skill of tailoring their approaches at the doorstep, which ultimately leads to higher success rates. Additionally, there might be a self-selection effect, where less successful interviewers are likely to leave the workforce more quickly (Jäckle, et al. 2013). However, there are also numerous studies that report curvilinear, negative, or null relationships between interviewer experience and response (e.g., Blom, de Leeuw, and Hox 2011; Couper and Groves 1992; Durrant, et al. 2010). This may be explained by two related mechanisms: experienced interviewers often receive the most difficult cases (e.g., Blom, de Leeuw, and Hox 2011) and are assigned higher workloads (e.g., Japec 2008; Loosveldt, Carton, and Pickery 1998), both of which may lead to lower response rates.

In recent years, there has been an increase in research on interviewer effects on unit non-response related to interviewer attitudes and personality. In general, having a positive atti-tude and being more persuasion-oriented (de Leeuw, et al. 1998; Durrant, et al. 2010; Hox and de Leeuw 2002; Jäckle, et al. 2013; Maynard and Schaeffer 2002; Vassallo, et al. 2015), being confident (Blom, de Leeuw, and Hox 2011; Durrant, et al. 2010), and being extroverted (Jäckle, et al. 2013) are associated with higher cooperation rates. In contrast, interviewers who believe in stressing the voluntary nature of a survey tend to achieve lower coopera-tion rates (de Leeuw, et al. 1998).

Interviewer behavior can have both a passive and active effect on the sampled person's decision to participate. Passive influences might occur because interviewers are simply present, while interviewers' actual behavior can actively influence the target person's deci-sion to participate. However, as Jäckle, et al. (2013) summarize, interviewer behavior is not often predictive of cooperation. These authors provide three possible reasons for these null findings. First, studies examining this relationship generally have lower statistical power, given that the number of interviewers working on a project is usually limited. Second, there are issues with measurement, as interviewers can forget the "exact components of interaction" when they are asked to report them. Finally, the level of measurement can be problematic as interviewers are typically asked about their usual behavior, even though individualized interactions might be more relevant than a general pattern of how inter-viewers behave, according to Durrant, et al. (2010).

Some studies have used audio recordings to analyze interviewer behaviors during respondent recruitment. Overall, the findings of these studies suggest higher coopera-tion rates for interviewers who introduce themselves and tailor their interactions, whereas less successful interviewers give sample persons too much room to escape (Groves and McGonagle 2001; Morton-Williams 1993; Schaeffer, et al. 2013; Snijkers, Hox, and de Leeuw 1999).

Overall, the literature explaining interviewer effects on survey unit nonresponse shows great variability across studies in the significance and even direction of the predictors of interviewer effects (see West and Blom 2017). This diversity in the findings regarding the predictors of survey unit nonresponse might be related to survey characteristics and vary-ing explanatory variables available for analyses. For example, studies differ in the set of interviewers employed, the survey organizations managing the interviewers, the sampling frames used, and the populations and time periods observed. In addition, the explanatory variables available to the researchers examining interviewer effects on unit nonresponse tend to differ greatly across studies, and this may cause differences in the results of studies trying to explain interviewer effects. Few research projects have attempted to explain interviewer effects on survey unit nonresponse across different types of surveys, while keeping the survey organization, interviewer population, explanatory characteristics, and models for estimating interviewer variance constant (Blom, de Leeuw, and Hox 2011; Jäckle, et al. 2013).

We aim to address these weaknesses in the literature examining interviewer effects on survey unit nonresponse by analyzing interviewer effects during the recruitment phases of four different surveys while harmonizing measurements and analytical strategies. In our study, we use the same measures of sample composition, interviewer socio-demographic characteristics, interviewer experience, and interviewer attitudes, behaviors, and expectations about the survey in each of the four surveys. We also estimate the same models to explain interviewer effects during the recruitment phase. With this approach, we compare the interviewer effects during the recruitment phase across the four surveys and aim to identify interviewer characteristics influencing contact and cooperation consistently over the observed studies. Such findings could have important implications for interviewer selection, fieldwork monitoring, and future interviewer training sessions.

14.2 Data

We conduct a comparison of interviewer effects on unit nonresponse in four interviewer-mediated data collections using face-to-face recruitment. These studies were implemented between 2011 and 2014 in Germany by the same survey agency, and included the 2011 German implementation of the Programme for the International Assessment of Adult Competencies (PIAAC),[*] the 2012 and 2014 face-to-face recruitment interviews of the German Internet Panel (GIP),[†] and the 2013 German refresher sample of the Survey of Health, Ageing, and Retirement in Europe (SHARE; see Table 14.1 for an overview).[‡]

PIAAC is a multi-cycle cross-sectional survey for assessing basic competencies of the adult population aged 16 to 65 years that is carried out every 10 years in over 40 countries (for an overview of the German implementation, see Zabal, et al. 2014). For PIAAC, we use data from the German implementation. SHARE is a multidisciplinary and cross-national panel measuring the health, socio-economic status, and social and family networks of

[*] This paper uses data from the German Implementation of PIAAC 2012: Rammstedt, B., Martin, S., Zabal, A., Massing, N., Ackermann-Piek, D., Helmschrott, S., … Maehler, B. D. (2014). Programme for the International Assessment of Adult Competencies (PIAAC), Germany – Master Dataset. Unpublished data. Mannheim: GESIS - Leibniz Institute for the Social Science.

[†] This paper uses data from the German Internet Panel recruitment interviews. A study description can be found in (Blom, Gathmann, and Krieger 2015). The German Internet Panel is funded by the Deutsche Forschungsgemeinschaft (DFG, German Research Foundation) – Project-ID 139943784 – Collaborative Research Center 884 "Political Economy of Reforms" (SFB 884).

[‡] The SHARE data collection has been funded by the European Commission through FP5 (QLK6-CT-2001-00360), FP6 (SHARE-I3: RII-CT-2006-062193, COMPARE: CIT5-CT-2005-028857, SHARELIFE: CIT4-CT-2006-028812), FP7 (SHARE-PREP: GA N°211909, SHARE-LEAP: GA N°227822, SHARE M4: GA N°261982) and Horizon 2020 (SHARE-DEV3: GA N°676536, SERISS: GA N°654221) and by DG Employment, Social Affairs & Inclusion. Additional funding from the German Ministry of Education and Research, the Max Planck Society for the Advancement of Science, the U.S. National Institute on Aging (U01_AG09740-13S2, P01_AG005842, P01_AG08291, P30_AG12815, R21_AG025169, Y1-AG-4553-01, IAG_BSR06-11, OGHA_04-064, HHSN271201300071C) and from various national funding sources is gratefully acknowledged (see www.share-project.org).

TABLE 14.1

Overview of the Data Collections

	Fieldwork Period	Sampling Frame	Survey Organization
PIAAC	August 2011–March 2012	Local community registers of individuals	TNS Infratest Sozialforschung[a]
GIP 2012	May 2012–August 2012	Database of areas, listing of households in areas	TNS Infratest Sozialforschung[a]
SHARE	February–September 2013	Local community registers of individuals	TNS Infratest Sozialforschung[a]
GIP 2014	April 2014–August 2014	Database of areas, listing of households in areas	TNS Infratest Sozialforschung[a]

Note: PIAAC = Programme for the International Assessment of Adult Competencies. GIP = German Internet Panel. SHARE = Survey of Health, Ageing, and Retirement in Europe.

[a] TNS Infratest Sozialforschung since 2017 Kantar Deutschland GmbH.

more than 120,000 individuals aged 50 or older. SHARE covers 27 European countries and Israel and is conducted every second year (Börsch-Supan, et al. 2013). For SHARE, we use data from the German refreshment sample of wave five so that no panel cases are included in our analyses. The GIP is an infrastructure project at the Collaborative Research Center "Political Economy of Reforms" (SFB 884). The GIP is a longitudinal online panel survey, where online and offline respondents between the ages of 16 and 75 in Germany are first recruited offline by interviewers and then asked questions about political and economic attitudes and reform preferences on a bi-monthly basis (Blom, Gathmann, and Krieger 2015). In PIAAC and SHARE, interviewers had to make contact and gain cooperation from sample persons who were preselected from the German register. The GIP 2012 and GIP 2014, in contrast, are surveys of individuals in households that had been selected by a random route procedure. The target respondent in the household could be any household member between the ages of 16 and 75.

For each survey, we combined three data sources: first, the survey's gross sample; second, small area information provided by the Federal Institute for Research on Building Urban Affairs and Spatial Development (2015); and third, data from interviewer surveys. Gross sample files provide information on survey participation and some basic information on respondents' characteristics. Small area information describes the immediate neighborhood of the sampled household in terms of the relative share of Germans, of single-person households and of unemployed persons.

Interviewer questionnaire data are from ex-ante harmonized interviewer surveys implemented in all four surveys. This questionnaire collected information on each interviewer's socio-demographic characteristics, experience, attitudes, behaviors, and expectations about the survey, and was based on Blom and Korbmacher (2013, see Online Appendix 14A for the original German version and English translation of the interviewer questionnaire). The descriptive summaries of interviewers' socio-demographic characteristics, work experience, and working hours for each of the four surveys separately are presented in Online Appendix 14B. Overall, the interviewers seem to be rather similar with regard to most of their socio-demographic characteristics, work experience, and working hours across the four surveys. This is not surprising, given that all surveys were implemented by the same survey agency with the same pool of interviewers, and confirms the assumption that similar interviewers were working for all four studies. The interviewer ID provided allows us to identify individual interviewers in each survey, but unfortunately, it does not allow identification of interviewers across surveys.

For our analyses, we reduced the number of items from the interviewer survey by calculating indicators based on factor analyses for the long item lists on the interviewer's attitudes, behaviors, and expectations (similar indicators are used by Ackermann-Piek 2018). The factor matrices as well as an overview of the indicators and single items included in the analyses are presented in Online Appendix 14C. The responses of interviewers to questions from three thematic blocks were included in this analysis: questions on standardized interviewing techniques (8 items, 4-point agree/disagree scale), questions on how to achieve response (12 items, 4-point agree/disagree scale), and questions on the reasons to work as an interviewer (10 items, 7-point important/not important scale). See Table A14A.1 for details.

We calculated three indicators measuring interviewers' reported behavior with regard to following standardized interviewing techniques: *tailor content* describes interviewers' deviations from standardized interviewing techniques due to respondents' difficulties with questions, *tailor questions to shorten survey* refers to statements that interviewers do not follow standardized interviewing techniques to shorten the interviewing time, and *tailor to adapt to respondent* refers to items that describe interviewers' deviations from standardized interviewing techniques to help respondents, such as speaking slower or speaking the dialect of the respondent. Additionally, our analyses include one single item: *stick to instructions*.

In addition, we include a section of indicators based on statements referring to strategies on how to achieve response. In this section, the following indicators were calculated: success via *hard work* consists of two items that refer to strategies to achieve response by working hard, *acceptance of refusal* is calculated with three items regarding statements that refusals of a sample person should be accepted by interviewers, and *respect towards voluntariness* includes items that emphasize that taking part in a survey is voluntary. In addition, from this section, we used the following two statements in their original version: *caught at right time most will cooperate* and *concerned about data protection*.

Regarding reasons for interviewers to work in this position, we computed three different indicators: science, people, and formal conditions. The indicator *science* refers to two items about reasons for working as an interviewer due to being involved in scientific research that serves society. The indicator *people* refers to reasons for working as an interviewer due to the ability to interact and socialize with people. The indicator *formal conditions* refers to reasons for working as an interviewer due to more formal conditions, like payment and flexible working hours.

Each of these data sources is affected by a small amount of missing information (see Table 14.2 for an overview). Depending on the extent and nature of the missingness,

TABLE 14.2

Sample Sizes

	PIAAC	GIP 2012	SHARE	GIP 2014
Gross sample of eligible sample units	9,369	4,871	8,873	10,200
Number of contacted sample units	7,704	4,185	8,668	8,187
Number of interviewed sample units	4,595	2,121	3,026	4,426
Number of interviewers working on the survey	129	133	172	189
Number of interviewers completing the interviewer survey	115	131	146	141
Number of sample units in the contact models	7,989	3,898	7,175	6,253
Number of sample units in the cooperation models	7,504	3,038	7,058	5,767

Note: PIAAC = Programme for the International Assessment of Adult Competencies. GIP = German Internet Panel. SHARE = Survey of Health, Ageing, and Retirement in Europe.

TABLE 14.3

Contact and Cooperation Rates

	PIAAC	GIP 2012	SHARE[a]	GIP 2014
Contact rate (%)	82.2	85.9	97.7	80.3
Cooperation rate (%)	59.6	50.7	34.9	54.1

Note: PIAAC = Programme for the International Assessment of Adult
         Competencies. GIP = German Internet Panel. SHARE = Survey of
         Health, Ageing, and Retirement in Europe.
[a] Malter and Börsch-Supan 2015; Kneip, Malther, and Sand 2015.

we either conducted a complete case analysis or imputed the missing information. Specifically, we list-wise deleted interviewers when the complete interviewer questionnaire was unanswered (PIAAC: 11.6%, GIP 2012: 1.5%, SHARE: 17.3%, GIP 2014: 25.4%). For the interviewer survey, item missing data (about 3% across all variables) were imputed once with either linear regression, ordered logistic regression, or multinomial logistic regression, depending on the level of measurement of the variable imputed, and all variables in the survey were included as predictors in the imputation models. Similar to the interviewers' socio-demographic characteristics, work experience, and working hours, the interviewers were quite similar across the four surveys with respect to their attitudes, reported behaviors, and expectations (see Online Appendix 14D for descriptive statistics for these variables, for all surveys together and for each separate survey).

The dependent variables in this chapter are indicators of making contact with target respondents and convincing target respondents to cooperate, conditional on making contact. Response by a specific sample unit in a face-to-face recruitment is obtained when the sample unit is successfully contacted and, subsequently, cooperates with the survey interview request. The respondent and interviewer characteristics leading to successful contact can be different from those leading to successful cooperation (see, e.g., Lynn and Clarke 2002). For this reason, we model contact and cooperation separately.

Following the guidelines of the American Association for Public Opinion Research (AAPOR 2016), a sampled person was defined as contacted if the interviewer had succeeded in contacting any household member in an eligible household. Cooperation was defined conditional upon successful contact (i.e., non-contacted sample units were excluded from the models of cooperation). If the sampled person started the interview, even if the interview was broken off at a later point in time, that person was deemed cooperative. Based on these definitions, Table 14.3 displays contact and cooperation rates achieved in each survey. Overall, PIAAC, GIP 2012, and GIP 2014 had similar contact rates, ranging between 80% and 86%. SHARE interviewers achieved a much higher contact rate of about 98%. Cooperation rates in the four surveys were much lower than the contact rates. Again, PIAAC, GIP 2012, and GIP 2014 had similar cooperation rates, ranging between 51% and 60%, whereas SHARE interviewers achieved a much lower overall cooperation rate of about 35%.

Survey organizations tend to re-assign difficult cases to their most productive and experienced interviewers, a practice that may bias the results of analyses of interviewer effects. For this reason, sample units that were approached by more than one interviewer during the contact and cooperation phase were excluded entirely from the analyses (PIAAC: 3.9%, GIP 2012: 5.3%, SHARE: 2.2%, GIP 2014: 9.1%).

14.3 Methods

Multilevel analysis is now a standard approach for the analysis of interviewer effects on nonresponse (Blom, et al. 2011). We started our analysis by fitting null models, more specifically three-level regressions (PIAAC) and two-level regressions (GIP 2012, SHARE, and GIP 2014), to identify the amount of raw variance at the interviewer level (Hox 2010). Due to differences in sampling methods and gross sample sizes, the PIAAC survey allocated at least two primary sampling units (PSUs) to each interviewer for most of the sample, while interviewers in the GIP 2012 survey, the GIP 2014 survey, and the SHARE survey worked in one PSU only. For this reason, three-level logistic regression models with sample units nested within PSUs and PSUs nested within interviewers were estimated for PIAAC, while two-level logistic regression models with sample units nested within interviewers were estimated for GIP 2012, GIP 2014, and SHARE.

The four surveys were implemented without an interpenetrated sample design that would enable us to separate interviewer effects from area effects. Studies of interviewer effects in surveys employing face-to-face recruitment often suffer from this problem. Following the guidance of Schaeffer, Dykema, and Maynard (2010), we specify multilevel models that include fixed effects for variables describing the sample compositions of the interviewers' assignments for all models, including the size of the sample point in terms of number of inhabitants, the region with four categories (north, south, west, east), the share of Germans, the share of single-person households, and the share of unemployed persons in the immediate neighborhood of the sampled addresses.

Taking the random effects at the interviewer level into account, we next specified a model for each survey outcome (contact and cooperation) that included all available interviewer characteristics from the interviewer questionnaire described above, in addition to the sample composition control variables. Models were estimated using the xtmelogit procedure in Stata 13. We tested odds ratios for significance using a two-sided Wald test, while interviewer variance components were tested against zero using an appropriate likelihood-ratio test (Rabe-Hesketh and Skrondal 2012).

14.4 Results

Our results show that for GIP 2012, PIAAC, and GIP 2014, around 20% of the variance in a sampled person's propensity to be successfully contacted is due to interviewer effects (GIP 2012: 18.5%, PIAAC: 20.7%, GIP 2014: 17.2%). The percentage of variance in a sampled person's propensity to be successfully contacted that can be explained by interviewer effects is much higher for SHARE (60.6%). These notable interviewer effects on contact in SHARE are likely arising from the fact that most of the interviewers in the SHARE sample have a contact rate of 100%, and only some interviewers are found to be "outliers" with a very low contact rate. Thus, we focus on explaining interviewer effects on contact for GIP 2012, PIAAC, and GIP 2014. (Regression tables are presented in Online Appendix 14E.)

In contrast, when examining a sampled person's propensity to participate in each of the respective surveys, the fraction of the raw variance due to interviewer effects is rather small for both PIAAC (2.1%) and SHARE (5.2%). The percentages of variance in propensity to participate due to interviewer effects for GIP 2012 and GIP 2014 are slightly higher

(GIP 2012: 12.8%, GIP 2014: 17.2%). As the fraction of the variance in a sample person's willingness to participate in PIAAC due to interviewer effects is so small in PIAAC, we do not conduct further analysis using interviewer characteristics. Thus, we focus on explaining interviewer effects on cooperation for GIP 2012, SHARE, and GIP 2014 (see regression tables in Online Appendix 14E).

The results from fitting the three-level logistic regression model for PIAAC and the two-level logistic regression models for GIP 2012 and GIP 2014 that added the fixed effects of interviewer characteristics to explain interviewer effects on a sample person's propensity to be successfully contacted are presented in Figure 14.1. The results from the models for a sampled person's propensity to participate in the survey are presented in Figure 14.2. In the plot, the markers are odds ratios, and the horizontal lines are confidence intervals.

Looking at Figure 14.1, it becomes obvious that none of the available interviewer characteristics have a significant relationship with a sampled person's propensity to be successfully contacted in each of the four surveys (see Online Appendix 14E).

For GIP 2012, we find that sample persons have a significantly higher propensity to be successfully contacted when interviewers report that they do stick to the interviewer instructions during the implementation of the questionnaire. The interviewer variance explained by including all interviewers' characteristics in one model is about 19%.

For PIAAC, we find that sample persons have a significantly lower propensity to be successfully contacted when interviewers report a higher tendency to tailor questions in order to shorten the questionnaire. Persons have a significantly *higher* propensity to be successfully contacted when interviewers report that they either deviate from standardized interviewing techniques to help respondents, such as speaking slower or speaking the dialect of the respondent, or report they stick to interviewer instructions. The latter two significant results are contradictory, as deviating from standardized techniques and sticking to instructions are seemingly opposite types of interviewer behaviors. Also, sample persons have a significantly higher propensity to be successfully contacted when interviewers say that refusals should not be accepted when respondents are reluctant. The interviewer variance explained by including fixed effects of all the interviewers' characteristics in one model is about 56%.

For GIP 2014, we find that sample persons have a significantly lower propensity to be successfully contacted when interviewers report a lower agreeableness towards the voluntariness of participation. The interviewer variance explained by including all interviewers' characteristics in one model is about 30%.

In Figure 14.2, we present estimated odds ratios from the two-level logistic regression models for cooperation in GIP 2012, SHARE, and GIP 2014. Similar to the results for the contact models, we do not find any interviewer characteristic that has a significant relationship with cooperation in each of the four surveys in our study.

For GIP 2012, our results indicated a significantly higher propensity to cooperate when female interviewers tried to gain the cooperation of sampled persons. Furthermore, we found a significant positive effect of tailoring content: sampled persons approached by interviewers who report that they tailor content less frequently have a higher propensity to cooperate. Sampled persons had a significantly *lower* propensity to participate when interviewers report a higher tendency to tailor questions in order to shorten the questionnaire. Furthermore, sampled persons' propensity to cooperate is significantly lower when interviewers report being concerned about data protection issues themselves. The interviewer variance explained by including the fixed effects of all the interviewers' characteristics in one model is about 45%.

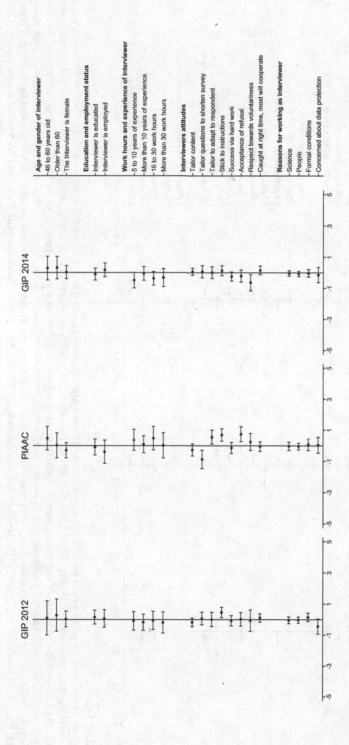

FIGURE 14.1

Estimated odds ratios for predictors of successful contact, multilevel logistic regression, across surveys. *Note:* Parameter estimates with 95% confidence intervals from three-level logistic regression for PIAAC and from two-level logistic regression for GIP 2012 and GIP 2014 for the dependent variable of successful contact. Significant = p < 0.05. Model controls for sample composition characteristics (share of Germans, share of single-person households, share of unemployed persons in the immediate neighborhood of the sampled household, PSU size, and region). Reference age: ≤ 45 years. Reference gender: the interviewer is male. Reference education: interviewer has no Abitur (A-levels). Reference employment status: interviewer is not full- or part-time employed. Reference experience: ≤ 5 years of experience. Reference working hours: ≤ 15 hours per week. Number of interviewers GIP 2012 = 131. Number of interviewers PIAAC = 115. Number of interviewers GIP 2014 = 141. Number of PSUs PIAAC = 251. Number of sample persons PIAAC = 3,898. Number of sample persons GIP 2012 = 3,898. Number of sample persons GIP 2014 = 6,253. PIAAC = Programme for the International Assessment of Adult Competencies. GIP = German Internet Panel. The full models can be found in Online Appendix 14E.

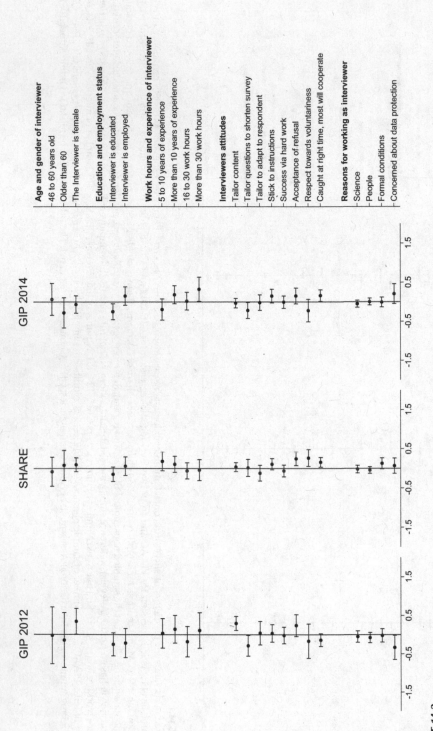

FIGURE 14.2

Estimated odds ratios for predictors of successful cooperation, multilevel logistic regression, across surveys. Parameter estimates with 95% confidence intervals from two-level logistic regression for GIP 2012, SHARE, and GIP 2014 for the dependent variable of successful cooperation. Significant = p < 0.05. Model control for sample composition characteristics (share of Germans, share of single-person households, share of unemployed persons in the immediate neighborhood of the sampled household, PSU size, and region). Reference age: ≤ 45 years. Reference gender: the interviewer is male. Reference education: interviewer has no Abitur (A-levels). Reference experience: ≤ 5 years of experience. Reference working hours: ≤ 15 hours per week. Number of interviewers GIP 2012 = 131. Number of interviewers SHARE = 142. Number of interviewers GIP 2014 = 141. Number of sample persons GIP 2012 = 3,038. Number of sample persons SHARE = 7,058. Number of sample persons GIP 2014 = 5,767. GIP = German Internet Panel. SHARE = Survey of Health, Ageing, and Retirement in Europe. The full models can be found in Online Appendix 14E.

For SHARE, we found three interviewer characteristics that had significant relationships with cooperation propensity, and all three were indicators measuring how to successfully achieve response. Persons had a significantly higher propensity to participate in SHARE when interviewers reported fewer acceptances of respondents' refusals, when interviewers reported higher agreeableness towards the voluntariness of participation, and when interviewers reported higher agreeableness towards the statement that people will cooperate when they are contacted at the right time. The interviewer variance explained by including fixed effects of all the interviewers' characteristics in one model is about 36%.

For GIP 2014, sampled persons had a significantly lower propensity to participate when interviewers with an Abitur (A-levels) education tried to gain their cooperation, compared to interviewers who have a lower educational degree. In addition, sampled persons approached by interviewers who reported that they tailored less frequently had a lower propensity to cooperate, and sampled persons approached by interviewers who reported a higher agreeableness towards the statement that people will cooperate when they are contacted at the right time had a higher propensity to cooperate. Also, sampled persons' propensity to cooperate is significantly higher when they are approached by interviewers working more than 30 hours per week, compared to interviewers working less than 30 hours per week. The interviewer variance explained by including the fixed effects of all the interviewers' characteristics in one model is about 35%.

14.5 Summary and Discussion

The literature explaining interviewer effects on survey unit nonresponse across different surveys has thus far reported a great deal of variability in the most important predictors of unit nonresponse. We reasoned that this is due to the findings in the literature being based on diverse surveys with a high level of inconsistency in terms of key survey characteristics. In this chapter, we attempted to explain interviewer effects on survey unit nonresponse in a more orchestrated way. We based our analyses on four surveys using face-to-face recruitment (GIP 2012, PIAAC, SHARE, and GIP 2014), each of which was conducted in Germany in approximately the same time period by the same survey agency with overlapping interviewers. Across the four surveys, we estimated the same multilevel models for two aspects of survey unit nonresponse – successful contact and cooperation – including the same explanatory variables.

Our study revealed two main findings. First, using multilevel modeling, we analyzed the amount of variance at the interviewer level in a sampled person's propensity to be successfully contacted and a sampled person's propensity to cooperate. To account for the non-random assignment of sampled persons to interviewers and the associated potential that area effects could be confounded with the interviewer effects, we adjusted for numerous sample composition variables in all models. Our results revealed a rather similar interviewer effect on a sampled person's propensity to be successfully contacted in the 2012 GIP, the PIAAC, and the 2014 GIP, with interviewer variance representing around 20% of the total variance. We assume this to be the case because the contact strategies used in the three surveys are rather similar. We did not estimate interviewer effects for SHARE as the majority of the interviewers have a contact rate of 100%.

For a sampled person's propensity to participate in each survey, the percentage of variance due to interviewer effects was rather small for PIAAC (2.1%) and for SHARE (5.2%).

The percentages of total variance due to interviewer effects for the 2012 and 2014 GIP were much higher (GIP 2012: 12.8%, GIP 2014: 17.2%). The reason for this difference is unknown. We can speculate that this is due to the fact that the procedures used to identify the household member eligible for the interview in the GIP surveys differ from those used for SHARE and PIAAC. GIP interviewers were allowed to select any household member within the eligible age range, whereas for PIAAC and SHARE the target person was pre-specified (see Blom, et al. 2015 for details). Interviewer traits may have a bigger influence on cooperation in the GIP field protocols.

The second finding, which is at the same time the main finding of our chapter, concerns the explanatory power of interviewers' characteristics across the four surveys. Although there was a high level of consistency in the designs of the four studies and the interviewers employed in the four studies were quite similar with regard to most of their characteristics collected via the interviewer questionnaire, there was no consistency regarding the significant predictors of survey unit nonresponse across the studies.

There were some differences between the four surveys that might influence the respondent's decision to participate more than the interviewer. For example, the survey topic, the amount of information provided in the cover letter, the age of the target population, the amount of interviewer training, the sponsor, and the research team differed between the surveys examined in this chapter. However, the differences in explanatory power of interviewers' characteristics between the GIP 2012 and GIP 2014 cannot be explained by differences in the survey design, as all the above-mentioned factors were kept constant between the two survey designs. Still, we found no consistent interviewer effects over these two surveys. We have to conclude that other factors unobserved in our study affect the interviewer's success in gaining contact and cooperation.

We can only speculate about the reasons for differences in the explanatory power of interviewer characteristics across the surveys. Even with a high level of consistency across the four studies, our results are consistent with the current literature: interviewer effects during the recruitment phase seem to be study-specific, and need to be analyzed, monitored, and treated accordingly. Thus, different surveys need to take steps to minimize variance in these recruitment outcomes through careful interviewer selection, training, and active fieldwork monitoring (e.g., re-training interviewers found to have unusually low contact and cooperation rates during data collection).

References

Ackermann-Piek, D. 2018. *Interviewer Effects in PIAAC Germany 2012*. Mannheim: Doctoral dissertation.

American Association for Public Opinion Research (AAPOR). 2016. *Standard Definitions: Final Dispositions of Case Codes and Outcome Rates for Surveys* (9th ed.), AAPOR.

Blom, A. G., E. D. de Leeuw, and J. J. Hox. 2011. Interviewer effects on nonresponse in the European Social Survey. *Journal of Official Statistics* 27(2):359–377.

Blom, A. G., C. Gathmann, and U. Krieger. 2015. The German Internet Panel: Method and results. *Field Methods* 27(4):391–408.

Blom, A. G., and J. M. Korbmacher. 2013. Measuring interviewer characteristics pertinent to social surveys: A conceptual framework. *Survey Methods – Insights from the Field* 1(1):1–19.

Börsch-Supan, A., M. Brandt, C. Hunkler, T. Kneip, J. Korbmacher, F. Malter, B. Schaan, S. Stuck, S. Zuber, and SHARE Central Coordination Team. 2013. Data resource profile: The Survey of Health, Ageing and Retirement in Europe (SHARE). *International Journal of Epidemiology* 42(4):992–1001.

Campanelli, P., and C. O'Muircheartaigh. 1999. Interviewers, interviewer continuity, and panel survey nonresponse. *Quality and Quantity* 33(1):59–76.

Couper, M. P., and R. M. Groves. 1992. The role of the interviewer in survey participation. *Survey Methodology* 18(2):263–277.

de Leeuw, E. D., J. J. Hox, G. Snijkers, and W. de Heer. 1998. Interviewer opinions, attitudes, and strategies regarding survey participation and their effect on response. In: *Nonresponse in Survey Research*, ed. A. Koch, and R. Porst, 239–262. Mannheim: ZUMA Nachrichten SPEZIAL 4.

Durbin, J., and A. Stuart. 1951. Differences in response rates of experienced and inexperienced interviewers. *Journal of the Royal Statistical Society: Series A (General)* 114(2):163–206.

Durrant, G. B., R. M. Groves, L. Staetsky, and F. Steele. 2010. Effects of interviewer attitudes and behaviors on refusal in household surveys. *Public Opinion Quarterly* 74(1):1–36.

Federal Institute for Research on Building Urban Affairs and Spatial Development. 2015. *Indicators and Maps on Territorial and Urban Development*. Distributed by BBSR. http://inkar.de/ (Accessed December 22, 2019).

Groves, R. M., and M. P. Couper. 1998. *Nonresponse in Household Interview Surveys*. New York: Wiley.

Groves, R. M., F. J. Fowler, M. P. Couper, J. M. Lepkowski, E. Singer, and R. Tourangeau. 2009. *Survey Methodology* (2nd ed.), Hoboken: John Wiley & Sons.

Groves, R. M., and K. A. McGonagle. 2001. A theory-guided interviewer training protocol regarding survey participation. *Journal of Official Statistics* 17(2):249–265.

Hox, J. J. 2010. *Multilevel Analysis - Techniques and Applications* (2nd ed.), New York: Routledge.

Hox, J. J., and E. D. de Leeuw. 2002. The influence of interviewers' attitude and behavior on household survey nonresponse: An international comparison. In: *Survey Nonresponse*, ed. R. M. Groves, D. A. Dillman, J. L. Eltinge, and R. J. A. Little, 103–119. New York: John Wiley & Sons.

Jäckle, A., P. Lynn, J. Sinibaldi, and S. Tipping. 2013. The effect of interviewer experience, attitudes, personality and skills on respondent co-operation with face-to-face surveys. *Survey Research Methods* 7(1):1–15.

Japec, L. 2008. Interviewer error and interviewer burden. In: *Advances in Telephone Survey Methodology*, ed. J. M. Lepkowski, C. Tucker, J. M. Brick, E. D. de Leeuw, L. Japec, P. J. Lavrakas, M. W. Link, and R. L. Sangster, 187–211. Hoboken: John Wiley & Sons.

Kneip, T., F. Malter, and G. Sand. 2015. Fieldwork monitoring and survey participation in the fifth wave of SHARE. In: *SHARE Wave 5: Innovations & Methodology*, ed. F. Malter, and A. Börsch-Supan, 101–157. Munich: Munich Center for the Economics of Aging (MEA).

Loosveldt, G., A. Carton, and J. Pickery. 1998. The effect of interviewer and respondent characteristics on refusals in a panel survey. In: *Nonresponse in Survey Research: Proceedings of the Eighth International Workshop on Household Survey Nonresponse, 24–16 September 1997*, ed. A. Koch, and R. Porst, 249–262. Mannheim: Zentrum für Umfragen, Methoden und Analysen - ZUMA (ZUMA-Nachrichten Spezial 4).

Lord, V. B., P. C. Friday, and P. K. Brennan 2005. The effects of interviewer characteristics on arrestees' responses to drug-related questions. *Applied Psychology in Criminal Justice* 1(1):36–55.

Lynn, P., and P. Clarke. 2002. Separating refusal bias and non-contact bias: Evidence from UK national surveys. *Journal of the Royal Statistical Society: Series D (The Statistician)* 51(3):319–333.

Malter, F., and A. Börsch-Supan. 2015. *SHARE Wave 5: Innovations & Methodology*. Munich: Munich Center for the Economics of Aging (MEA).

Maynard, D. W., and N. C. Schaeffer. 2002. Standardization and its discontents. In: *Standardization and Tacit Knowledge: Interaction and Practice in the Survey Interview*, ed. R. M. Groves, G. Kalton, J. N. K. Rao, N. Schwarz, and C. Skinner, 3–45. New York: John Wiley & Sons.

Moorman, P. G., B. Newman, R. C. Millikan, C.-K. Tse, and D. P. Sandler. 1999. Participation rates in a case-control study: The impact of age, race, and race of interviewer. *Annals of Epidemiology* 9(3):188–195.

Morton-Williams, J. 1993. *Interviewer Approaches*. Brookfield: Dartmouth Publishing Company.

Rabe-Hesketh, S., and A. Skrondal. 2012. *Multilevel and Longitudinal Modeling Using Stata* (3rd ed.), College Station: Stata Press.

Schaeffer, N. C., J. Dykema, and D. W. Maynard. 2010. Interviewers and interviewing. In: *Handbook of Survey Research*, ed. P. V. Marsden, and J. D. Wright, 437–470. Bingley: Emerald Group Publishing Limited.

Schaeffer, N. C., D. Garbarski, J. Freese, and D. W. Maynard. 2013. An interactional model of the call for survey participation: Actions and reactions in the survey recruitment call. *Public Opinion Quarterly* 77(1):323–351.

Snijkers, G., J. J. Hox, and E. D. de Leeuw. 1999. Interviewers' tactics for fighting survey nonresponse. *Journal of Official Statistics* 15(2):185–198.

Vassallo, R., G. B. Durrant, P. W. F. Smith, and H. Goldstein. 2015. Interviewer effects on non-response propensity in longitudinal surveys: A multilevel modeling approach. *Journal of the Royal Statistical Society: Series A (Statistics in Society)* 178(1):83–99.

Webster, C. 1996. Hispanic and Anglo interviewer and respondent ethnicity and gender: The impact on survey response quality. *Journal of Marketing Research* 33(1):62–72.

West, B. T., and A. G. Blom. 2017. Explaining interviewer effects: A research synthesis. *Journal of Survey Statistics and Methodology* 5(2):175–211.

West, B. T., M. R. Elliott, Z. N. Mneimneh, J. Wagner, A. Peytchev, and M. Trappmann. 2019. An examination of an interviewer-respondent matching protocol in a longitudinal CATI study. *Journal of Survey Statistics and Methodology* 0:1–21.

Zabal, A., S. Martin, N. Massing, D. Ackermann, S. Helmschrott, I. Barkow, and B. Rammstedt. 2014. *PIAAC Germany 2012: Technical Report*. Münster: Waxmann.

15

Comparing Two Methods for Managing Telephone Interview Cases

Jamie Wescott

CONTENTS

15.1 Introduction

In an interviewer-administered survey, interviewers can have a notable effect on data quality (West and Blom 2017). Studies have shown that there is an association between unit noncontact and nonresponse rates and interviewer characteristics and behaviors, such as experience levels and approaches to making contact (Blom 2012; Groves and Couper 1998; Purdon, Campanelli, and Sturgis 1999).

A standard practice for reducing nonresponse rates in interviewer-administered surveys is to vary the times of day and days of week of contact attempts, thereby maximizing the chances of making contact (Groves and Couper 1998). In most telephone surveys, automated call scheduling systems are used to help mitigate potential impacts of interviewers' decision-making on contact and cooperation. These systems use algorithms based on information about prior contacts, the time elapsed since the last call attempt, the time zone in which cases are located, scheduled appointments, and the overall status of each case to provide interviewers with a sequence of cases to dial. They also dictate when to leave a voice message and how often to schedule callbacks. As such, they control the timing, frequency, and number of times each case is dialed to ensure that cases receive a consistent number of calls at optimal time points.

When an automated call scheduler is not available, a case ownership model is an alternative approach for case management. In a case ownership model, telephone interviewers are each assigned a set of individual sample units (cases) and encouraged to use their judgment and experience to determine how best to work the cases. Interviewers are responsible for all aspects of managing their assigned cases, including reviewing case histories, deciding which of their assigned cases and phone numbers to call at any given time, making judgments about how often to attempt contact with a particular case, scheduling callbacks, and conducting interviews.

This chapter will describe our experiences implementing a case ownership approach for part of a national study of young adults. We also will compare the case ownership approach to the automated call scheduler approach used for a later stage of the study. Specifically, we compare the number of call attempts per case, variability in the timing of call attempts between the two approaches, and number of calls per interviewer hour worked. We identify strengths and weaknesses of the case ownership approach vis-à-vis the automated call scheduler approach, provide lessons learned for other researchers planning to utilize this approach, and identify areas for future research. As described in further detail below, the decision to use the case ownership approach was made for pragmatic reasons; this study is observational, not experimental. However, the similarity of the two data collections within the same larger study involving the same study cohort provides a unique opportunity to compare the two case management approaches.

15.2 Background of the Study and Design of the Case Management Approaches

The larger study in which these two case management approaches were used is conducted for the U.S. Department of Education's Institute of Education Sciences and is focused on the experiences and outcomes of youth with disabilities as they transition into adulthood. The study follows a cohort of nearly 22,000 youth, who were first sampled as 7th to 12th graders in late 2011, as they progress through high school and into postsecondary education and the workforce. The sample is composed of three subgroups of analytical importance: (1) students who have an individualized education plan (IEP) and are eligible to receive special education services under the Individuals with Disabilities Education Act (IDEA); (2) students who have not been identified as needing special education services but who have a condition that qualifies them for accommodations under Section 504 of the Vocational Rehabilitation Act of 1973; and (3) students with no identified disability who have neither an IEP nor a Section 504 plan.

The study was divided into two phases. In Phase I, sampled youth and their parents, teachers, and other secondary school staff were interviewed as part of the base-year data collection in 2012. At that time, in anticipation of Phase II, sample members and their parents were asked for their consent to collect their high school transcripts in the future and to link their data to other administrative records. The goal of the Phase II collection is to link sample members' transcript, administrative data, and Phase I interview data to create a rich data set of students' high school experiences, postsecondary education, financial aid, vocational rehabilitation agency services, and employment outcomes. Data collection for Phase II is carried out by RTI International, an independent, nonprofit research organization. Phase II transcript collection began in early 2017 and a second round of collection to obtain updated transcripts has been planned for early 2020.

In Phase I, there were 2,312 cases that refused consent to the collection of high school transcripts (10.5% of the sample) and 2,909 additional cases that refused consent for linking to administrative data (13.2% of the sample). Without their consent, these sample members could not be included in the Phase II transcript collection or administrative matching. The lack of consent for these cases was concerning for two reasons. First, while high response rates provide no guarantee of high data quality, high unit nonresponse rates, caused in part by excluding cases that did not consent to either the transcript collection or administrative data linking, would increase the *risk* for nonresponse bias in the final data (Groves, et al. 2009). If these cases did not differ from consenting cases, their nonresponse would have little effect on final estimates. However, given the potential for consent missingness to be linked to respondent characteristics (e.g., whether the student has an IEP or a 504 plan) or outcomes of interest (e.g., whether the student completed high school, enrolled in postsecondary education, or is employed), the absence of these cases could bias the final data. Previous research has shown an association between education and linkage consent, with college graduates being more likely to consent to data linkage than those who did not complete high school (Sakshaug, et al. 2012), raising the specter of non-consent bias in the current study. Second, the study cohort includes several subgroups of interest (e.g., students with specific disabilities) for policymakers and advocates for students with disabilities; subgroup analyses are important to the overall goals of the study. Therefore, maximizing response rates was critical to ensuring that the responding sample was large enough to perform these analyses with sufficient statistical power.

Thus, to maximize the final response rate and sample size, the study team conducted two consent conversion efforts. The first effort was conducted in late 2016, immediately prior to the start of transcript collection in early 2017, and focused on gaining consent to collect high school transcripts from the 2,312 sample members who did not provide consent during the Phase I interview. The second effort was conducted in the summer of 2019 and focused on gaining consent for administrative data matching from the 2,909 sample members who did not provide consent during the Phase I interview.

Both consent collections used similar data collection protocols:

- Sample members and their parents or guardians received hardcopy announcement mailings, a hardcopy reminder postcard, an email announcement, and an email reminder asking them to complete the consent form.

- Respondents were offered three options for providing consent: completing the self-administered web consent form, returning a paper consent form by mail, or providing consent over the phone.

- Respondents were offered a $10 promised incentive as a token of appreciation for the time needed to complete the consent form.

- Telephone interviewers made outbound prompting calls to obtain consent over the phone, assist the respondent with completing the self-administered web-based consent form, or remind the respondent to return the paper consent form. Interviewers were located in a centralized call center; standard dialing hours were 9:00 am to 11:00 pm (EST) Monday through Saturday and 1:00 to 9:00 pm (EST) on Sundays.

While there were many similarities between the consent collections for transcripts and administrative matching, they differed in the time and resources available to conduct the work. Scheduling for the transcript consent effort in 2016 was constrained by the time required to conduct an advance pilot study (testing the effectiveness of the incentive and prompting calls using approximately 650 cases from the transcript consent group) and the impending start of transcript collection in 2017. Between these two efforts, there were only nine weeks available to conduct the transcript consent collection for the remaining 1,648 cases. In early October 2016, announcement mailings and e-mails were sent to sample members and their parents inviting them to complete the self-administered consent form. Outbound prompting calls to obtain this consent began after two weeks and were conducted over seven weeks from October 2016 to December 2016. This tight window and cost considerations made it infeasible to develop a fully programmed automated computer-assisted telephone interview (CATI) case management system (CMS). Thus, the 2016 transcript consent collection effort used an interviewer case ownership model, which could be designed and deployed in about a quarter of the time required to design and deploy a CATI call scheduling system.

The administrative matching collection in 2019 had fewer schedule constraints and thus a longer collection period of 13 weeks, from May 2019 to August 2019. With more time and resources, this effort was able to use a programmed, automated CATI scheduler to manage the 2,909 sample cases. In this chapter, we compare productivity measures for these two different case management systems. For comparison purposes, all analyses presented in this chapter compare results from all seven weeks of prompting for transcript consent and the first seven weeks of prompting for administrative consent.

While circumstances dictated the use of the case ownership approach for the 2016 transcript consent collection, there were anticipated drawbacks. The study team had previously used this approach for in-person field interviews, institution-based administrative record collections, and small-scale telephone re-interviews (in which a small subset of respondents are interviewed again to test the performance of interview items over time), but had not previously used it as the primary method of managing telephone data collection. Because interviewers were free to schedule and manage calls based on their experience and judgment, there was a risk of less consistent case management practices than would occur in a study with automated call scheduling. While the project provided interviewers with training on case management best practices, training is not a guarantee of success; researchers have documented instances in which field interviewers have not fully complied with data collection protocols despite training (Walejko and Wagner 2015). Thus, on case ownership studies, additional effort may be required for supervisors to review cases to ensure that case management guidelines are followed consistently.

While the ownership and management of cases differed, once contact had been made, interviewers working on both consent collection efforts administered a web-based consent form over the phone (which took approximately five minutes to complete), prompted the respondent to complete the self-administered web form, or reminded the respondent to return the paper form. Interviewers were also expected to document the results of their call attempts in the case management system and report any particularly challenging situations to supervisory staff. Of the 1,648 total cases included in the 2016 transcript consent collection effort, 19% provided consent (59% over the phone, 17% by mail, and 24% by web). Of the 2,909 cases included in the 2019 administrative matching consent effort, 14% provided consent in the first seven weeks of data collection (53% over the phone, 18% by mail, and 29% by web). While it is interesting to note that the automated CATI scheduler collection resulted in lower consent rates during the first seven weeks, the 2016 and 2019 consent collections asked different subsets of sample members to provide consent for different activities, thus we cannot attribute differences in consent rates solely to differences in contacting approach.

15.2.1 Case Management Systems

The CMS used for the 2016 transcript consent collection was a custom-built web-based system similar to those used for institution-based data collections, during which a sampled institution is assigned to a specific staff member who serves as the case manager for the duration of the study. For this study, sample members were jointly assigned to a pair of interviewers who worked complementary shifts throughout the week. Interviewers were paired such that their combined shifts covered as much of the week as possible. The cases were initially distributed among interviewer pairs using random assignment, although adjustments to case assignments were made throughout data collection based on staff workload. Each interviewer was assigned approximately 250 cases on average, although the specific number of cases assigned varied based on the number of hours the interviewer worked per week.

The CMS included a case management report that was tailored for each interviewer (see online supplemental materials); this report included a list of all cases assigned to the interviewer; the date, time, and results of the most recent call attempt; the next appointment for the case; the time zone in which the case was located; and the overall status of the case. The CMS also included a list of all known contacts associated with each case (e.g., parents) and their contact information, an appointment feature that allowed interviewers to schedule callbacks, a contact history page that included a record of every contact attempt associated with the sample member, answers to frequently asked questions about the study, a link to complete the web-based consent form, and suggested call scripts (see online supplemental materials). The appointment feature allowed interviewers to schedule call attempts during any shift, not merely their own, so that other interviewers could call the case when the assigned interviewers were not scheduled to work.

For the 2019 administrative consent collection, cases were managed using a proprietary CATI-CMS. This CMS included a call scheduler algorithm that automatically provided interviewers with cases to call based on the number and results of previous call attempts, the status of the case, and any scheduled appointments. As interviewers completed each call attempt, the next case to dial appeared on the CATI-CMS screen. Similar to the case ownership CMS, the CATI-CMS included an event history of all previous call attempts, a list of all contact information associated with a sample member, an appointment feature, a link to the web-based consent form, and call scripts. Unlike the case ownership CMS, in which interviewers had to manually follow printed skip logic to navigate through the call scripts, the CATI-CMS call scripts were divided into separate screens with routing logic based on responses recorded by interviewers.

15.2.2 Interviewer Training

Six experienced interviewers who had worked on prior projects with young adult sample populations and were proficient in administering complex survey instruments were trained to use the case ownership approach for the 2016 transcript consent collection. Training consisted of eight hours of in-class instruction conducted over two days, providing information about the background of the study, strategies for gaining cooperation, answers to frequently asked questions, and how to launch and complete the web-based consent form (see online supplemental materials for a sample training agenda). None of the interviewers had previously used the case ownership model, so training emphasized case management skills such as reviewing case histories to determine which cases to call, best practices for timing call attempts, and how to schedule callback attempts using the CMS. Interviewers were instructed to allow an average of five days between call attempts, but to use their experience and judgment to refine their approach for each individual case. Interviewers were also taught to vary the timing of their callbacks (rather than continuing to call at unproductive times), to only call during standard interviewing hours for the sample members' time zone (9:00 am to 9:00 pm Monday through Saturday, 1:00 pm to 9:00 pm Sunday), and to set appointments during other interviewers' shifts, if necessary. Training included hands-on sessions in which interviewers acted out scenarios and practiced gaining cooperation without reading from a screen. Throughout the collection period, interviewers participated in weekly check-in meetings to ask questions and share successful strategies with their colleagues.

Fourteen interviewers were trained for the 2019 administrative consent collection. Training was conducted over three days and consisted of four in-class hours of general training on using the CATI CMS and call center onboarding procedures and eight in-class hours of project-specific training. The interviewers were a mix of new interviewers and experienced staff who had worked on other projects with young adult sample populations, but none of the interviewers had worked on the 2016 consent collection effort. Training included the same modules noted above about the background of the study, strategies for gaining cooperation, answers to frequently asked questions, and how to launch and complete the web-based consent form, as well as hands-on sessions for navigating the CATI CMS. Similar to the 2016 collection, interviewers participated in weekly check-in meetings throughout data collection.

15.2.3 Interviewer Monitoring and Case Review

All outbound telephone calls were audio-recorded and subject to quality review. For both the 2016 and 2019 consent collections, approximately 10% of all outbound calls were monitored by a supervisor, using either live or recorded audio, and assessed for quality. After each monitoring session, the supervisor provided the interviewer with feedback on their performance during the call and, if necessary, guidance to improve their case management, form administration, and refusal aversion skills. Monitoring was conducted continuously throughout both consent collections. Common challenges that arose in monitoring sessions were addressed in weekly check-in meetings with the entire interviewing staff.

The supervisory team also conducted case review throughout both consent collection periods. For the 2016 collection, case review focused on monitoring the productivity of each interviewer, verifying that the number of call attempts per case was consistent with best practice guidelines presented in interviewer training, and verifying that appointments were met. When case review efforts revealed underworked cases – cases that were outliers based on the number of call attempts or the amount of time elapsed since the last call attempt – the supervisory team provided interviewers with feedback and lists of cases that needed attention. Because interviewers were not automatically prompted to call appointments at the scheduled time, a deliberate effort was made to review appointments daily and remind interviewers to check their appointment schedule at the beginning of each shift. For the 2019 collection, case review was intended to verify that the CATI system applied the correct status codes and the call scheduler placed cases in the appropriate dialing queue.

15.3 Results

We compared the effectiveness of the case ownership approach against the automated CATI scheduler approach using three metrics: the number of call attempts per case, variability in the timing of call attempts, and the number of calls per interviewer hour worked.

15.3.1 Number of Call Attempts Per Case

First, we reviewed the number of call attempts per case across both the 2016 transcript consent and 2019 administrative matching consent samples. Consistency in the number of call attempts across cases is desirable because it indicates that data collection effort has been applied equally across the sampled cases. During interviewer training for the case ownership approach, interviewers were instructed that each case should receive a call attempt every five to seven days, on average, which means that non-responding cases should receive seven to ten call attempts during the seven-week outbound calling period. Non-responding cases that received fewer than seven call attempts in the data collection period would be considered "under-worked" cases. Figure 15.1 shows the distribution of call counts by data collection round for all outbound calling cases and non-responding cases. For the 2016 transcript consent collection using the case ownership approach, calls per case ranged from 1 to 17, with most sample members (nearly 61%) receiving between 3 and 6 calls and only about 24% receiving 7 or more calls. For the 2019 administrative consent collection using the automated CATI scheduler, calls per case ranged from 1 to 17, similar to the case ownership approach, but the sample generally received more call attempts than the 2016 cases; here only about 19% of the cases received 3–6 calls and about 73% received 7 or more calls.

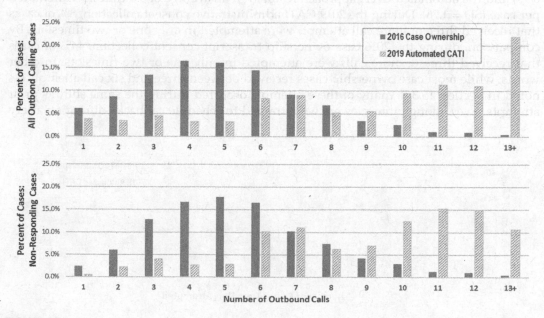

FIGURE 15.1

Outbound call count distribution by data collection round.

One key point demonstrated by Figure 15.1 is that there is a small but noteworthy group of nonrespondents who received very few calls. These low call count frequencies were observed in both rounds of data collection, but affected a larger proportion of cases in the case ownership round. In the case ownership transcript consent collection, 283 non-respondents (21.2% of nonrespondents) received three or fewer calls; 33 of these were no longer being dialed because they were dead-ended (e.g., final refusals, disconnected phone numbers), leaving 250 or 18.8% of nonresponse cases receiving too few calls. For the CATI administrative consent collection, 131 nonrespondents (6.7% of nonrespondents) received three or fewer calls; 10 of these (13.1%) were dead-ended cases, leaving 121 or 6.1% receiving too few calls. These cases represent opportunities for improvement in our scheduler algorithm and case review procedures – had these cases received more call attempts, it is possible that more sample members would have provided consent.

15.3.2 Variability in Timing of Call Attempts

Next, we examined the extent to which interviewers varied the timing of their call attempts. Variation of call timing is desirable because attempting contact at differing days of the week or times of the day increases the likelihood of reaching sample members (while continuing to call sample members at the same unsuccessful timeslots tends to result in lower contact rates). For both consent collections, there were 12 possible calling slots: dayshift (calls before 5:00 pm) and nightshift (calls after 5:00 pm) for Monday through Friday (10 total slots for weekdays), Saturday, and Sunday. Figure 15.2 shows the distribution of cases by number of timeslots attempted for all cases that received three or more outbound calls. The automated CATI scheduler resulted in greater variety in the timing of call attempts and more variance in the number of timeslots attempted when compared to the case ownership approach. Among cases that received three or more calls, the case ownership approach resulted in an average of 2.72 timeslots attempted per case (SD = 0.95) and the automated CATI approach resulted in an average of 5.5 timeslots attempted per case (SD = 1.76). During the 2019 CATI administrative consent collection, 5% of cases that received three or more call attempts were attempted in only one or two timeslots. By comparison, during the 2016 case ownership transcript consent collection, 44% of cases that received three or more calls were attempted in only one or two timeslots. In other words, while most case ownership cases received between three and six call attempts (as noted in Section 15.3.1), many of these attempts occurred during the same shift as prior attempts. Even though interviewers were trained to schedule callbacks during a variety

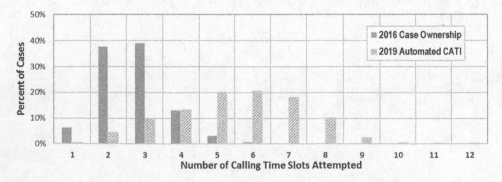

FIGURE 15.2
Number of calling timeslots attempted for cases that received three or more outbound calls.

of timeslots (including timeslots that fell outside of their shift), they tended to schedule callbacks during the same timeslots. Additional analysis is needed to examine how these call timing patterns intersect with interviewers' dialing shifts and to quantify the impact of call attempt variability on the likelihood of successfully contacting a case.

To further compare the variability in call timing across call attempts, we created a simple variability ratio for all cases that received three or more calls:

$$\text{Variability ratio} = \frac{\text{Number of timeslots dialed}}{\text{Number of total call attempts}} \qquad (15.1)$$

Cases with values approaching 0.0 received call attempts with the lowest variability of timing; that is, all or most attempts were made in the same timeslot (because all cases in this analysis received at least three attempts, it is not possible for the variability score to equal 0.0). Cases with values closest to 1.00 are those that received call attempts during the widest range of timeslots (for cases with values exactly equal to 1.00, every call attempt occurred in a unique timeslot). Figure 15.3 shows the distribution of this ratio for the two data collections. The 2016 case ownership approach resulted in lower variation of call attempts, as demonstrated by a higher percentage of cases in the lower ratios and a lower percentage of cases in the higher ratios, compared to the 2019 CATI scheduler approach.

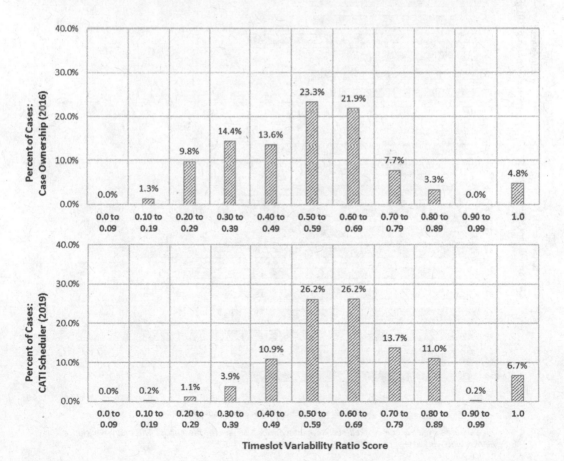

FIGURE 15.3
Timeslot variability of outbound calls.

For the case ownership approach, 61% of cases had scores of 0.50 or above, meaning that the timeslot of outbound calls was duplicated for no more than 50% of the call attempts for that case. For the automated CATI scheduler approach, 84% of cases had scores of 0.50 or above.

15.3.3 Number of Call Attempts Per Interviewer Hour

Finally, we examined the productivity of each interviewer, as measured by the number of call attempts made per hour. Figure 15.4 shows the average number of calls made per hour by each interviewer who completed at least 80 hours of work during the 2016 and 2019 data collections. Interviewers are listed in decreasing order of hours worked on the project, such that interviewer 1 worked the highest number of hours. While there are too few interviewers to analyze statistical correlations, there does not appear to be any relationship between interviewers' number of hours worked on the project and number of calls made per hour. For both approaches, there was wide variation in the average number of calls per hour across interviewers: the average number of calls per hour ranged from

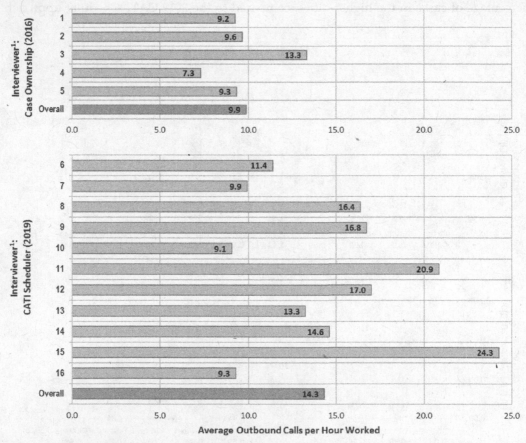

[1]*Interviewers are sorted by hours worked, such that Interviewer 1 worked the most hours. This analysis excludes interviewers who worked fewer than 80 hours on the project.*

FIGURE 15.4
Average number of outbound calls per hour worked, by interviewer.

7.3 to 13.3 (9.9 calls per hour overall) for the 2016 case ownership interviewers and from 9.1 to 24.3 (14.3 calls per hour overall) for the 2019 automated CATI scheduler interviewers. Although it had wide variation in the average number of calls per hour across interviewers, the automated CATI scheduler resulted in more calls per hour overall than the case ownership approach. The higher number of calls per hour is likely due to the CATI system's immediate delivery of cases to call, allowing interviewers to dial quickly with little downtime between calls. In the case ownership approach, interviewers had to spend time between calls reviewing and selecting which case to call next. However, the CATI scheduler did not prompt or force interviewers to make a set number of calls per hour, so we still observe variation across staff.

15.3.4 Operational Observations

At the end of each collection, interviewers participated in debriefing meetings in which they reflected on their experiences through informal discussions. In these meetings, interviewers' subjective assessment of the case ownership approach was positive. Interviewers directly expressed that they appreciated being able to work cases as they saw fit and enjoyed the challenge of case management. One interviewer noted that she enjoyed being able to "use her experience" to decide how to work a case. Supervisory staff also noted that interviewers were enthusiastic in weekly check-in meetings and were eager to share their approaches with the rest of the team.

Live and recorded monitoring sessions revealed that both sets of interviewers were able to establish rapport with sample members using the sample scripts and talking points provided in training. Supervisory staff working on the case ownership collection reported that additional time was spent conducting case review and providing feedback to interviewers to ensure that interviewers were managing their caseloads using the best practices covered in training. However, labor hours were not tracked for specific activities on the project, so there is no data available to evaluate these subjective perceptions. Similarly, missed appointments were rare but did occur during both collections; however, it is not possible to compare the frequency of missed appointments across the two collections because data for missed appointments was not recorded by the case ownership CMS.

15.4 Discussion

While the case ownership approach used in the 2016 transcript consent collection contributed to the project meeting budget, schedule, and response rate targets, it resulted in fewer calls per interviewing hour, fewer call attempts per case, and less variability in the timing of call attempts compared to the 2019 CATI approach. These results revealed inefficiencies in interviewers' case management skills despite the efforts of the project team to train interviewers on case management best practices, review cases, and intervene throughout data collection by reassigning cases and providing feedback to interviewers. Thus, the results suggest that with respect to these outcomes, more work may be needed to improve interviewers' case management skills and practices.

This analysis was a retrospective observational study, and as such there are limitations to the conclusions that can be drawn regarding the impact of the case ownership approach. While the data collection protocols were similar, the consent being requested differed between data collections. The collections also requested consent from different subsets of the study cohort. Either of these factors may explain the difference in consent rates between collections. In addition, in this study there were no interviewers who worked on both the 2016 and 2019 collections, so it is not possible to compare how an individual interviewer's performance and perceptions varied across the data collections. Future research should utilize experimental designs to evaluate the effect of these two case management approaches on consent rates overall and by important subgroups. Such an assessment will require a robust methodological data set, including case reassignments (including the reason for reassignments), the number of cases assigned to interviewers on each day of data collection, missed appointments, and clustering of callbacks within interviewers' work schedules. These studies should also administer a standardized questionnaire to interviewers and should assess the relationship of interviewer characteristics and perceptions with the effectiveness of each approach.

Data collection budgets are not limitless, and data collection managers are tasked with balancing resources to maximize survey quality (Biemer 2010). Because these data collections were carried out primarily for production purposes rather than methodological research, labor data was not recorded in sufficient detail to examine specific tasks such as the time spent programming and testing, conducting case review, providing interviewer feedback, monitoring, and other non-interviewing tasks; therefore, it was not possible in this study to compare the two approaches in terms of efforts and costs associated with each specific task. Future comparisons of these case management approaches should collect and analyze such data. It is likely that case ownership approaches are most feasible for studies with small sample sizes, in which the labor costs associated with intensive interviewer training and case review are small compared to those required for programming complex CATI systems, or for establishment surveys where a dedicated interviewer is needed to build rapport with the establishment's gatekeepers. For larger studies, the labor costs associated with the training, review, and manual intervention required for the case ownership approach may outweigh system development savings; it may be more cost-effective to program the CATI system. In the 2016 collection, the cost savings from programming allowed for a robust tracing, locating, and mailing effort that would not have been possible if a CATI system had been developed. It is possible that these efforts had more positive impact on the consent rate than would have been achieved through automated call scheduling. Future research is needed to estimate the relative cost and benefit of these activities when compared to automated call scheduling and to determine the inflection point at which the benefits of an automated CATI system outweigh the development costs.

The case ownership approach can be implemented successfully for telephone interviewing through rigorous analysis of paradata throughout data collection and interventions to ensure that cases are worked consistently. Researchers planning the use of a case ownership approach should consider the following recommendations:

1. Train interviewers on case management skills, including factors interviewers should consider when deciding when to contact each case and expectations for variation of contact attempts.

2. Build a specialized interviewing team of experienced staff who are adept at implementing the case ownership approach.

3. Provide interviewers with sample call scripts and conduct role play exercises for interviewers to practice using the call scripts. We found that interviewers were

successful in establishing rapport and gaining cooperation without verbatim call scripts, but that they needed the call scripts as a starting point.

4. Set up real-time reporting to monitor the number of call attempts received by each case, the variation in timeslots attempted, and interviewer productivity from the start of data collection.

5. Adjust interviewer case assignments throughout data collection to ensure that cases receive a consistent level of effort and sufficient variation of call attempts.

References

Biemer, P. 2010. Total survey error: Design, implementation, and evaluation. *Public Opinion Quarterly* 74(5):817–848.

Blom, A. G. 2012. Explaining cross-country differences in survey contact rates: Application of decomposition methods. *Journal of the Royal Statistical Society: Series A* 175(1):217–242.

Groves, R. M., and M. P. Couper. 1998. *Nonresponse in Household Interview Surveys*. New York: Wiley.

Groves, R. M., F. J. Fowler, M. P. Couper, J. M. Lepkowski, E. Singer, and R. Tourangeau. 2009. *Survey Methodology*. New York: Wiley.

Purdon, S., P. Campanelli, and P. Sturgis. 1999. Interviewers' calling strategies on face-to-face interview surveys. *Journal of Official Statistics* 15(2):199–216.

Sakshaug, J. W., M. P. Couper, M. B. Ofstedal, and D. R. Weir. 2012. Linking survey and administrative records: Mechanisms of consent. *Sociological Methods & Research* 41(4):535–569.

Walejko, G., and J. Wagner. 2015. The challenges associated with relying on CAPI interviewers to implement novel field procedures. Paper presented at the 70th Annual Conference of the American Association of Public Opinion Research, Hollywood, FL.

West, B. T., and A. G. Blom. 2017. Explaining interviewer effects: A research synthesis. *Journal of Survey Statistics & Methodology* 5(2):175–211.

16

Investigating the Use of Nurse Paradata in Understanding Nonresponse to Biological Data Collection

Fiona Pashazadeh, Alexandru Cernat, and Joseph W. Sakshaug

CONTENTS

16.1 Introduction

The collection of biological data in social surveys has led to new possibilities for biosocial research by enabling a more comprehensive understanding of the interplay between the different factors affecting health and behavioral outcomes. This combination of survey and biological data has already yielded some significant discoveries in the social and health sciences (e.g., Banks, et al. 2006; Kim, et al. 2013; Puterman, et al. 2016). The biological measurements collected in social surveys include blood and saliva specimens, which can be used to assess markers of disease (e.g., hemoglobin A1c). For example, the Wisconsin Longitudinal Study collects salivary DNA samples from respondents by mail (Dykema, et al. 2017). The Survey of Health, Retirement, and Aging in Europe collects dried blood spots in several European countries plus Israel. Lay interviewers are used to collect the blood spots in addition to carrying out the interview (Weiss, Sakshaug, and Börsch-Supan 2018). Other examples include Understanding Society: the UK Household Longitudinal Study (UKHLS) and the English Longitudinal Study of Ageing (ELSA), which use nurses to collect venous blood samples from respondents (Benzeval, et al. 2014; Steptoe, et al. 2012). Recent analyses of the UKHLS show evidence of links between informal caregiving and adverse metabolic markers (Lacey, McMunn, and Webb 2018), and higher levels of chronic stress biomarkers for those reemployed in poorer quality jobs (Chandola and Zhang 2017).

The collection of biological data in conjunction with survey data comes with a unique set of challenges, one of which is nonresponse. There are at least three stages at which nonresponse can occur. The first stage is the decision of the respondent to take part in the biological collection component of the interview. Respondents could decide to forego the biological component entirely because of time constraints, general unwillingness, or physical limitations. This would result in nonresponse for all possible measurements, and this could lead to nonresponse bias if those refusing to participate in the biological component are significantly different from those who participate (Groves 2006). This initial stage of participation is likely to have different effects depending on whether the biological data collection occurs at the same time as the interview (as in the lay interviewer collection model), or if there is a gap between the interview and the biological data collection (as in the nurse collection model). In the case of the UKHLS, the nurse visit took place five months after the interview. By this time, respondents may have reconsidered their decision to participate in the survey, or they may be harder to reach due to increasing work, family, or other competing demands. The 1970 British Birth Cohort tried using nurses to conduct interviews and collect biological data all in one visit, but these efforts were not successful, as nurses did not like making repeat visits and persuading reluctant respondents to participate (Brown 2018).

For respondents who do decide to proceed with the biological component, some may only be willing to consent to a subset of possible measurements. This introduces a second stage of participation. For example, a respondent may be unwilling to consent to the collection of blood, but willing to consent to a less invasive collection, such as saliva. This situation is typical in biosocial surveys where cooperation rates are generally lower for the most invasive measurements (Dykema, et al. 2017; Jaszczak, Lundeen, and Smith 2009; Weiss, Sakshaug, and Börsch-Supan 2018). In the third stage of participation, respondents who consent to specific measurements may be unwilling, or physically unable, to complete the measurement once it begins. For example, it may not be possible to extract sufficient amounts of blood or saliva from the respondent, or the respondent may experience physical discomfort during the collection procedure that prevents them from finishing the collection. Even if a sufficient amount of sample is collected, nonresponse can still occur if the collected sample is mishandled or damaged during the shipment or processing stage.

While samples selected for surveys are designed to be representative of the target population, the multiple stages at which nonresponse can occur during the biological data collection can lead to bias in statistical analyses of the collected data (Cernat, et al. 2018; Korbmacher 2014; Sakshaug, Couper, and Ofstedal 2010). The situation is likely to be more severe when a separate nurse visit is required following the interview and/or more invasive biological measures, such as blood, are collected (McFall, et al. 2014). In order to address this nonresponse, survey researchers typically create and apply sampling weights based on response propensity models (Brick and Kalton 1996). These models may also be useful in adapting or targeting approaches to increase contact and cooperation during fieldwork (e.g., Lynn 2017). However, variables included in the models must be available for both respondents and nonrespondents in the sample. Furthermore, to be effective at reducing bias, the variables used in response propensity models must be related both to the response outcome and the analysis variables of interest (Little and Vartivarian 2005).

Auxiliary variables used in response propensity models include different types of data collected during the survey process, known as paradata (for a collection of examples, see Kreuter 2013; for the use of paradata in nonresponse adjustment, see Olson 2013). Paradata can include variables such as call histories, response timings, and interviewer observations, and may provide a cost-effective way to address nonresponse in analyses which use survey data. Such data have been shown to be related to response and improved the fits of response propensity models, though in many studies these data tend not to be strongly related to the survey variables of interest (Diez Roux 2001; Kreuter and Kohler 2009; Kreuter, et al. 2010; Lin and Schaeffer 1995; Peytchev and Olson 2007; Sakshaug and Kreuter 2011; West, Kreuter, and Trappmann 2014). Paradata can have other quality issues, such as missingness and measurement error, which can vary across sources and variables (Sinibaldi, Durrant, and Kreuter 2013; Smith 2011; Stoop, et al. 2010; West 2013).

The use of paradata to study nonresponse to biological data collection and adjust for its occurrence is an understudied area of research, particularly when the collection involves the employment of trained nurses to carry out the fieldwork in a separate visit following the interview, such as in the UKHLS and the ELSA studies. There is evidence that nurses can systematically affect response at each stage of participation (Cernat, et al. 2018). Although separate weights are constructed to adjust for nonresponse to biological data collection in these large-scale surveys, the auxiliary data used in the adjustment mostly come from the sampling frame and the questionnaire. Paradata collected from the attempted nurse visit may provide further useful information to better predict nonresponse and improve the response propensity models fitted after biological data collection, especially since there is some evidence that nurses may differ from interviewers in terms of how they approach and carry out survey tasks related to the contact and cooperation of sample members (Brown 2018). However, nurses are not used to making follow-up visits and collecting paradata (e.g., doorstep interactions) and may be more prone to errors, especially for difficult cases. Therefore, the range and quality of survey paradata collected by nurses is likely to vary.

The aim of this chapter is to explore the potential for using nurse visit paradata to address nonresponse to biological data collection. We carry out an investigation of the quality of paradata variables collected during the nurse visit phase of the UKHLS. Through this analysis, we determine whether such data can help to predict response across the different stages of biological data collection. This determination will inform our conclusion about whether paradata collected by nurses during biological data collection should be incorporated into nonresponse adjustment procedures, or at least disseminated to data users who can then decide for themselves whether to use these data as control variables in their analysis.

In this chapter, we aim to address the following three research questions: (1) What types of paradata are available at each stage of biological data collection where nonresponse might occur? (2) What is the quality of the available paradata – for example, how much variation exists in the variables, and how much data are missing? (3) Can paradata variables improve models of nonresponse for each of the stages of biological data collection? Can these variables still be useful even when there are significant concerns about their quality?

16.2 Data

The data sets used here are from the UKHLS, which provides a rich and readily available source of auxiliary variables collected during fieldwork. In waves 2 and 3, biological data collection was included for two different subsets of the full sample. This involved re-establishing contact with the eligible sample members to arrange a separate visit by a trained nurse, approximately five months after the main survey interview. In wave 2, the biological data sample was from the General Population Sample in England, and in wave 3, the sample was from the predecessor to the UKHLS, the British Household Panel Survey (BHPS).

Nurses were required to make contact with sample members by telephone to make an appointment for a personal visit. This direct contact from the nurse followed an advance letter to sample members introducing the process for biological data collection and a telephone call from a central call center. During the attempts to contact the sample members and the visit itself, nurses were instructed to collect and record paradata. The paradata variables made available with the data set include: (1) the number of calls to each household (including both telephone calls and personal visits), (2) status (e.g., no reply, contact made, appointment made) and length of each call, (3) timings for each module of biological data collection during the nurse visit (once sample members had agreed to participate), and (4) reasons for missing measurements (e.g., refusal, illness, afraid of needles). The recording of paradata at each call introduces possibilities for including the sequences of contact attempts as predictors in the response propensity models, especially as these variables are available for both respondents and nonrespondents. Such variables have been found to be predictive of nonresponse, although their inclusion is not necessarily straightforward due to the complex nature of the data (Durrant, D'Arrigo, and Steele 2013; Hanly, Clarke, and Steele 2016; Kreuter and Kohler 2009).

In the UKHLS, call record paradata are available for 11,580 households in the wave 2 biological data sample and 3,398 households in wave 3. Descriptive statistics for the available call records (available in Online Appendix) illustrate the nature of these paradata and provide insights into their quality. The greatest variation in call length is for calls that included interviewing, as might be expected, compared to calls that established contact or made appointments. There are also some outliers, in particular an interview visit that is recorded as lasting 691 minutes; this may indicate issues with the accuracy of the paradata in some cases. Another potential concern with these data is that the total number of calls per household recorded for both waves 2 and 3 indicate that the large majority of available call histories for households contain just one call rather than a longer sequence. Nurses were expected to record all telephone contact attempts as well as personal visits, so this is unexpected and again raises questions about the quality and utility of the paradata.

An additional quality issue emerged when matching the call records to the nurse visit sample data sets for individuals who were deemed eligible during the pre-visit telephone interview and were subsequently assigned a nurse. Here, we find that there are missing call records for 2,753 and 825 individuals in waves 2 and 3, respectively. Table 16.1 shows that the missing call records are mainly present at stage 1 of the nurse visit, i.e., where all eligible sample members are included. This indicates that call records are more likely to be present for those who agreed to participate in the nurse visit compared to those who did not agree.

16.3 Methods

Given the initial analysis in Section 16.2, the quality of the available paradata from the biological data collection component of the UKHLS is called into question in a way that

TABLE 16.1

Call Record Data Availability for Individuals at Each Stage of Biological Data Collection in Waves 2 and 3

Stage of Biological Data Collection	Number of Eligible Individuals	Number Participating in the Stage	Number with Call Record Data	Number with Missing Call Record Data	Percentage with Call Record Data
Wave 2					
(1) Nurse visit	21,161	15,608	18,408	2,753	86.99%
(2) Consent to the blood sample	14,264	11,018	14,196	68	99.52%
(3) Obtaining blood sample given consent	11,018	9,905	10,965	53	99.52%
Wave 3					
(1) Nurse visit	6,604	5,046	5,779	825	87.51%
(2) Consent to the blood sample	4,857	3,741	4,845	12	99.75%
(3) Obtaining blood sample given consent	3,741	3,342	3,732	9	99.76%

cannot be ignored. In particular, the descriptive analysis of the paradata shows that call records for the UKHLS nurse visit are disproportionately available for individuals who had agreed to participate in the nurse visit. This would cause bias if these data were used in subsequent models of response without taking this into account, i.e., in a complete case analysis.

Given these concerns about quality, we investigate whether these paradata can still be useful by considering two different approaches to including these variables in models of nonresponse to the stages of biological data collection. The first approach involves using the available paradata as explanatory variables with a separate category for those with missing values, to evaluate the predictive power of both the variables and the missingness. This also allows us to evaluate possible bias if missingness tends to be associated with significantly different expectations of response propensity. The second approach is to use the missingness of the paradata to create "nurse performance" indicators that could be used in subsequent models of response to biological data collection, as an alternative to including the paradata variables directly. This approach tests whether the performance of nurses in recording the required call information is in turn associated with nurse performance at other interviewer tasks, such as gaining cooperation for biological data collection from sample members. The inclusion of the nurse performance indicators in the response propensity models precludes the use of the actual paradata variables as the information on the missingness would be duplicated, creating problems of multicollinearity.

The models of response to the stages of biological data collection included the explanatory variables described in Table 16.2. These are based on the findings of studies reported by Groves and Couper (1998) and Watson and Wooden (2009), which regard the characteristics of respondents to longitudinal surveys and include variables describing the individual sample members as well as their households and geographical locations. In addition, health variables were included due to their likely importance in biological data collection, as well as indicators of survey reluctance (Sakshaug, Couper, and Ofstedal 2010).

The UKHLS involves sampling by geographical area, which may lead to clustering in responses and therefore incorrect conclusions from statistical analyses using standard

TABLE 16.2

Details of the Respondent and Household-Level Explanatory Variables for the Models of Response to the UKHLS Nurse Visits

Category	Variables
Personal characteristics	Sex, age, marital status, employment status, highest qualification, ethnicity (White/non-White)
Household characteristics	Number of adults, children, housing tenure, Government Office Region (London or elsewhere), urban/rural
Health status	Long-standing illness, self-rated health
Indicators of survey reluctance	Number of calls for the main interview, interviewer ratings of cooperation, and suspicion following the main interview

techniques (Goldstein 2011). The way that the sample is assigned to nurses may also introduce clustering, and nurses may affect nonresponse in the same way that interviewers do (e.g., Cernat, et al. 2018; West and Blom 2017). In the UKHLS biological data collection fieldwork for waves 2 and 3, nurses were allocated to an average of 16.41 and 9.72 postcode sectors, respectively. Postcode sectors were covered by an average of 1.15 nurses in wave 2 and 3.19 nurses in wave 3. In order to account for and explore these effects, cross-classified multilevel logistic regression models were used in this analysis to accommodate the non-hierarchical nature of the sample, with crossed random effects for geographical area (postcode sector) and nurse (Campanelli and O'Muircheartaigh 1999; Pickery, Loosveldt, and Carton 2001; Vassallo, et al. 2015).

Recall that there are three distinct stages of the nurse visit where nonresponse can occur: agreeing to the nurse visit, consenting to the blood sample, and obtaining the blood sample. Each stage was modeled separately to incorporate and estimate the different nonresponse mechanisms and associations. Responses for each stage were coded as binary variables with agreement/consent/obtain equal to one and other outcomes equal to zero.

16.3.1 Paradata Variables

The characteristics of the nurses and a selection of paradata variables from each of the stages of biological data collection, as well as the main survey interview, were included in the extended response propensity models, as described in Table 16.3. As many of the paradata variables from the call records included repeated information, we selected "Outcome

TABLE 16.3

Details of the Paradata Variables for the Models of Response to the UKHLS Nurse Visits

Variable	Description
Nurse characteristics	Nurse age
	Nurse years of experience working for NatCen
Outcome of first call to the household	"No reply," "Contact made," "Appointment made," "Any interviewing done," "Any other status," "Missing"
Main interview length	Time in minutes of main survey interview
Length of nurse visit excluding the blood sample (stages 2 and 3)	Time in minutes to complete the nurse visit excluding all blood sample components
Missing contact info	We give a value of 1 for all the cases that have missing contact information and 0 otherwise

of first call to the household" and the lengths of the nurse visit and interview. Missing values were coded as a separate category for the call outcome variables. Any negative times were set to zero for the interview and nurse visit lengths.

Response propensity models were first estimated for each stage of biological data collection in waves 2 and 3 and included the explanatory variables in Table 16.2. The nurse characteristics and the paradata variables were then added to the models, with likelihood ratio tests carried out to assess whether any of the fixed effects associated with these additional predictors were statistically significant (with a p-value of 0.05 or less). This would indicate that the variables are improving the models and are useful additions. Effect sizes were also evaluated to see if the variables contributed in a substantively meaningful way to the analysis of response to each stage.

16.3.2 Nurse Performance Indicators

The models for response at each stage of the wave 3 nurse visit were then re-estimated using the nurse performance indicators derived from wave 2, as an alternative to using the paradata variables directly. The same pool of nurses worked on waves 2 and 3, so it was possible to match the indicators across these waves. The nurse performance indicators were calculated using cross-classified multilevel logistic regression models for the log-odds of an individual sample member having call records at wave 2. Here, $\pi_{1i(jk)}$ is the probability of having call records for individual i with nurse j in area k, and $\mathbf{X}'_{i(jk)}\boldsymbol{\beta}$ is a linear combination of coefficients and covariates for nurse characteristics (age and experience). The nurse-level residuals are denoted by u_j where $u_j \sim N\left(0, \sigma_u^2\right)$ and the area-level residuals are denoted by v_k where $v_k \sim N(0, \sigma_v^2)$

$$\log\left(\frac{\pi_{1i(jk)}}{1 - \pi_{1i(jk)}}\right) = \mathbf{X}'_{i(jk)}\boldsymbol{\beta} + u_j + v_k$$

As with the response propensity models, the cross-classified model accounts for the clustering by area in the UKHLS sample while also allowing for clustering by nurses, depending on the likelihood that call records are available for the sample members in their assignment, conditional on their age and experience (experience is included as number of years working for the survey organization). Predictions (or more specifically, Empirical Best Linear Unbiased Predictions, also known as EBLUPs) of the nurse-level residuals from this model were used to create three categories of nurse performance indicators: "average," "above average," and "below average." To create the categories, we use the interval estimates for the EBLUPs. If the EBLUP is negative and the interval estimate does not contain 0, we code the nurses as performing below average. If they have a positive EBLUP and their interval estimate does not include 0, we code this as above average. All the rest of the nurses are in the average performance category.

The nurse performance indicators were then matched to sample members in wave 3 and included with the other covariates in the response propensity models for the three stages of biological data collection in wave 3 (overall participation in the nurse visit, consent to the blood sample, and obtaining of the blood sample). Again, likelihood ratio tests were performed to determine whether the nurse performance indicators had improved the response propensity models. We use procedures in the R software for all of the analyses, and our code is readily available upon request.

16.4 Results

16.4.1 Approach 1: Response Propensity Models Including Nurse Characteristics and Paradata Variables

Table 16.4 shows the key results from the models for response across the wave 2 nurse visit stages (full results are available in the online supplemental materials). The addition of the nurse characteristics in model 2 for stage 1 shows that the age of the nurse increases the likelihood of participating, but experience does not. Extending the model with the paradata variables in model 3 shows that the outcome of the first call is significant, with appointments and any interviewing carried out at this stage increasing the log-odds of positive responses (by 2.48 and 3.90 units, respectively), as would be expected. Missingness of call records has a strong negative association with response propensity, which matches with the observations from the data description regarding the disproportionate availability of call records for those who participated in biological data collection. Likelihood ratio tests indicate that the extended model with all the additional variables is an improvement over the basic model.

The extended models for stage 2 in Table 16.4 show that the nurse characteristics are now not significant. This is also the case for the outcome of the first call, apart from the missing category. However, the lengths of the main interview and the pre-blood sample components of the nurse visit increase the likelihood of consenting to the blood sample, when adjusting for the indicator of missing call records. Here, the full extended model is shown to be an improvement over the basic model using likelihood ratio tests. Similarly, for stage 3, the likelihood of obtaining the blood sample is not associated with nurse age or experience. The only paradata variable that shows a significant, and negative, relationship with the likelihood of obtaining the blood sample is the length of the main interview, although the effect size is relatively small.

Table 16.5 shows the key results from the wave 3 nurse visit response propensity models. The extended models indicate that the age of the nurse again increases the likelihood of participating in the nurse visit. The outcome of the first call variable also shows the same associations as for wave 2. However, for stages 2 and 3, the additional variables are not statistically significant and the extended models are not an improvement over the basic models.

16.4.2 Approach 2: Response Propensity Models Including Nurse Performance Indicators

The results from the cross-classified logistic model of sample members having call records in wave 2 show that, of the 155 nurses at wave 2, 10 had an above-average likelihood of recording calls and 39 had a below-average likelihood, when conditioning on their age and experience. This is also confirmed by the quantile–quantile plot of nurse residuals (available in the online supplemental materials).

In wave 3, there were 131 nurses. All of these nurses had also worked on wave 2 and therefore were able to be matched to the calculated performance indicators. Table 16.6 shows the key results of the response propensity models for individuals at stage 1 of the wave 3 nurse visit. These results suggest that the nurse performance indicators are strongly significant and also have the expected effects – being assigned a nurse with above-average performance increases the likelihood that an individual will participate in the nurse visit and a nurse with below-average performance decreases the likelihood. The performance

TABLE 16.4

Results from Cross-Classified Multilevel Logistic Regression Models for the Log-Odds of Participation Outcomes in the Wave 2 Nurse Visit Using Nurse Characteristics and Paradata Variables[a]

	Participation in Nurse Visit			Consenting to the Blood Sample			Obtaining the Blood Sample		
	Model 1	Model 2	Model 3	Model 1	Model 2	Model 3	Model 1	Model 2	Model 3
Nurse age (mean-centered)/100		2.00*** (0.52)	1.35* (0.59)		0.40 (0.59)	0.15 (0.60)		0.71 (0.85)	0.62 (0.86)
Nurse experience (mean-centered)/100		1.14 (0.95)	−0.53 (1.06)		−1.07 (1.06)	−0.22 (1.09)		2.17 (1.56)	2.64 (1.58)
Outcome of first call (ref = no reply)									
Contact made			0.02 (0.13)			−0.38 (0.22)			−0.10 (0.36)
Appointment made			2.48*** (0.19)			−0.10 (0.22)			−0.01 (0.35)
Any interviewing done			3.90*** (0.10)			0.16 (0.14)			0.20 (0.22)
Any other status			0.01 (0.16)			0.20 (0.31)			0.35 (0.45)
Missing contact info			−3.65*** (0.19)			1.65* (0.65)			−0.35 (0.54)
Main interview length in minutes/100			−0.03 (0.03)			0.06* (0.03)			−0.11** (0.03)
Non-blood visit length in minutes/100						0.91*** (0.22)			0.58 (0.31)
Log likelihood	−10833.91	−10823.43	−5371.75	−7003.33	−7002.77	−6982.23	−3352.05	−3350.09	−3342.67
Area-level variance	0.46	0.46	0.43	0.19	0.19	0.20	0.04	0.04	0.04
Nurse-level variance	0.22	0.17	0.17	0.24	0.23	0.24	0.51	0.49	0.49

***p < 0.001, **p < 0.01, *p < 0.05

[a] Variables also included in the models: intercept, sex, age, marital status, employment status, highest qualification, ethnicity (White/non-White), number of adults, children, housing tenure, Government Office Region (London or elsewhere), urban/rural, long-standing illness, self-rated health, number of calls at main interview, and interviewer observations of survey reluctance. Full results available in Online Appendix.

TABLE 16.5

Results from Cross-Classified Logistic Regression Models of the Log-Odds of Participation Outcomes in the Wave 3 Nurse Visit Using Nurse Characteristics and Paradata Variables[a]

	Participation in Nurse Visit			Consenting to the Blood Sample			Obtaining the Blood Sample		
	Model 1	Model 2	Model 3	Model 1	Model 2	Model 3	Model 1	Model 2	Model 3
Nurse age (mean-centered/100)		2.72** (0.92)	−0.01 (1.02)		1.23 (0.92)	1.19 (0.93)		−0.59 (1.11)	−0.41 (1.13)
Nurse experience (mean-centered/100)		0.32 (1.57)	1.30 (1.73)		0.74 (1.58)	0.79 (1.61)		2.76 (1.94)	2.42 (1.98)
Outcome of first call (ref=no reply)									
Contact made			0.09 (0.28)			0.14 (0.46)			−0.49 (0.69)
Appointment made			2.28*** (0.42)			0.17 (0.48)			−0.30 (0.71)
Any interviewing done			3.38*** (0.20)			0.11 (0.29)			0.03 (0.48)
Any other status			−0.19 (0.29)			0.03 (0.48)			0.18 (0.80)
Missing contact info			−4.92*** (0.49)			0.42 (1.20)			12.71 (181.08)[b]
Main interview length in minutes/100			0.01 (0.07)			0.04 (0.06)			−0.03 (0.07)
Non-blood visit length in minutes/100						0.03 (0.41)			−0.48 (0.60)
Log likelihood	−3120.68	−3115.67	−1580.80	−2347.41	−2346.03	−2345.58	−1157.19	−1156.13	−1154.62
Area-level variance	0.28	0.29	0.17	0.10	0.10	0.10	0.12	0.15	0.14
Nurse-level variance	0.45	0.40	0.32	0.36	0.35	0.35	0.30	0.31	0.31

*** p < 0.001, ** p < 0.01, * p < 0.05.

[a] Variables also included in the models: intercept, sex, age, marital status, employment status, highest qualification, ethnicity (White/non-White), number of adults, children, housing tenure, Government Office Region (London or elsewhere), urban/rural, long-standing illness, self-rated health, number of calls at main interview, and interviewer observations of survey reluctance. Full results available in Online Appendix.

[b] This very large coefficient could be due to the small sample size in this group. All other coefficients in the model indicate that the model has converged.

TABLE 16.6

Results from Cross-Classified Logistic Regression Models of the Log-Odds of Participation Outcomes in the Wave 3 Nurse Visit Using Nurse Performance Indicators[a]

	Participation in Nurse Visit		Consent to Blood Sample		Obtaining Blood Sample	
	Model 1	Model 2	Model 1	Model 2	Model 1	Model 2
"Above-average performance"[b]		0.63** (0.22)		−0.22 (0.23)		−0.27 (0.26)
"Below-average performance"[b]		−0.71*** (0.16)		−0.22 (0.18)		−0.18 (0.22)
Log likelihood	−3120.68	−3106.05	−2347.41	−2346.40	−1157.19	−1156.43
Area-level variance	0.28	0.28	0.10	0.10	0.12	0.13
Nurse-level variance	0.45	0.28	0.36	0.35	0.30	0.31

***p < 0.001, **p < 0.01, *p < 0.05

[a] Variables also included in the models: intercept, sex, age, marital status, employment status, highest qualification, ethnicity (White/non-White), number of adults, children, housing tenure, Government Office Region (London or elsewhere), urban/rural, long-standing illness, self-rated health, number of calls at main interview, and interviewer observations of survey reluctance. Full results available in Online Appendix.

[b] The reference category is average performance.

indicators also reduce the estimated nurse-level variance component from 0.45 to 0.28, suggesting that some of the variation in the likelihood of responding to the wave 3 nurse visit attributed to clustering by nurse can be explained by these variables. Table 16.6 also shows that these results are not replicated in stages 2 and 3 of biological data collection as the performance indicators are not statistically significant and do not improve the response propensity models.

16.5 Discussion

There are mixed results from the addition of paradata variables to the models of response to various stages of the UKHLS biological data collection at waves 2 and 3. There are also clear differences not only between the stages of biological data collection but also between the two waves, and this may be attributed to the nature of each subsample (wave 3 contains long-standing BHPS respondents). The nurse performance indicators and characteristics, along with the paradata variables, appear to be generally more useful in models of initial participation rather than in models of obtaining the subsequent blood sample.

The absence of call records from a large number of cases precluded the use of some potentially useful paradata variables, such as the number and results of all the call attempts made to each household by the nurses. Instead, we were able to investigate the role of the assigned nurses in their new roles as "interviewers" responsible for making contact with and gaining cooperation and consent from sample members, as well as collecting the biological data samples.

The results suggest that, even with concerns about the quality of the paradata available (in particular measurement error and missingness; e.g., Durrant, D'Arrigo, and Steele

2013), these new forms of data may still contribute to models of response to biological data collection. These data also allow us to gain insights into the fieldwork practices of those not in a traditional survey interviewer role, and this may help to inform future training carried out by survey organizations, e.g., to increase the likelihood that contact protocols are followed.

However, a clear limitation is that we do not know whether the lack of call records in the case of the UKHLS is an indicator of nurse performance alone – it may be that there are some other effects being detected. These could include, for example, a geographical area effect related to the nurse not being able or willing to record the call details that also affects the likelihood of an individual to respond to biological data collection, such as concerns about the safety of the neighborhood.

This chapter provides a starting point for investigating the use of paradata in the study of response to biological data collection in large-scale social surveys. The next steps include further study of the usefulness of these variables in post-survey adjustments, in particular looking at their associations with outcome measures of interest in biosocial research, and whether "interviewer" performance indicators would also be informative here.

From our initial work, survey organizations should aim to record and make available a range of paradata during biological data collection. Nevertheless, any extra burden to record additional data on those carrying out the fieldwork in addition to their central tasks (in this case, the collection of a wide range of biological data that requires professional training) would have to be weighed against the overall quality of the data produced and also staff retention and continuity.

References

Banks, J., M. Marmot, Z. Oldfield, and J. P. Smith. 2006. Disease and disadvantage in the United States and in England. *Journal of the American Medical Association* 295(17):2037–2045.

Benzeval, M., A. Davillas, M. Kumari, and P. Lynn. 2014. *Understanding Society: The UK Household Longitudinal Study Biomarker User Guide and Glossary.* Essex: Institute for Social and Economic Research, University of Essex.

Brick, J., and G. Kalton. 1996. Handling missing data in survey research. *Statistical Methods in Medical Research* 5(3):215–238.

Brown, M. 2018. We all see the value in combining social and biomedical data collection, so let's ensure we can do it effectively. *Closer.* https://www.closer.ac.uk/news-opinion/blog/social-biomedical-data-collection-matt-brown/ (Accessed December 22, 2019).

Campanelli, P., and C. O'Muircheartaigh. 1999. Interviewers, interviewer continuity, and panel survey nonresponse. *Quality and Quantity* 33(1):59–76.

Cernat, A., J. W. Sakshaug, T. Chandola, J. Nazroo, and N. Shlomo. 2018. Nurse effects on nonresponse in survey-based biomeasures. Paper submitted.

Chandola, T., and N. Zhang. 2017. Re-employment, job quality, health and allostatic load biomarkers: Prospective evidence from the UK Household Longitudinal Study. *International Journal of Epidemiology* 47(1):47–57.

Diez Roux, A. V. 2001. Investigating neighborhood and area effects on health. *American Journal of Public Health* 91(11):1783–1789.

Durrant, G. B., J. D'Arrigo, and F. Steele. 2013. Analysing interviewer call record data by using a multilevel discrete time event history modelling approach. *Journal of the Royal Statistical Society: Series A, (Statistics in Society)* 176(1):251–269.

Dykema, J., K. DiLoreto, K. D. Croes, D. Garbarski, and J. Beach. 2017. Factors associated with participation in the collection of saliva samples by mail in a survey of older adults. *Public Opinion Quarterly* 81(1):57–85.

Goldstein, H. 2011. *Multilevel Statistical Models,* 4th ed. Chichester: John Wiley & Sons, Inc.

Groves, R. M. 2006. Nonresponse rates and nonresponse bias in household surveys. *Public Opinion Quarterly* 70(5):646–675.

Groves, R. M., and M. P. Couper. 1998. *Nonresponse in Household Interview Surveys.* Chichester: John Wiley & Sons, Inc.

Hanly, M., P. Clarke, and F. Steele. 2016. Sequence analysis of call record data: Exploring the role of different cost settings. *Journal of the Royal Statistical Society: Series A, (Statistics in Society)* 179(3):793–808.

Jaszczak, A., K. Lundeen, and S. Smith. 2009. Using nonmedically trained interviewers to collect biomeasures in a national in-home survey. *Field Methods* 21(1):26–48.

Kim, E. S., J. K. Sun, N. Park, and C. Peterson. 2013. Purpose in life and reduced incidence of stroke in older adults: The Health and Retirement Study. *Journal of Psychosomatic Research* 74(5):427–432.

Korbmacher, J. M. 2014. *Interviewer Effects on Respondents' Willingness to Provide Blood Samples in SHARE.* SHARE Working Paper Series 20.

Kreuter, F. 2013. *Improving Surveys with Paradata: Analytic Uses of Process Information.* Hoboken: John Wiley & Sons, Inc.

Kreuter, F., and U. Kohler. 2009. Analyzing contact sequences in call record data. Potential and limitations of sequence indicators for nonresponse adjustments in the European Social Survey. *Journal of Official Statistics* 25(2):203–226.

Kreuter, F., K. Olson, J. Wagner, T. Yan, T. M. Ezzati-Rice, C. Casas-Cordero, M. Lemay, A. Peytchev, R. M. Groves, and T. E. Raghunathan. 2010. Using proxy measures and other correlates of survey outcomes to adjust for non-response: Examples from multiple surveys. *Journal of the Royal Statistical Society: Series A, (Statistics in Society)* 173(2):389–407.

Lacey, R. E., A. McMunn, and E. A. Webb. 2018. Informal caregiving and metabolic markers in the UK Household Longitudinal Study. *Maturitas* 109:97–103.

Lin, I.-F., and N. C. Schaeffer. 1995. Using survey participants to estimate the impact of nonparticipation. *Public Opinion Quarterly* 59(2):236–258.

Little, R. J., and S. Vartivarian. 2005. Does weighting for nonresponse increase the variance of survey means? *Survey Methodology* 31(2):161–168.

Lynn, P. 2017. From standardised to targeted survey procedures for tackling non-response and attrition. *Survey Research Methods* 11(1):93–103.

McFall, S. L., J. Petersen, O. Kaminska, and P. Lynn. 2014. *Understanding Society–UK Household Longitudinal Study: Waves 2 and 3 Nurse Health Assessment, 2010–2012, Guide to Nurse Health Assessment.* Colchester: University of Essex.

Olson, K. 2013. Paradata for nonresponse adjustment. *The Annals of the American Academy of Political and Social Science* 645(1):142–170.

Peytchev, A., and K. Olson. 2007. Using interviewer observations to improve nonresponse adjustments: NES 2004. *Proceedings of the American Statistical Association, Survey Research Methods Section.*

Pickery, J., G. Loosveldt, and A. Carton. 2001. The effects of interviewer and respondent characteristics on response behavior in panel surveys: A multilevel approach. *Sociological Methods and Research* 29(4):509–523.

Puterman, E., A. Gemmill, D. Karasek, D. Weir, N. E. Adler, A. A. Prather, and E. S. Epel. 2016. Lifespan adversity and later adulthood telomere length in the nationally representative US Health and Retirement Study. *Proceedings of the National Academy of Sciences of the United States of America* 113(42):E6335–E6342.

Sakshaug, J. W., and F. Kreuter. 2011. Using paradata and other auxiliary data to examine mode switch nonresponse in a recruit-and-switch "telephone survey." *Journal of Official Statistics* 27(2):339.

Sakshaug, J. W., M. P. Couper, and M. B. Ofstedal. 2010. Characteristics of physical measurement consent in a population-based survey of older adults. *Medical Care* 48(1):64–71.

Sinibaldi, J., G. B. Durrant, and F. Kreuter. 2013. Evaluating the measurement error of interviewer observed paradata. *Public Opinion Quarterly* 77(S1):173–293.

Smith, T. W. 2011. The report of the international workshop on using multi-level data from sample frames, auxiliary databases, paradata and related sources to detect and adjust for nonresponse bias in surveys. *International Journal of Public Opinion Research* 23(3):389–402.

Steptoe, A., E. Breeze, J. Banks, and J. Nazroo. 2012. Cohort profile: The English longitudinal study of ageing. *International Journal of Epidemiology* 42(6):1640–1648.

Stoop, I., H. Matsuo, A. Koch, and J. Billiet. 2010. Paradata in the European social survey: Studying nonresponse and adjusting for bias. *Paper Presented at the JSM Proceedings*, Vancouver.

Vassallo, R., G. B. Durrant, P. W. F. Smith, and H. Goldstein. 2015. Interviewer effects on non-response propensity in longitudinal surveys: A multilevel modelling approach. *Journal of the Royal Statistical Society: Series A, (Statistics in Society)* 178(1):83–99.

Watson, N., and M. Wooden. 2009. Identifying factors affecting longitudinal survey response. In: *Methodology of Longitudinal Surveys*, ed. P. Lynn, 157–182. Hoboken: John Wiley & Sons, Inc.

Weiss, L. M., J. W. Sakshaug, and A. Börsch-Supan. 2018. Collection of biomeasures in a cross-national setting: Experiences in SHARE. In: *Advances in Comparative Survey Methods: Multinational, Multiregional, and Multicultural Contexts (3MC)*, ed. T. Johnson, B.-E. Pennell, I. Stoop, and B. Dorer, 623–641. Hoboken: John Wiley & Sons, Inc.

West, B. T. 2013. The effects of error in paradata on weighting class adjustments: A simulation study. In: *Improving Surveys with Paradata: Analytic Uses of Process Information*, ed. F. Kreuter, 361–388. Hoboken: John Wiley & Sons, Inc.

West, B. T., and A. G. Blom. 2017. Explaining interviewer effects: A research synthesis. *Journal of Survey Statistics and Methodology* 5(2):175–211.

West, B. T., F. Kreuter, and M. Trappmann. 2014. Is the collection of interviewer observations worthwhile in an economic panel survey? New evidence from the German Labor Market and Social Security (PASS) Study. *Journal of Survey Statistics and Methodology* 2(2):159–181.

Section VI

Interview Pace and Behaviors

17

Exploring the Antecedents and Consequences of Interviewer Reading Speed (IRS) at the Question Level

Allyson L. Holbrook, Timothy P. Johnson, Evgenia Kapousouz, and Young Ik Cho

CONTENTS

17.1 Introduction

The pace of interviews (or how quickly interviewers and respondents move through a survey) may vary considerably across interviewers (e.g., Cannell, Miller, and Oksenberg 1981). Researchers have extensively examined the effect of interviewers' pace during survey introductions (which are often unscripted) on participation (e.g., Groves, et al. 2008; Oksenberg and Cannell 1988; Oksenberg, Coleman, and Cannell 1986). Although research investigating the effect of pace within the survey itself is more sparse, there is some evidence that faster pace is associated with lower data quality (Fowler and Mangione 1990; Vandenplas, et al. 2018). Although standardized interviewer training typically includes instructions to read questions slowly based on the assumption that doing so maximizes data quality (e.g., Alcser, et al. 2016; Fowler and Mangione 1990), much research examining interview pace during interviews has operationalized pace using completion times for modules or entire interviews (e.g., Loosveldt and Beullens 2013; Olson and Peytchev 2007; Vandenplas, et al. 2018, Olson and Smyth, Chapter 20, this volume). Overall interview completion time is affected by interviewer reading speed (IRS), but it may also be affected by other factors (e.g., interviewer reading errors and respondent questions).

Our research extends this work to examine the antecedents and consequences of IRS at the question level across a range of questions. Using data from in-person interviews in a laboratory setting, we test hypotheses about the effects of interviewer experience, question length and sensitivity, and location in the questionnaire on IRS. We also examine the effects of other question characteristics on IRS, including degree of question abstraction and response format. Finally, we examine the impact of respondent and interviewer race and ethnicity (interviewers were matched to respondents on race/ethnicity) on IRS.

Although understanding the antecedents of IRS is important, IRS is also of interest to researchers because it may influence the process by which respondents answer questions. We test the impact of IRS on three measures of the respondent's response process: response latencies and respondent behaviors that may indicate comprehension and mapping difficulties.

17.1.1 Background

One of the key goals of standardized interviewing is to minimize interviewer-related sources of measurement error, including the speed with which interviews are completed (Fowler and Mangione 1990; Loosveldt and Beullens 2017). Total completion times vary across interviewers (Couper and Kreuter 2013; Loosveldt and Beullens, 2013; Vandenplas, et al. 2018), but few studies have specifically examined the speed with which interviewers read questions (independent of other behaviors that might affect overall interview length). Interviewer reading speed is of particular concern because reading speeds that exceed typical conversational norms may compromise respondent comprehension and contribute to measurement error. Subsequently, further investigation of interviewer pace, and specifically IRS, is needed.

Researchers have examined the effects of respondent, interviewer, question, and study design features on interview pace. Older respondents and those with fewer cognitive skills tend to have longer interview times (Couper and Kreuter 2013; Loosveldt and Beullens 2013; Olson and Smyth 2015). Employment status and computer use have also been found to be associated with interview time (Olson and Smyth 2015).

Overall, evidence suggests interviewers have a greater influence on pace than respondents (Loosveldt and Beullens 2013). Studies have found that greater interviewer experience (often operationalized as experience interviewing on a particular survey) is associated with faster interview pace (Bergmann and Bristle 2016; Loosveldt and Beullens 2013; Olson and Peytchev 2007, Garbarski, et al., Chapter 18, this volume). When examining pace at the item level, Olson and Smyth (2015) found that greater interviewer experience was associated with faster response times (including the whole interaction between the interviewer and respondent) for yes/no questions, but slower response times for open-ended questions. Other question characteristics are also known to be associated with interview speed, including question length (Couper and Kreuter 2013; Olson and Smyth 2015), question sensitivity, and question type (Olson and Smyth 2015).

There is also some evidence that at least one element of study implementation – piecemeal payment of interviewers, rather than on an hourly basis – increases interview pace (Bergmann and Bristle 2016; Loosveldt and Beullens 2013). Interviewers may be less motivated to carefully follow standardized protocols and have a strong incentive to finish interviews faster when they are paid by the interview than by the hour (Bergmann and Bristle 2016; Cannell 1977), and payment per completed interview may also compromise data quality and lead to falsification.

Faster survey interviews have also been found to be detrimental to survey quality, increasing both straight-lining (Fowler and Mangione 1990; Vandenplas, et al. 2018) and the likelihood of providing "don't know" responses (Vandenplas, et al. 2018). Although empirical research is currently unavailable, common sense suggests that faster interviewer reading of questions may also lead to respondent difficulties with question processing.

17.1.2 Hypotheses

The current research builds on past evidence to examine the antecedents and consequences of IRS specifically. Available findings suggest that interviewer experience is a significant predictor of overall interview length, such that interviews conducted by experienced interviews are shorter (Olson and Peytchev 2007). This leads to our first hypothesis:

> H1: Interviewer experience will be positively associated with IRS.

Interviewer and respondent behavior may also change over the course of an interview. Respondents may become more comfortable with the survey process and interviewers may feel comfortable speeding up in response. Interviewers (and respondents) may become impatient as a long interview progresses and engage in strategies to get through the questions more quickly. For respondents, this may result in behaviors like survey satisficing (Krosnick 1991; Krosnick, et al. 2002). For interviewers, this may be reflected in faster IRS. As a result, we hypothesize that:

> H2: IRS will increase as an interview progresses.

Like any conversation, surveys involve turn taking and there are rules and norms that govern turn taking in order to facilitate communication (e.g., Wiemann and Knapp 1975). The survey interview involves fairly frequent hand-offs in turn taking and relatively short periods of talking from both parties, which also establishes conversation-specific norms for turn taking. Talking for a long time without giving the other person a chance to speak may be seen as violating norms (Wiemann and Knapp 1975). Furthermore, interviewers are trying to build rapport with respondents, and talking for too long may undermine this goal (e.g., Leech 2002). As a result, interviewers may be uncomfortable with longer questions, and may want to get through them quickly so that they can give respondents an opportunity to talk. This leads to our next hypothesis:

> H3: IRS will be positively associated with question length.

We know that respondents experience discomfort when being asked sensitive questions (Krumpal 2013). We argue that interviewers may be as uncomfortable asking sensitive questions as respondents are answering them and they may therefore read these questions more quickly to minimize the length of their discomfort. This leads us to hypothesize that:

> H4: Interviewers will read sensitive questions faster than nonsensitive questions.

Our remaining hypotheses address the potential consequences of interviewer pace. First, we examine whether interviewer pace is associated with response latencies, or the amount of time it takes respondents to answer a question. With regard to interviewer pace, some researchers have argued that interviewers communicate their desired pace of the conversation by the speed at which they speak, and respondents match their speed to the interviewers (e.g., Holbrook, Green, and Krosnick 2003). Consistent with this argument, Fowler (1966) found that faster interviewer pace led respondents to perceive that answering quickly was more important than answering completely or accurately. This argument leads to the following hypothesis:

> H5a: Faster IRS will be associated with shorter response latencies.

However, one could also make the counterargument that faster interviewer pace may increase respondents' cognitive burden and reduce the speed with which they are able to answer. Faster IRS may make the task of listening to and comprehending the survey question more difficult, reducing respondents' capacity to start the process of retrieving relevant information from memory or constructing a response. As a result, faster IRS may slow respondents' answering process, leading to the following competing hypothesis:

H5b: Faster IRS will be associated with longer response latencies.

One of the key reasons why researchers have studied interviewer pace is because it may be associated with data quality. There is some evidence from other contexts that a fast speaking speed can lower a listener's comprehension (e.g., Conrad 1989; Griffiths 1990). Following this logic, we hypothesize that:

H6: Faster IRS will be associated with more comprehension difficulties.

Faster IRS could also increase the extent to which respondents show mapping difficulties, at least in part because it affects comprehension. It could do so by affecting respondent ability to properly recall all of the response categories when forming judgments. Therefore, we hypothesize that:

H7: Faster IRS will be associated with more mapping difficulties.

The data we analyzed included questions that varied along a number of dimensions (see Figure A17.1 in the online supplemental materials). We controlled for these characteristics in our analyses predicting IRS. Because question characteristics can influence the response process (see Tourangeau, Rips, and Rasinski 2000), we also included these variables in our analyses exploring the potential consequences of IRS.

17.2 Methods

Respondents in the survey were 405 adults 18 years or older living in the Chicago metropolitan area, including 103 non-Hispanic Whites, 100 non-Hispanic Blacks, 102 Mexican-Americans, and 100 Korean-Americans. Fifty-two of the Mexican-American respondents were interviewed in English and 50 were interviewed in Spanish. Forty-one of the Korean-American respondents were interviewed in English and 59 were interviewed in Korean.

Interviews took place between August 13, 2008 and April 10, 2010 and were conducted by the Survey Research Laboratory (SRL) of the University of Illinois at Chicago (UIC). Respondents were recruited using Random Digit Dial (RDD) sampling procedures. Recruitment efforts targeted geographic areas with high proportions of residents in the targeted race/ethnic groups, and areas that were accessible to UIC. The oldest/youngest male/female approach was used to randomly select a household member between the ages of 18 and 70. The sample was limited to respondents in the four targeted racial and ethnic groups who lived in Chicago and spoke one of the targeted languages (English for Whites and Blacks, Spanish or English for Mexican-Americans, and Korean or English for Korean-Americans). To recruit Korean-Americans, we also purchased samples of and contacted those in households with Asian and Korean surnames (e.g., Kim), expanded the targeted geographic area to all of Cook County, sent advance letters (in English and Korean) to likely Korean households, and recruited alternative household members if the initially selected respondent was unavailable.

We cannot estimate a standard response rate for this study because we used a quota sampling strategy and discontinued recruitment when the targeted number of cases was reached. In the non-Korean sample, appointments were scheduled for 10.0% of the sample and interviews were completed with 61.1% of those scheduled. Among the Korean sample, these numbers were 1.8% and 77.6%, respectively. Mexican- and Korean-Americans were allowed to choose their interview language; those who were bilingual and had no language preference were randomly assigned to interview language until quotas were met.

Eligible respondents were interviewed at SRL. They completed an initial paper-and-pencil self-interview (PAPI) that included several measures not employed in this analysis, followed by a computer-assisted personal interview (CAPI; average length of 67 minutes), followed by a second PAPI that included respondents' demographic characteristics. Interviews and respondents were matched on race and ethnicity. Bilingual interviewers conducted the interviews with Mexican- and Korean-Americans. Respondents received $40 for participating. All study procedures were approved by the UIC Institutional Review Board.

We measured reading time and response latencies for 150 CAPI questions. The core of the questionnaire included 90 items that varied along a number of dimensions (see Appendix 17C of the online materials for question wording). Questions varied in the type of judgment they asked respondents to make: subjective judgments (e.g., attitudes, beliefs, or values), self-relevant knowledge (behaviors or characteristics), or factual knowledge. Within each of these sets of questions, half of the questions required time-qualified judgments (that required the respondent to recall whether something happened within a specific period of time) and half required judgments that were not time-qualified. The response format of questions was also manipulated. Six different response formats were used: (1) yes/no; (2) categorical; (3) fully labeled unipolar scales; (4) fully labeled bipolar scales with a midpoint; (5) fully labeled bipolar scales without a midpoint; and (6) numerical open ends. The remaining questions included semantic differential scales, feeling thermometer questions about groups in society, agree/disagree questions, questions that explicitly offered or omitted a "no-opinion" option or filter, and questions that asked respondents about the process of answering questions in the survey. A number of questions were deliberately designed to be problematic (as denoted in Appendix 17C of the online supplemental materials).

The CAPI interview was separated into four modules. Half of the respondents were randomly assigned to be asked Modules I and II before Modules III and IV. The other half of respondents were randomly assigned to receive Modules III and IV before Modules I and II. All respondents received Module V questions (demographics) last.

Behavior coding was used to obtain information about the validity of timing measures (see below) and indicators of comprehension and mapping difficulties. Interviews were video and audio recorded if respondents gave permission, and trained staff coded interviewer and respondent behaviors from the video recordings. A total of 398 of the 405 interviews were able to be coded. Validation was done for 77 cases across all three languages, which included 29,281 behavior codes. Coder agreement for these validated cases was high (95.8% overall agreement).

Question reading time was measured as the amount of time it took interviewers to read the question, as captured by a timer that started when the question screen was loaded and ended when interviewers hit "enter" to indicate they finished reading the question and were ready for the response screen. *Response latencies* were measured as the amount of time it took respondents to answer the question, as measured by a timer that started when the response screen was loaded and ended when a response was entered. Response latencies were transformed by taking the square root of this measure. IRS was calculated in words per minute for each question using question length and question reading time. For more information on this time measurement procedure, which was modified from one suggested by Bassili and Fletcher (1991), see Appendix 17B. Respondent and interviewer behaviors rendered both response and question latencies invalid. Figures 17.A2 and 17.A3 show the distribution of invalidated question and response latencies, respectively. Eliminating these questions resulted in a total of 31,520 valid question–answer sequences included in our analyses.

Variables were coded to assess each of our hypotheses about the predictors of IRS. For each interview, *interviewer experience* on the project was calculated as the number of successive interviews (e.g., an interviewer's first interview was coded 1, their second interview was coded 2, and so on). For each question, the *number of previous questions* asked of each respondent was coded. *Question length* was operationalized as the number of words in the question.

Question sensitivity was rated independently by two of the investigators. Each question was coded as not at all sensitive, somewhat sensitive, or very sensitive. Concordance between the two coders was over 80%, and all disagreements were discussed and consensus reached on a final code. Dummy variables were created for somewhat and very sensitive questions (with not at all sensitive questions as the comparison group).

Behavior coding data were used to estimate two variables. A variable for *comprehension difficulty* was coded 1 (versus 0) if behavior coding indicated that a respondent showed any of the behaviors listed in the top panel of Figure 17.1 for a given question. A variable for *mapping difficulty* was coded 1 (versus 0) if behavior coding indicated that a respondent showed any of the behaviors listed in the bottom panel of Figure 17.1 for a given question.

A number of control variables were used to control for question characteristics. A variable to represent whether a question was *time-qualified* or not was coded 1 for questions that requested time-qualified judgments and 0 for those that did not. *Judgment type* was coded using two dummy variables for subjective judgments and factual knowledge (self-relevant knowledge questions were the comparison group). Questions that used a *show card* were coded 1 and those that did not were coded 0. *Question abstraction* was rated independently by two of the investigators as not at all abstract (i.e., about a concrete or physical construct), somewhat abstract (e.g., about objective or physical construct but responding requires some interpretation of terms or ideas; e.g., "corruption in government"), or very abstract (e.g., not objective or grounded in physical world; e.g., "obedience"). Concordance between the two coders was over 80%; all disagreements were discussed and consensus reached on a final code. Dummy variables were created for somewhat and very abstract questions (not at all abstract questions as the comparison group). Several *deliberately problematic questions* were included in the survey instrument including those that: (1) asked about nonexistent policies or objects, (2) had mismatched question stem and response options, (3) had non-mutually exclusive or non-exhaustive response options, (4) requested information that was too specific, and (5) were double barreled. A dummy variable was coded 1 for deliberately problematic questions and 0 for all other questions. Two dummy variables captured the inclusion of an *explicitly offered "no-opinion" response option* and the use of *"no-opinion" filter* questions. For the first variable, each question was given a code of 1 if it explicitly offered respondents a no-opinion response option and a 0 otherwise. For the second variable, each question was coded 1 if it was preceded by a no-opinion filter question and 0 otherwise.

Question comprehension codes

(30)   **Clarification (Unspecified):** Respondent indicates uncertainty about question, but it is unclear as to whether the problem is related to the construct or the context. (e.g., "What is the question asking?" or "What?")

(31)   **Clarification (Construct/Statement):** Respondent makes a statement indicating uncertainty about question meaning. (e.g., "I'm not sure what 'depressed' means.")

(32)   **Clarification (Construct/Question):** Respondent asks for clarification of question meaning. (e.g., "What do you mean by 'depressed'?" or "Depressed?")

(33)   **Clarification (Context):** Respondent indicates s/he understands the meaning of the construct, but indicates uncertainty about question meaning within the context of the question as stated. (e.g., "What do you want to know about being depressed?"; "How often do you pay with cash at restaurants?" Response: "Does that include debit cards?")

(34)   **Clarification (Time frame):** Respondent indicates uncertainty about the question's time frame.

(35)   **Rewording (Question):** Respondent rephrases/repeats the question, or part of the question, before answering. This is not an indication of needing question clarification.

(36)   **Respondent asks for repeat of question:** Respondent asks interviewer to repeat the question, or part of the question. (e.g., "What did you say?", "What was that?", or "Who?" when asking a single word question, such as "Jews?")

(37)   **Difficulty using show card:** Respondent indicates difficulty understanding how to use show card or asks for help in using show card. This should be used instead of 51 if a show card is being used on the question.

(38)   **Clarification (Not enough information):** Respondent indicates that there is not enough information given in the question to answer. (Key phrases include "It depends on the situation.", "It is case by case.", and "I don't have enough information.")

Response mapping codes

(50)   **Inadequate answer (General):** Respondent gives answer that does not meet question objective. Response usually prompts probing. (No other response mapping codes are appropriate.)

(51)   **Clarification (Response format):** Respondent indicates uncertainty about the format for responding. (e.g., "I'm not sure how to answer that.", "What else, is that all you are offering me?", or "Are you asking for a percentage?")

(52)   **Respondent asks for repeat of response options:** Respondent asks interviewer to repeat the response options, or some of the response options, only.

(53)   **Clarification (Response option meaning):** Respondent asks for clarification of a response option meaning. (e.g., "What is the meaning of 'sometimes'?")

(55)   **Rewording (Response options):** Respondent rephrases/repeats the response options, or some of the response options, before answering. This is not an indication of needing response option clarification.

(60)   **Imprecise response (General):** Respondent gives answer that only partially meets question objective. This code is typically used on an open-ended question when the response is not specific enough or is qualified. Response usually prompts probing. (e.g., "Well over 10 times."; "At least twice."; "George Bush is very energetic." when asking about his intelligence and energy; "Five minutes" or "Half an hour." when question asks for hours.)

(61)   **Imprecise response (Different response option):** Respondent gives answer that does not use the response options provided with the question. Response usually prompts probing. (e.g., "Not so good health" instead of excellent, very good, good, fair or poor; "Agree" instead of strongly agree, somewhat agree, somewhat disagree or strongly disagree.)

(62)   **Imprecise response (Inferred answer):** Respondent gives answer that does not use the response options provided for the question. However, sufficient information is provided by the respondent to infer the correct answer. Response can prompt probing or confirmation. (e.g., "Do you know what stage of breast cancer you have?" Response: "Stage 1" rather than "yes."; "Do you favor the death penalty?" Response: "Favor" rather than "yes."; "Uh-huh" rather than "yes.")

(63)   **Imprecise response (Range):** Respondent answers question with a range rather than a single number. Response usually prompts probing.

(64)   **Multiple answers:** Respondent answers question with more than one response categories. Response usually prompts probing. (e.g., "All of the above." or "Both.")

FIGURE 17.1
Respondent behavior codes used to identify comprehension and mapping problems.

Question format was captured with a set of dummy variables for questions with (1) agree/disagree response options, (2) yes/no response options, (3) categorical response options, (4) fully labeled unipolar response scales, (5) fully labeled bipolar response scales with a midpoint, (6) fully labeled bipolar response scales without a midpoint, (7) semantic differentials, and (8) feeling thermometers (numerical open-ended response format was the comparison group).

Respondent demographics were also measured. *Education* was measured using a series of dummy variables for highest degree earned, including less than a high school degree (as the comparison group), high school graduate, some college, college graduate (four-year degree), and college graduate (advanced degree). Respondents' *age* was calculated from reports of birth year and was coded as a continuous variable ranging from 0 (age 18) to 1 (age 70). *Gender* was observed by the interviewer and coded 0 for females and 1 for males. Household *income* was coded into five categories (coded 0 = less than \$10,000; .2 = \$10,001–\$20,000; .4 = \$20,001–\$30,000; .6 = \$30,001–\$50,000; .8 = \$50,001–\$70,000; and 1 = \$70,001 or more). Respondent *race/ethnicity* was obtained via the initial RDD screening and confirmed at the beginning of the CAPI survey. Race/ethnicity was coded using three dummy variables (with non-Hispanic White as the comparison group) to indicate non-Hispanic Black, Mexican-American, and Korean-American. *Language of interviews* was coded using two dummy variables, representing Spanish and Korean interviews. English language interviews were the comparison group.

To examine the antecedents and consequences of IRS, we conducted a series of generalized linear models. The models were fitted with the identity link function [E(*y*) = *x*β, *y* ~ Normal] for continuous outcomes of interviewer pace and response latencies, and the logit link function [logit{E(*y*)} = *x*β, *y* ~ Bernoulli] for dichotomous outcomes of any comprehension and any mapping difficulty. All models were fitted with a cluster-robust sandwich variance estimator (Hardin and Hilbe 2018) to adjust for cross-classified clusters of questions and interviewers in which IRS was nested. Only questions with valid response and question latencies were included (see Online Appendix 17B for more details about this process). All analyses were conducted using Stata 14 (StataCorp 2015) with unweighted data.

17.3 Results

We first examine the antecedents of interviewer pace (see Table 17.1). Model 1 includes the focal independent variables: interviewer experience, number of previous questions, question length, and question sensitivity. IRS became faster as interviewer experience increased (H1; B = .565, SE = .040, p<.001), as the number of previous questions increased (H2; B = .048, SE = .009, p<.001), and as question length increased (H3; B = .879, SE = .042, p<.001). IRS is also significantly faster for somewhat (B = 7.811, SE = 1.291, p<.001) and very sensitive (B = 5.220, SE = 2.202, p = .018) questions than for not at all sensitive questions (H4).

When question characteristics are entered into the model (see Model 2), the substantive conclusions about H1–H3 remain unchanged (see rows 2–4 of Model 2 in Table 17.1). However, the effect of question sensitivity becomes nonsignificant, suggesting that the effect observed in Model 1 is due to these question characteristics (see rows 5 and 6 of Model 2 in Table 17.1).

Some of these other question characteristics are also associated with IRS. Questions that were preceded by a don't know filter question, used a show card, and were deliberately

TABLE 17.1

Predicting Interviewer Reading Speed in Words per Minute (n = 31,520)

Predictor	Model 1: Focal IVs			Model 2: Focal IVs and Question Characteristics			Model 3: Focal IVs, Question Characteristics, and Stratum		
	B	SE	p	B	SE	p	B	SE	p
Intercept	125.942	1.477	<0.001	141.800	2.296	<0.001	162.843	2.100	<0.001
Interviewer experience	0.565	0.040	<0.001	0.582	0.037	<0.001	0.425	0.030	<0.001
Number of previous questions	0.048	0.009	<0.001	0.061	0.008	<0.001	0.056	0.007	<0.001
Number of words in question	0.879	0.042	<0.001	0.992	0.057	<0.001	0.708	0.049	<0.001
Sensitivity (ref: not at all sensitive)									
Somewhat sensitive	7.811	1.291	<0.001	1.145	1.327	0.388	1.452	1.059	0.170
Very sensitive	5.220	2.202	0.018	-0.586	2.556	0.819	0.283	2.012	0.888
Time-qualified judgment				-1.550	1.179	0.189	-1.101	0.993	0.267
Includes an explicit "Don't Know" option				-6.743	4.742	0.155	-4.537	3.939	0.249
Preceded by a don't know filter				-12.371	5.341	0.021	-5.283	7.248	0.466
Used a show card				-21.974	7.905	0.005	-20.010	5.452	<0.001
Deliberately problematic question				-2.238	0.578	<0.001	-2.539	0.527	<0.001
Type of judgment (ref: self-knowledge)									
Subjective (e.g., attitude)				1.061	1.563	0.497	2.708	1.225	0.027
Factual knowledge				-19.697	1.802	<0.001	-19.366	1.490	<0.001
Abstraction (ref: not at all abstract)									
Somewhat abstract				-9.032	1.497	<0.001	-8.597	1.267	<0.001
Very abstract				-10.471	2.065	<0.001	-9.376	1.671	<0.001
Format (ref: open-ended numeric)									
Agree/disagree				-10.505	3.758	0.005	-11.162	2.793	<0.001
Yes/no				-6.596	2.269	0.004	-7.446	1.995	<0.001
Feeling thermometer				13.479	8.547	0.115	7.175	6.056	0.236
Categorical				-14.331	2.333	<0.001	-11.701	1.968	<0.001
Unipolar scale				-2.838	2.189	0.195	-0.748	1.847	0.686
Bipolar scale with a midpoint				-5.230	2.448	0.033	-3.047	1.914	0.111

(Continued)

TABLE 17.1 (CONTINUED)

Predicting Interviewer Reading Speed in Words per Minute (n = 31,520)

Predictor	Model 1: Focal IVs	Model 2: Focal IVs and Question Characteristics		Model 3: Focal IVs, Question Characteristics, and Stratum	
Bipolar scale without a midpoint	−0.023	2.577	0.993	0.397	1.924	0.837
Semantic differential	−26.594	9.194	0.004	−14.523	7.166	0.043
Stratum (ref: non-Hispanic White)						
Korean-American (Asian Ints)				−21.721	1.611	<0.001
Mexican-American (Latino/a Ints)				−0.930	1.424	0.514
Non-Hispanic African American				−14.997	1.257	<0.001
Language (ref: English)						
Spanish				−33.047	1.106	<0.001
Korean				−37.603	1.797	<0.001
AIC	10.110	10.053		9.819		

problematic are associated with slower interviewer pace, as were more abstract questions (relative to those rated as not at all abstract). Factual knowledge questions were read more slowly than those asking about self-relevant knowledge. Interviewer pace also varied across response formats.

Model 3 includes respondent-level control variables of race/ethnicity (using stratum since interviewers and respondents were matched on race/ethnicity) and the language of interview. The inclusion of these variables does not change any of the effects of the focal IVs from those found in Model 2 (see rows 2–6 of Model 4 in Table 17.1). IRS in both the Korean-American and African-American strata is significantly slower than in the non-Hispanic White stratum, but the interviews in the Mexican-American stratum are not (when controlling for language). IRS is slower in the Spanish and Korean language interviews than in the English language interviews.

Next, we examine the consequences of IRS for three measures of the response process: response latencies, comprehension difficulties, and mapping difficulties (controlling for all the variables used as predictors of IRS as well as additional respondent demographic characteristics; see Table 17.2). IRS is negatively and significantly associated with response latencies (B = −.019, SE = .002, p<.001), such that faster IRS was associated with faster responding (H5a). Contrary to H6, IRS is negatively and significantly associated with comprehension difficulties (B = −.004, SE =.001, p<.001). Comprehension difficulties are *less* likely when interviewer reading pace was faster. However, if faster interviewer pace indicates to respondents that it is more important for them to answer quickly than to answer completely or accurately (as is suggested by the findings regarding response latencies), respondents might be less likely to actively engage in the kinds of behaviors that indicate comprehension problems (e.g., requests for clarification). IRS is not associated with mapping difficulties (H7; B = −.002, SE = .001, p = .159).

17.4 Discussion and Conclusions

Very little research has examined IRS at the question level across a broad range of questions, and much prior research has used overall measures of pace (e.g., interview length) that are affected by more than simply IRS (e.g., interviewer errors, respondent requests for clarification, etc.). Our analyses uncovered important relationships that highlight the usefulness of measuring interviewer pace at the question level.

We found a number of theoretically sensible predictors of interviewer pace, including the interviewer's experience, position of the question in the interview, question length, and a number of question characteristics. We also found that interviewer pace was associated with several indicators of the response process: evidence that faster IRS results in shorter response latencies supports the argument that faster interviewer pace may communicate to respondents that it is important to answer quickly, consistent with past research (Fowler 1966; Holbrook, Green, and Krosnick 2003).

Our research, however, has several important limitations. Our study was conducted as an in-person survey in a laboratory setting, which may limit the external validity of our findings (e.g., would our findings replicate to in-person interviews conducted in respondents' homes?). Interviewers were matched to respondents by race/ethnicity, which precluded us from fully examining the effect of interviewer and respondent characteristics.

TABLE 17.2

Consequence of Interviewer Pace (Words per Minute; n = 29,275)

Predictor	Model 1: Response Latencies			Model 2: Comprehension Difficulties			Model 3: Mapping Difficulties		
	B	SE	p	B	SE	p	B	SE	p
Intercept	25.637	0.606	<0.001	-2.829	0.292	<0.001	-1.953	0.309	<0.001
Mapping difficulties				0.585	0.123	<0.001			
Comprehension difficulties							0.530	0.121	<0.001
Words per minute	-0.019	0.002	<0.001	-0.004	0.001	<0.001	-0.002	0.001	0.159
Interviewer experience	-0.019	0.005	<0.001	-0.005	0.002	0.024	-0.015	0.003	<0.001
Number of previous questions	-0.008	0.001	<0.001	-0.001	0.001	0.652	-0.003	0.001	<0.001
Number of words in question	0.022	0.007	0.004	-0.022	0.006	<0.001	0.010	0.004	0.005
Sensitivity (ref: not at all sensitive)									
Somewhat sensitive	-0.067	0.180	0.709	-0.237	0.098	0.016	0.184	0.096	0.055
Very sensitive	-1.393	0.351	<0.001	-0.590	0.173	0.001	0.105	0.193	0.586
Time-qualified judgment	0.764	0.185	<0.001	0.978	0.086	<0.001	-0.338	0.094	<0.001
Includes an explicit "Don't Know" option	-0.071	0.402	0.860	-0.624	0.351	0.076	-1.918	0.519	<0.001
Preceded by a don't know filter	2.601	1.408	0.065	1.194	0.481	0.013	-0.007	0.414	0.987
Used a show card	2.759	1.201	0.022	0.077	1.230	0.95	0.952	1.001	0.341
Intentionally difficult question	1.254	0.191	<0.001	0.120	0.043	0.005	0.149	0.042	<0.001
Type of judgment (ref: self-knowledge)									
Subjective (e.g., attitude)	1.688	0.191	<0.001	0.256	0.121	0.034	0.516	0.104	<0.001
Factual knowledge	3.332	0.298	<0.001	0.743	0.123	<0.001	-0.649	0.135	<0.001
Abstraction (ref: not at all abstract)									
Somewhat abstract	-0.262	0.200	0.191	-0.387	0.100	<0.001	-0.554	0.114	<0.001
Very abstract	0.269	0.226	0.234	-0.386	0.140	0.006	-0.671	0.143	<0.001
Format (ref: open-ended numeric)									
Agree/disagree	-5.046	0.539	<0.001	0.197	0.218	0.366	0.804	0.195	<0.001
Yes/no	-8.833	0.393	<0.001	-0.853	0.115	<0.001	-0.609	0.131	<0.001
Feeling thermometer	-7.270	1.287	<0.001	-0.191	1.243	0.878	-3.939	1.039	<0.001
Categorical	-7.225	0.411	<0.001	-1.066	0.163	<0.001	-1.009	0.170	<0.001

(Continued)

TABLE 17.2 (CONTIUED)

Consequence of Interviewer Pace (Words per Minute; n = 29,275)

Predictor	Model 1: Response Latencies			Model 2: Comprehension Difficulties			Model 3: Mapping Difficulties		
	B	SE	p	B	SE	p	B	SE	p
Unipolar scale	-8.653	0.397	<0.001	-0.657	0.133	<0.001	-1.472	0.152	<0.001
Bipolar scale with a midpoint	-7.868	0.436	<0.001	-1.357	0.182	<0.001	-0.890	0.175	<0.001
Bipolar scale without a midpoint	-6.386	0.434	<0.001	-0.564	0.152	<0.001	-0.203	0.148	0.170
Semantic differential	-6.372	1.057	<0.001	0.101	1.222	0.934	-0.757	1.089	0.487
Stratum (ref: non-Hispanic White)									
Korean-American (Asian Ints)	1.454	0.275	<0.001	0.765	0.114	<0.001	0.593	0.133	<0.001
Mexican-American (Latino/a Ints)	-0.542	0.217	0.013	-0.509	0.132	<0.001	-0.234	0.144	0.103
Non-Hispanic African American	0.289	0.209	0.166	-0.082	0.104	0.428	-0.157	0.123	0.202
Language (ref: English)									
Spanish	-0.142	0.174	0.415	-0.355	0.163	0.029	-0.023	0.143	0.871
Korean	-1.722	0.309	<0.001	-0.659	0.131	<0.001	-0.212	0.158	0.181
Demographic background									
Male	-0.440	0.098	<0.001	0.048	0.050	0.337	-0.059	0.063	0.353
Age	2.151	0.183	<0.001	1.071	0.102	<0.001	0.842	0.119	<0.001
Income	-0.026	0.037	0.480	0.050	0.021	0.017	-0.010	0.024	0.664
Education (ref: <high school)									
High school graduates	0.551	0.193	0.004	0.534	0.153	<0.001	-0.112	0.130	0.388
Some college	0.127	0.211	0.548	0.400	0.162	0.014	-0.045	0.142	0.751
College graduates	0.182	0.227	0.421	0.540	0.174	0.002	-0.373	0.149	0.013
Advanced degree	0.757	0.237	0.001	0.629	0.173	<0.001	-0.513	0.153	0.001
AIC	6.959			0.400			0.300		

In addition, although our approach to analyzing behavior coding, response latencies, and interviewer pace was innovative because it allowed us to separate interviewer reading time from response latencies at the question level, we were only able to conduct these analyses for paradigmatic sequences of interactions between interviewers and respondents. Many nonparadigmatic sequences were excluded from our analyses (see Appendices Figures 17.A2 and 17.A3), and it is unclear whether our results would generalize to those types of interactions.

This research largely supports the current best practices in interviewer training to instruct interviewers to avoid speaking too fast because doing so can communicate to respondents that it is more important to answer quickly than to answer carefully and completely. It also suggests that faster IRS may reduce the extent to which respondents feel like they can ask for clarification or assistance. It suggests that interviewer training might particularly focus on this for sensitive questions and that interviewers be vigilant about not speeding up as the interview progresses. Finally, it suggests that training regarding pace should be part of re-training or continuing training for more experienced interviewers (since those interviewers read at the fastest pace).

There are a number of ways that future research could build on or extend these findings. For example, one direction for future research would be to examine pace in instances where question or response latencies are indicated as invalid using the approach here. This would be labor-intensive, however, as it would require analysis of individual recordings to assess pace when interviews deviate from the standardized script. Another direction for future research could be examining potential moderators or mediators of the effects observed here. For example, it may be that some of the observed relationships hold only for specific types of questions or types of respondents. Alternatively, one could examine more directly whether respondents' perceptions of the importance of answering quickly (versus carefully and accurately) are affected by pace. Future work is needed to assess the association of IRS with other indicators of data quality (e.g., behaviors that indicate survey satisficing). Finally, more work could be done to assess the effect of context on question-level interactions. For example, how does the pattern of interviewer or respondent behaviors affect IRS or respondents' reaction to IRS (e.g., do respondent behaviors in previous questions lead interviewers to read faster or slower?).

The current research is one of the first explorations of the antecedents and consequences of interview pace measured as IRS at the question level across a broad range of question types, including many designed to be deliberately problematic. Our research was also unique in that it involved an examination of questions designed to systematically vary on important dimensions (e.g., judgment type, response format, time qualification, etc.). The approach we used to assess IRS (adapted from procedures initially developed to measure response latencies) was feasible for assessing IRS. Our measure of IRS showed theoretically sensible associations with hypothesized antecedents and consequences of IRS.

Acknowledgments

This work was supported by the National Science Foundation [Grant #0648539] and the National Institutes of Health [R01HD053636-01A1].

References

Alcser, K., J. Clemens, L. Holland, H. Guyer, and M. Hu. 2016. Interviewer recruitment, selection, and training. In: *Guidelines for Best Practice in Cross-Cultural Surveys*. Ann Arbor: Survey Research Center, Institute for Social Research, University of Michigan. http://www.ccsg.isr.umich.edu/ (accessed January 4, 2019).

Bassili, J. N., and J. F. Fletcher. 1991. Response-time measurement in survey research: A method for CATI and a new look at nonattitudes. *Public Opinion Quarterly* 55(3):331–346.

Bergmann, M., and J. Bristle. 2016. Do interviewers' reading behaviors influence survey outcomes? Evidence from a cross-national setting. *SHARE Working Paper Series 23- 2016*. http://www.share-project.org/uploads/tx_sharepublications/WP_Series_23_2016_Bristle_Bergmann_01.pdf (accessed January 12, 2019).

Cannell, C. F. 1977. *A Summary of Studies of Interviewing Methodology*. Rockville: U.S. Dept. of Health, Education, and Welfare, Public Health Service, Health Resources Administration. National Center for Health Statistics.

Cannell, C., P. Miller, and L. Oksenberg. 1981. Research in interviewing techniques. *Sociological Methodology* 12:389–437.

Conrad, L. 1989. The effects of time-compressed speech on native and EFL listening comprehension. *Studies in Second Language Acquisition* 11(1):1–16.

Couper, M. P., and F. Kreuter. 2013. Using paradata to explore item level response times in surveys. *Journal of the Royal Statistical Society: Series A* 176(1):271–286.

Fowler, F. J. 1966. *Education, Interaction, and Interview Performance*. Ph.D. dissertation. University of Michigan.

Fowler, F. J. Jr., and T. W. Mangione. 1990. *Standardized Survey Interviewing: Minimizing Interviewer-Related Error*. Beverly Hills: Sage Publications.

Griffiths, R. T. 1990. Speech rate and NNS comprehension: A preliminary study in time-benefit analysis. *Language Learning* 40(3):311–336.

Groves, R. M., B. C. O'Hare, D. Gould-Smith, J. Benki, P. Maher, and S. Hansen. 2008. Telephone interviewer voice characteristics and the survey participation decision. In: *Telephone Survey Methodology*, ed. J. Lepkowski, C. Tucker, J. M. Brick, E. D., de Leeuw, L. Japec, P. J. Lavrakas, M. W. Link, and R. Sangster, 385–400. New York: John Wiley.

Hardin, J. W., and J. M. Hilbe. 2018. *Generalized Linear Models and Extensions*. 4th ed. College Station: Stata Press.

Holbrook, A. L., M. C. Green, and J. A. Krosnick. 2003. Telephone versus face-to-face interviewing of national probability samples with long questionnaires: Comparisons of respondent satisficing and social desirability response bias. *Public Opinion Quarterly* 67(1):79–125.

Krosnick, J. A. 1991. Response strategies for coping with the cognitive demands of attitude measures in surveys. *Applied Cognitive Psychology* 5(3):213–236.

Krosnick, J. A., A. L. Holbrook, M. K. Berent, R. T. Carson, M. W. Hanemann, R. J. Kopp, R. C. Mitchell, S. Presser, P. A. Ruud, V. K. Smith, W. R. Moody, M. C. Green, and M. Conaway. 2002. The impact of "no opinion" response options on data quality: Non-attitude reduction or an invitation to satisfice? *Public Opinion Quarterly* 66:371–403.

Krumpal, I. 2013. Determinants of social desirability bias in sensitive surveys: A literature review. *Quality and Quantity* 47(4):2025–2047.

Leech, B. L. 2002. Asking questions: Techniques for semistructured interviews. *PS: Political Science and Politics* 35(4):665–668.

Loosveldt, G., and K. Beullens. 2013. The impact of respondents and interviewers on interview speed in face-to-face interviews. *Social Science Research* 42(6):1422–1430.

Loosveldt, G., and K. Beullens. 2017. Interviewer effects on non-differentiation and straightlining in the European Social Survey. *Journal of Official Statistics* 33(2):409–426.

Oksenberg, L., and C. Cannell. 1988. Effects of interviewer vocal characteristics on nonresponse. In: *Telephone Survey Methodology*, ed. R. M. Groves, P. P. Biemer, L. E. Lyberg, J. T. Massey, W. L. Nicholls II, and J. Waksberg, 257–269. New York: John Wiley & Sons.

Oksenberg, L., L. Coleman, and C. F. Cannell. 1986. Interviewers' voices and refusal rates in telephone surveys. *Public Opinion Quarterly* 50(1):97–111.

Olson, K., and A. Peytchev. 2007. Effect of interviewer experience on interview pace and interviewer attitudes. *Public Opinion Quarterly* 71(2):273–286.

Olson, K., and J. D. Smyth. 2015. The Effect of CATI questions, respondents, and interviewers on response time. *Journal of Survey Statistics and Methodology* 3(3):361–396.

StataCorp. 2015. *Stata Statistical Software: Release 14*. College Station: StataCorp.

Tourangeau, R., L. J. Rips, and K. Rasinski. 2000. *The Psychology of Survey Response*. Cambridge: Cambridge University Press.

Vandenplas, C., G. Loosveldt, K. Beullens, and K. Denies. 2018. Are interviewer effects on interview speed related to interviewer effects on straight-lining tendency in the European Social Survey? An interviewer-related analysis. *Journal of Survey Statistics and Methodology* 6(4):516–538.

Wiemann, J. M., and M. L. Knapp. 1975. Turn-taking in conversations. *Journal of Communication* 25(2):75–92.

18

Response Times as an Indicator of Data Quality: Associations with Question, Interviewer, and Respondent Characteristics in a Health Survey of Diverse Respondents

Dana Garbarski, Jennifer Dykema, Nora Cate Schaeffer, and Dorothy Farrar Edwards

CONTENTS

18.1 Introduction

Response time (RT) – the time elapsing from the beginning of question reading for a given question until the start of the next question – is a potentially important indicator of data quality that can be reliably measured for all questions in a computer-administered survey using a latent timer (i.e., triggered automatically by moving on to the next question).* In interviewer-administered surveys, RTs index data quality by capturing the entire length of time spent on a question–answer sequence, including interviewer question-asking behaviors and respondent question-answering behaviors. Consequently, longer RTs may indicate longer processing or interaction on the part of the interviewer, respondent, or both.

RTs are an indirect measure of data quality; they do not directly measure reliability or validity, and we do not directly observe what factors lengthen the administration time. In addition, either too long or too short RTs could signal a problem (Ehlen, Schober, and Conrad 2007). However, studies that link components of RTs (interviewers' question

---

* RTs are distinct from response latencies (RLs). RLs measure time from the end of question reading to the respondent's answer. RLs have been shown to be associated with, for example, response accuracy (Draisma and Dijkstra 2004) and task difficulty (Garbarski, Schaeffer, and Dykema 2011).

reading and response latencies) to interviewer and respondent behaviors that index data quality strengthen the claim that RTs indicate data quality (Bergmann and Bristle 2019; Draisma and Dijkstra 2004; Olson, Smyth, and Kirchner 2019). In general, researchers tend to consider longer RTs as signaling processing problems for the interviewer, respondent, or both (Couper and Kreuter 2013; Olson and Smyth 2015; Yan and Olson 2013; Yan and Tourangeau 2008).

Previous work demonstrates that RTs are associated with various characteristics of interviewers (where applicable), questions, and respondents in web, telephone, and face-to-face interviews (e.g., Couper and Kreuter 2013; Olson and Smyth 2015; Yan and Tourangeau 2008). We replicate and extend this research by examining how RTs are associated with various question characteristics and several established tools for evaluating questions. We also examine whether increased interviewer experience in the study shortens RTs for questions with characteristics that impact the complexity of the interviewer's task (i.e., interviewer instructions and parenthetical phrases). We examine these relationships in the context of a sample of racially diverse respondents who answered questions about participation in medical research and their health.

18.1.1 Response Times and Question Characteristics

Questions vary in many ways, including their structural features (e.g., number of words or clauses), difficulty (e.g., readability level), response format (e.g., yes/no, ordinal rating scale, open response), topic, and content (Dykema, et al. 2019). RTs have been shown to be related to several question characteristics, including question type (e.g., events and behaviors vs. evaluations), question length, response format, inclusion of instructions, presence of ambiguous terms, and use of fully vs. partially labeled response categories (e.g., Couper and Kreuter 2013; Olson and Smyth 2015; Yan and Tourangeau 2008). Studies of RTs and question characteristics are largely based on observational approaches (see review in Dykema, et al. 2019) in which researchers make use of a survey conducted for another purpose, code specific characteristics of the questions in the survey, and examine the association of those characteristics with RTs. The characteristics examined vary across studies as a function of the types of questions available in the questionnaire and researcher interests. Replication across surveys, topics, and populations is critically important, given that many question characteristics are study-specific and collinear (Schaeffer and Dykema forthcoming).

In this chapter, we examine the association between RTs and question characteristics available in our own observational study. Table 18.1 provides the list of question characteristics and hypotheses. We base our hypotheses on relationships demonstrated in previous research and expectations about whether the characteristic is likely to increase the cognitive processing burden of the respondent, interviewer, or both. Some hypotheses are evident; others require explication. See Online Appendix 18A for background and justification regarding H1a–H1l. We formulate hypotheses under the assumption that other question characteristics are held constant.

In addition to the individual or "ad hoc" question characteristics described above, we also examine the association of several established question evaluation tools with RT, including the Flesch–Kincaid grade level, the Question Understanding Aid (QUAID; Graesser, et al. 2006), the Question Appraisal System (QAS; Willis 2005; Willis and Lessler 1999), and the Survey Quality Predictor (SQP; Saris and Gallhofer 2007) (see Online Appendix 18B). Each tool identifies multiple question characteristics that may be problematic for

TABLE 18.1

Hypotheses about the Effect of Question Characteristics on Response Times

Hypothesis	Question Characteristic	Effect on RTs
H1a	Number of words	+
H1b	Question order	−
H1c	Question type	Demographics < events/behaviors < subjective
H1d	Question form	Yes/no < unipolar ordinal, bipolar ordinal, nominal, discrete value
H1e	Definition in the question	+
H1f	List-item question	+
H1g	Sensitive question	−
H1h	Race-related question	+
H1i	Battery structure	First in battery > later; First in series > later
H1j	Emphasis in the question	−
H1k	Interviewer instructions	+
H1l	Parenthetical phrases	−
H2a, 3a	Flesch–Kincaid grade level	+
H2b, 3b	QUAID problem score	+
H2c, 3c	QAS problem score	+
H2d, 3d	SQP quality score	−
H4a	Interaction of number of interviews by interviewer instructions	−
H4b	Interaction of number of interviews by parenthetical phrases	−

Notes: H1a–H1l and H3a–H3d are net of the effects of other question characteristics; H2a–H2d are for bivariate relationships.

respondents or interviewers, and the tools can be used to code questions and characteristics from any type of survey. Although the tools differ in their implementation and scope, they can be used to produce a question-level "problem" or "quality" score that indicates the complexity of the question. We expect that more complex questions (as indicated by scores from the established tools) are associated with longer RTs because they are harder for interviewers to read and harder for respondents to answer (Table 18.1 H2a to H2d; H2d is negative because a higher SQP quality score indicates less complexity). Consistent with expectations, Olson and Smyth (2015) reported that questions with higher reading levels (harder to read) took longer to administer. We are not aware of studies that examine the relationship between the other tools and RTs. (Yan and Tourangeau [2008] examined the relationship between individual question characteristics and QUAID, but they did not include QUAID as a predictor of RT.)

Coding individual question characteristics and generating scores using the established tools is time-consuming and can be costly. Thus, whether the individual characteristics and scores from established tools each independently account for variance in RTs or are duplicative of each other is of interest. We evaluate this by examining whether scores from the established tools predict RTs net of individual question characteristics (H3a to H3d in Table 18.1): although some aspects of the characteristics that are coded to produce these scores overlap with individual question characteristics (e.g., question length), they also incorporate features beyond the individual characteristics with potential implications for RTs.

18.1.2 Response Times and Interviewers' Experience

An important interviewer characteristic to consider in predicting RTs is the interviewer's level of experience. Interviewers appear to increase their pace within an interview (as they gain experience with an individual respondent), within a study (as they gain experience with the particular questionnaire), and across studies (as they become more experienced in general). Their faster speed may be because they develop shortcuts (e.g., alter questions or decrease standardized practices), become more fluent, head-off problems, and so forth (Bergmann and Bristle 2019; Böhme and Stöhr 2014; Holbrook, et al. Chapter 17; Kirchner and Olson 2017; Olson and Peytchev 2007; Olson and Smyth Chapter 20).

In this chapter, we are primarily concerned with within-study experience (i.e., the number of interviews interviewers have conducted). Previous research indicates that the time to complete an entire interview (the aggregate of RTs) and interviewer reading times decrease with the number of interviews completed for a given study (Bergmann and Bristle 2019; Kirchner and Olson 2017; Loosveldt and Beullens 2013; Olson and Peytchev 2007), particularly for inexperienced interviewers (Olson and Peytchev 2007), and accounting for changes in the types of respondents interviewers encounter over the course of the field period (Kirchner and Olson 2017).

We propose that interviewer experience interacts with question characteristics that primarily impact interviewers' task complexity (Olson and Smyth 2015) in predicting RTs because these are the characteristics for which interviewers have the most discretion. In this study, these question characteristics include interviewer instructions and parenthetical phrases. As they complete interviews and become more familiar and comfortable with the questionnaire, we expect interviewers will decrease their attention to and reading of interviewer instructions and be less likely to incorporate discretionary parenthetical phrases. Thus, with increasing interviewer experience (more interviews completed), RTs will decline more rapidly for questions with instructions or parenthetical phrases than without them (H4a and 4b; Table 18.1).

18.2 Data and Methods

Data for this study are from the Voices Heard computer-assisted telephone interview (CATI) survey, which was designed to measure perceptions of barriers and facilitators to participating in medical research studies that collect biomarkers (e.g., saliva and blood) among respondents from various racial and ethnic groups (White, Black, Latino, and American Indian). We employed a quota sampling strategy because screening to identify members in non-White groups would have been prohibitively expensive. The quota sample consisted primarily of volunteers but also used a targeted list of names provided by a commercial vendor (see Online Appendix 18C for more detail). Interviewers conducted 410 usable interviews (in English only) with an average length of 25.21 minutes between October 2013 and March 2014. Respondents received a $20 cash incentive. The 96 questions included in the survey asked about: likelihood to participate in medical research based on the type of study (e.g., to collect tissue) and characteristics of requestor (e.g., "a member of your community"); things medical researchers do to encourage participation (e.g., provide results); concerns about participating in medical research; attitudes toward medical researchers; health status, health-related quality of life, health behaviors and conditions, and health care use; knowledge of research procedures; and social and demographic characteristics.

18.2.1 Measures

Dependent variable. RTs were collected by the CATI computer software as the amount of time (in seconds) spent on each question (mean 13.22 seconds, standard deviation 8.96, range 1–110). Values were top- and bottom-coded at the 99th and 1st percentiles within each item and log-transformed to correct for outliers and skew (Yan and Tourangeau 2008).

Individual question characteristics. Research assistants coded the previously identified individual question characteristics (H1a–H1l in Table 18.1) under the direction of the authors; no interrater reliability statistics were calculated, but codes were verified by the first author. Descriptive statistics for question characteristics are provided in the first column of Table 18.2.

Established tools for evaluating questions. We measured readability using the Flesch–Kincaid grade level. A higher level indicates the question's text is more difficult to read. For QUAID, we tallied the number of problems flagged by the online tool across five comprehension difficulty categories. QAS was coded by a member of the research team and operationalized as a composite sum of the number of problems identified out of 27 possible problems. SQP was coded by an undergraduate research assistant using SQP's online documentation. We use SQP's "quality estimate" (the product of a question's estimated reliability and validity) (see Online Appendix 18B).

Interviewer and respondent characteristics. The key interviewer characteristic of interest is within-study experience (number of interviews the interviewer completed up to the current interview). Other interviewer characteristics included as controls are: race (White, non-White [very few interviewers were Black, Latino, or Asian]), gender, age, and prior interviewing experience (less than one year or one year or more). Respondent characteristics included as controls are: race/ethnicity (Black, Latino, American Indian, and White), gender, age, and education (high school education or less, some college, and college or more). The last two characteristics are used in prior studies to examine or control for factors associated with response processing and cognitive ability (see Online Appendix 18D, Table A18.D1).

18.2.2 Analytic Strategy

The analytic sample includes 410 respondents asked 95 or 96[*] questions by one of 24 interviewers, yielding 39,052 question–answer sequences, which are the unit of analysis. We use cross-classified random-effects linear regression models to predict the log-transformed RTs using Stata 15.1. We use the *mixed* command with restricted maximum likelihood (*reml*) to analyze the data with a variance structure that uses crossed random effects to account for the fact that RT for each question is measured for each respondent and interviewer, and

---

[*] One question was a follow-up to a filter question that was not asked if respondents answered "yes" to the filter question.

TABLE 18.2

Descriptive Statistics and Regression Results of Response Times on Characteristics of Questions, Interviewers, and Respondents, Voices Heard Study

Question Characteristics	Descriptive Statistics				Regression		
	Mean or Percent	Std. Dev.	Min.	Max.	Coef.	Std. Err.	
Number of words	30.47	16.17	5.00	75.00	0.018	0.003	***
Question order	48.50	27.86	1.00	96.00	−0.002	0.002	
Question type							
Event or behavior (reference category)	57.3%						
Subjective	28.1%				0.211	0.147	
Demographic	14.6%				0.231	0.155	
Question form							
Yes/no (reference category)	30.2%						
Nominal	8.3%				0.208	0.123	
Discrete value	2.1%				0.328	0.188	
Bipolar ordinal	16.7%				1.067	0.170	***
Unipolar ordinal	42.7%				0.600	0.138	***
Definition in the question (vs. not)	5.2%				0.079	0.159	
List-item question (vs. not)	35.4%				0.027	0.063	
Sensitive question (vs. not)	10.4%				0.073	0.088	
Race-related question (vs. not)	9.4%				−0.017	0.100	
Battery structure							
First in battery	9.4%				0.102	0.114	
Later in battery (reference category)	44.8%						
First in series	6.3%				0.275	0.127	*
Later in series	31.3%				0.165	0.105	
Stand-alone	8.3%				0.242	0.139	
Emphasis in the question (vs. not)	19.8%				−0.316	0.102	**
Interviewer instructions (vs. not)	9.4%				0.201	0.112	
Parenthetical phrases (vs. not)	34.4%				−0.336	0.082	***
Flesch–Kincaid grade level	12.22	5.16	0.00	22.10	0.018	0.008	*
QUAID problem score	4.38	2.30	1.00	12.00	0.012	0.014	
QAS problem score	1.00	1.02	0.00	4.00	−0.033	0.045	
SQP quality score	0.50	0.05	0.44	0.67	−0.319	0.720	
Intercept					1.166	0.443	**
Random-effects parameters							
Interviewer-level variance					0.003	0.001	*
Question-level variance					0.045	0.007	***
Respondent-level variance					0.012	0.001	***
Residual variance					0.085	0.001	***
Wald chi-square					693.83	(df 35)	***
Log-restricted likelihood					−8,268.60		

Notes: Std. Dev. = standard deviation, Min. = minimum, Max. = maximum, Coef. = coefficient, Std. Err. = standard error. Descriptive statistics are calculated at the level of the question (N = 96) for question characteristics. Regression analysis is conducted at the level of the question–answer sequence (N = 39,052). Regression model also controls for respondent (race/ethnicity, gender, age, and education; N = 410) and interviewer characteristics (race/ethnicity, gender, age, prior interviewing experience, and study-specific experience; N = 24).

*p < 0.05, **p < 0.01, ***p < 0.001.

respondents are nested within interviewers. The base model predicting RT i for question j_1, respondent j_2, and interviewer k is $\ln(\text{Response time})_{i(j_1,j_2)k} = \beta_0 + u_{j_1} + u_{j_2} + v_k + e_{i(j_1,j_2)k}$. In this model, $u_{j_1} \sim N(0, \sigma^2_{u(1)})$, $u_{j_2} \sim N(0, \sigma^2_{u(2)})$, $v_k \sim N(0, \sigma^2_v)$, and $e_{i(j_1,j_2)k} \sim N(0, \sigma^2_e))$.

The full model predicting RT includes a series of fixed effects for questions, respondents, and interviewers:

$$\ln(\text{Response time})_{i(j_1,j_2)k} = \beta_0 + \sum_{b=1}^{B} \beta_b \text{Question characteristics}_{j_1 k}$$

$$+ \sum_{c=1}^{C} \beta_c \text{Respondent characteristics}_{j_2 k} + \sum_{d=1}^{D} \beta_d \text{Interviewer characteristics}_k$$

$$+ u_{j_1} + u_{j_2} + v_k + e_{i(j_1,j_2)k}$$

Because RTs are (natural) log-transformed, the coefficients can be interpreted in terms of percentage change, such that RTs change by 100 * [exp(β) − 1] percent for a one-unit increase in the independent variable, holding all other variables in the model constant.

18.3 Results

Table 18.2 presents a full model that regresses RTs on characteristics of questions, interviewers, and respondents (see Online Appendix 18D, Table A18.D2 for results from the partial models). Several of the significant effects of individual question characteristics align with our expectations (Table 18.1), net of the other characteristics. Each additional word in the question is associated with a 1.8% increase (i.e., 100 * [exp(.018) − 1]) in RTs (Table 18.2), consistent with H1a. Increasing question order is associated with a decrease in RT when the model does not control for scores from established tools for evaluating questions (Online Appendix 18D, Table A18.D2, Model 1), but this effect is not significant in the full model (Table 18.2), so H1b is not supported in the full model. Questions that have bipolar or unipolar ordered categories have longer RTs than yes/no questions (the reference group), but nominal and discrete-value questions are not significantly different from yes/no questions, so H1d is partially supported.* We find no evidence supporting H1c (question type), H1e (definition), H1f (list item), H1g (sensitive), H1h (race-related), and H1i (battery structure). However, when the question includes emphasis (i.e., bolded text), RTs are shorter, consistent with the expectation that emphasis aids in respondents' processing efficiency (H1j).

The hypotheses focused on question characteristics that impact the complexity of the interviewer's task are partially supported. The presence of an interviewer instruction is associated with increased RTs in the model that examines individual question characteristics

---

* The discrete-value questions ask respondents to report numerical answers: year of birth and number of days they drank alcohol in the past month. RTs are lower for all question forms compared to bipolar ordinal questions (p<.001) and lower for nominal questions compared to unipolar ordinal questions (p<.01) (not shown).

(Online Appendix 18D, Table A18.D2, Model 1), but it is not significant when controlling for scores from established tools for evaluating questions (Table 18.2), so H1k is not supported in the full model. The presence of a parenthetical phrase is also associated with decreased RTs, consistent with H1l and the expectation that, on average, interviewers read parenthetical phrases only when deemed necessary rather than with every question administration (Olson, Smyth, and Kirchner 2019).

When we examine the association between RTs and scores from the established tools to evaluate survey questions – important because investigators might only use one measure – we find that Flesch–Kincaid grade level (H2a) and QUAID problem score (H2b) are each positively associated with RTs (Online Appendix 18D, Table A18.D2, Models 2 and 3). Thus, hypotheses concerning bivariate relationships are supported for Flesch–Kincaid grade level and QUAID problem score, but not for QAS problem score or SQP quality score. When the individual question characteristics are included in the model with the established tools (Table 18.2), the effect of QUAID problem score is attenuated and not statistically significant, while the effect of grade level is attenuated but still significant (each additional grade level is associated with a $100 * [\exp(.018) − 1] = 1.8\%$ increase in RTs). Thus, H3 is only supported for the Flesch–Kincaid grade reading level (H3a).

The full model reduces the question-level variance relative to the base model by 87%, while the model with the individual question characteristics (that is, without the established tools) reduces the question-level variance by 86% (Online Appendix 18D, Table A18.D3). Thus, most of the question-level variation in RTs in these data is explained by this set of individual question characteristics. In the model that controls for the characteristics of respondents and interviewers, RTs are significantly different and longer for women, older respondents, and Latino respondents compared to other racial and ethnic groups (Online Appendix 18D, Table A18.D1); these effects remain largely unchanged in the models that also control for the question characteristics and established tools for evaluating questions (not shown).

Next we turn to our hypotheses that RTs will decrease faster with an increasing number of interviews completed in questions with interviewer instructions or parenthetical phrases than in questions without these characteristics. As predicted, there is a significant negative interaction of number of interviews completed with interviewer instructions and parenthetical phrases (results available upon request). Figures 18.1 and 18.2 show the predicted marginal means of log-transformed RTs with increasing numbers of interviews completed for questions with and without interviewer instructions and parenthetical phrases. Questions with interviewer instructions have longer RTs than those without interviewer instructions, but, as interviewers conduct more interviews, RTs decrease more rapidly for questions with instructions than without (Figure 18.1). The interaction effect is significant ($p<.05$) overall, yet its significance varies across the span of number of interviews completed: the marginal effect of having interviewer instructions as part of the question (vs. not) is statistically significant ($p = .046$) with the first interview and drops below statistical significance ($p = .052$) by the fourth interview (not shown). Thus, H4a is supported – for the first few interviews. Figure 18.2 shows that RTs are shorter, on average, when questions include parenthetical phrases compared to when they do not, and although RTs decrease for questions with and without parenthetical phrases over the number of interviews completed, the slope is steeper for questions that contain parenthetical phrases. The marginal effect is significant across the span of number of interviews completed (not shown). Thus, H4b is supported.

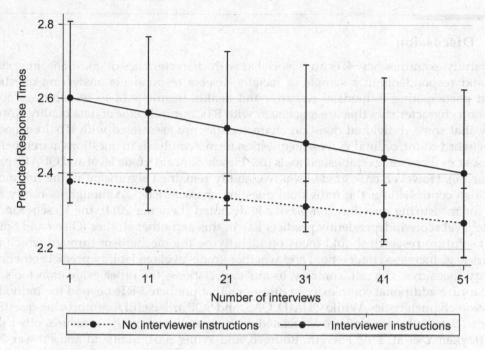

FIGURE 18.1
Predicted marginal means of (log-transformed) response times by the number of interviews completed and interviewer instructions in the question.

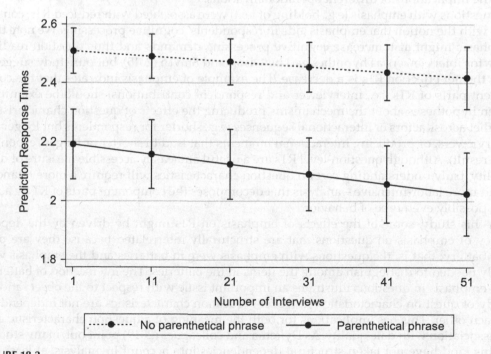

FIGURE 18.2
Predicted marginal means of (log-transformed) response times by the number of interviews completed and parenthetical phrases in the question.

18.4 Discussion

This study examines how RTs are associated with characteristics of questions, interviewers, and respondents in a sample of racially diverse respondents answering questions about participating in medical research and health. Results add to the findings about question characteristics that are associated with RTs, our indicator of data quality. Results show that some individual question characteristics are associated with RTs in expected ways (word count, ordinal vs. yes/no question forms, emphasis in questions, parenthetical phrases), as were some established tools (i.e., Flesch–Kincaid grade level and QUAID problem score). However, only grade-level readability remained significant in the full model that also controlled for the individual question characteristics. Although its utility as a tool for measuring question complexity is disputed (Lenzner 2014), the Flesch–Kincaid grade level score independently predicts RTs in this and other studies (Olson and Smyth 2015) – future research should focus on identifying the mechanism through which this occurs (e.g., increased interaction) and whether grade-level readability predicts other data quality measures. Overall, contrary to our expectations, the other evaluation tools did not capture additional complexity in questions that predicted RTs beyond the individual question characteristics. While QUAID, QAS, and SQP are useful for improving questions prior to data collection and are associated with several data quality measures other than RTs (Dykema, et al. 2019; Forsyth, Rothgeb, and Willis 2004; Maitland and Presser 2016; Olson 2010; Olson, Smyth, and Kirchner 2019; van der Zouwen and Smit 2004), they did not contribute to explaining variation in RT in this study. The methods researchers use to operationalize scores from these tools vary across studies. Future research should examine the implications of different operationalizations.

Questions with emphasis (e.g., bolding of text) were associated with reduced RTs, consistent with the notion that emphasis aids in respondents' cognitive processing. We note that emphasis might also increase cognitive processing demands and thus question reading time for interviewers as hypothesized by Olson and Smyth (2015), but our study suggests that the net effect on RTs is a decrease. The example of emphasis indicates that the component parts of RTs (i.e., interviewer and respondent contributions) should be examined when hypotheses about the mechanisms producing the effects of question characteristics conflict across actors or interactional sequences (e.g., shorter for respondents but longer for interviewers, or producing interactional moments that lead to shorter or longer responses as a result). Although question-level RTs are a useful and easily accessible measure of data quality, truly understanding certain question characteristics will require a more nuanced – and more labor-intensive – analysis that decomposes the component parts of RT by actor and possibly even type of behavior.

In this study, some of the effects of emphasis on RTs might be driven by the dependency of emphasis on questions that are structurally interrelated because they are part of a battery; that is, the questions with emphasis were in batteries and the emphasis was likely needed to distinguish among the items in the batteries. The intersection of batteries and emphasis in questions illustrates an important issue with respect to the observational study of question characteristics, however – question characteristics are not independent of each other. This has implications for both the meaning of a question characteristic and its association with data quality. As Dykema and colleagues (2019) point out, many studies of this kind have not taken structural dependencies into account in analysis, at least not systematically; that is the case in this study as well. In observational studies in particular, the joint distribution of question characteristics affects whether group sizes are sufficient to estimate main effects and interactions (Dykema, et al. 2016). The results of these types of studies may depend on which combinations of characteristics were accounted for in each particular analysis, which may contribute to the lack of replicability in the effects of certain question characteristics (individual characteristics or established tools that combine multiple characteristics into more comprehensive scores) across studies, complicating findings of which question characteristics are better predictors of data quality. Future research should include more study replications under different survey conditions and experimental designs to parse the dependencies where possible, especially for those characteristics for which existing findings are the least consistent.

Overall, interviewers' experience within the survey (i.e., number of interviews completed) is not associated with RTs. This may be due to the telephone mode, which is more monitored compared to face-to-face modes (Kirchner and Olson 2017). However, we found that interviewers' experience in the study interacted with key question characteristics – interviewer instructions and parenthetical phrases – that are used at or attended to with the interviewer's discretion. Specifically, questions with interviewer instructions have longer RTs for interviewers with fewer interviews and the slope decreases more rapidly for questions with interviewer instructions compared to questions without. However, the difference in slope is only significant for the first three interviews, indicating that interviewers no longer read or attend to these instructions after the first few interviews and may apply them from memory. In contrast, the inclusion of parenthetical phrases in questions served to decrease RTs at a significantly steeper rate with more interviews completed. This relationship may indicate that interviewers are treating parenthetical phrases as optional within interviews (as the unconditional effect of parentheticals indicated), and increasingly so as they complete more interviews during the study, with the result of lowering average RTs over time (that is, with more interviews completed).

If interviewers treat the parenthetical phrases as optional during question reading, this has implications for standardized survey administration, as all respondents are not hearing the same question. It raises the question of whether the parenthetical or non-parenthetical version of the item counts as the scripted administration (e.g., Olson, Smyth, and Kirchner 2019). However, as interviewers complete more interviews, they may be more adept at learning and facilitating small micro-adjustments to question asking, such as whether and when to omit parenthetical phrases or attend to instructions, that are aligned with the goals of standardization and keeping with conversational practices and maintaining rapport – if not necessarily aligned with the rules of standardization (Garbarski, Schaeffer, and Dykema 2016). This speaks to the notion of interviewers as pragmatists who work to complete the interview and learn the complexity of the interview task over time (Paul Beatty 2019, personal communication). As we do for respondents, future research must consider visual design (Dillman, Smyth, and Christian 2014) as integral for interviewers in terms of whether and how they read and attend to parts of the instrument; at least with interviewers involved, we can train on attention to various cues and retrain if standardized practices diminish over time. With respect to parenthetical phrases in particular, the evidence here and elsewhere is becoming clear: they are associated with indicators of lower data quality (Dykema, et al. 2016, 2019; Olson, et al. 2019).

With regard to limitations of the study, we note that RT is an indirect measure of data quality: we can presume that interviewers are choosing to not read parenthetical phrases, and increasingly so as they complete more interviews, but we are not directly observing behavior in this study. The strength of a measure of RT is that it is low-cost and easily obtained for every question–answer sequence in the data (with the correct programming capabilities to capture it). In terms of its validity, however, there is more work to be done to examine what actually underlies RT as a measure of data quality. For example, behavior coding could be used to examine what behaviors are associated with longer or shorter RTs.

Such studies would lend more credibility to using RTs as an indirect measure of data quality in more studies. Indeed, the results of any study depend on the quality of the criteria at hand, both the dependent and independent variables.

An additional limitation is that respondents were not recruited randomly due to cost and feasibility constraints, which limits the generalizability of our sample to a larger population of respondents. As with other observational studies of question characteristics, this issue exists at the level of the question (characteristics are not randomly sampled from the universe of all question characteristics but rather fit for the purpose of a given study) as well as at the level of the interviewer (who are employees at one particular survey organization, and not randomly assigned to cases but rather assigned due to proximity, shift, and so forth). Another factor that would be useful to know is respondents' status as English language speakers (e.g., is it their primary language spoken), for which we do not have information.

This research advances the field of survey methodology by examining RTs in the context of different question characteristics and established tools for evaluating questions as well as specific interviewer characteristics within a uniquely diverse sample of respondents. The results have direct implications for survey measurement, questionnaire design, interviewing methods, and interviewer training. The results expand our understanding of the joint influence of characteristics of questions, interviewers, and respondents – the first two of which may be modifiable in the course of survey research – and their application to the development of practical methods for improving the quality of survey data.

Acknowledgments

The collection of survey data for the Voices Heard Survey was funded by NIMHD grant P60MD003428 (PD: A. Adams). Project: Increasing Participation of Underrepresented Minorities in Biomarker Research (PI: D. Farrar Edwards). This study is based upon work supported by the National Science Foundation (grant number SES-1853094 to J. Dykema and D. Garbarski]. Project: Effects of Interviewers, Respondents, and Questions on Survey Measurement. Additional support was provided by the University of Wisconsin Survey Center (UWSC), which receives support from the College of Letters and Science at the University of Wisconsin-Madison; the facilities of the Social Science Computing Cooperative and the Center for Demography and Ecology (NICHD core grant P2C HD047873) at the University of Wisconsin-Madison; and the Graduate School and the Gannon Center for Women and Leadership at Loyola University Chicago. The authors thank the editors of this volume for their helpful comments on earlier drafts. Opinions expressed here are those of the authors and do not necessarily reflect those of the sponsors or related organizations.

References

Bergmann, M., and J. Bristle. 2019. Reading fast, reading slow: The effect of interviewers' speed in reading introductory texts on response behavior. *Journal of Survey Statistics and Methodology*. Advanced Access.

Böhme, M., and T. Stöhr. 2014. Household interview duration analysis in CAPI survey management. *Field Methods* 26(4):390–405.

Couper, M. P., and F. Kreuter. 2013. Using paradata to explore item level RTs in surveys. *Journal of the Royal Statistical Society: Series A, (Statistics in Society)* 176(1):271–286.

Dillman, D. A., J. D. Smyth, and L. M. Christian. 2014. *Internet, Phone, Mail, and Mixed-Mode Surveys: The Tailored Design Method*, 4th edition. Hoboken, NJ: Wiley & Sons Inc.

Draisma, S., and W. Dijkstra. 2004. Response latency and (para)linguistic expression as indicators of response error. In: *Methods for Testing and Evaluating Survey Questionnaires*, ed. S. Presser, J. M. Rothgeb, M. P. Couper, J. T. Lessler, E. Martin, J. Martin, and E. Singer, 131–148. New York: Springer-Verlag.

Dykema, J., N. C. Schaeffer, D. Garbarski, and M. Hout. 2019. The role of question characteristics in designing and evaluating survey questions. In: *Advances in Questionnaire Design, Development, Evaluation, and Testing*, ed. P. Beatty, D. Collins, L. Kaye, J. Padilla, G. Willis, and A. Wilmot, 119–152. Hoboken, NJ: Wiley.

Dykema, J., N. C. Schaeffer, D. Garbarski, E. V. Nordheim, M. Banghart, and K. Cyffka. 2016. The impact of parenthetical phrases on interviewers' and respondents' processing of survey questions. *Survey Practice* 9(2). https://www.surveypractice.org/article/2817-the-impact-of-parenthetical-phrases-on-interviewers-and-respondents-processing-of-survey-questions.

Ehlen, P., M. F. Schober, and F. G. Conrad. 2007. Modeling speech disfluency to predict conceptual misalignment in speech survey interfaces. *Discourse Processes* 44(3):245–265.

Forsyth, B., J. M. Rothgeb, and G. B. Willis. 2004. Does pretesting make a difference? An experimental test. In: *Methods for Testing and Evaluating Survey Questionnaires*, ed. S. Presser, J. M. Rothgeb, M. P. Couper, J. T. Lessler, E. Martin, J. Martin, and E. Singer, 525–546. New York: Springer-Verlag.

Garbarski, D., N. C. Schaeffer, and J. Dykema. 2011. Are interactional behaviors exhibited when the self-reported health question is asked associated with health status? *Social Science Research* 40(4):1025–1036.

Garbarski, D., N. C. Schaeffer, and J. Dykema. 2016. Interviewing practices, conversational practices, and rapport: Responsiveness and engagement in the standardized survey interview. *Sociological Methodology* 46(1):1–38.

Graesser, A. C., Z. Cai, M. M. Louwerse, and F. Daniel. 2006. Question Understanding AID (QUAID): A web facility that tests question comprehensibility. *Public Opinion Quarterly* 70(1):3–22.

Kirchner, A., and K. Olson. 2017. Examining changes of interview length over the course of the field period. *Journal of Survey Statistics and Methodology* 5:84–108.

Lenzner, T. 2014. Are readability formulas valid tools for assessing survey question difficulty? *Sociological Methods and Research* 43(4):677–698.

Loosveldt, G., and K. Beullens. 2013. The impact of respondents and interviewers on interview speed in face-to-face interviews. *Social Science Research* 42(6):1422–1430.

Maitland, A., and S. Presser. 2016. How accurately do different evaluation methods predict the reliability of survey questions? *Journal of Survey Statistics and Methodology* 4(3):362–381.

Olson, K. 2010. An examination of questionnaire evaluation by expert reviewers. *Field Methods* 22(4):295–318.

Olson, K., and A. Peytchev. 2007. Effect of interviewer experience on interview pace and interviewer attitudes. *Public Opinion Quarterly* 71(2):273–286.

Olson, K., and J. D. Smyth. 2015. The effect of CATI questions, respondents, and interviewers on RT. *Journal of Survey Statistics and Methodology* 3(3):361–396.

Olson, K., J. D. Smyth, and A. Kirchner. 2019. The effect of question characteristics on question reading behaviors in telephone surveys. *Journal of Survey Statistics and Methodology*. Advanced Access.

Saris, W. E., and I. N. Gallhofer. 2007. *Design, Evaluation, and Analysis of Questionnaires for Survey Research*. New York: Wiley.

Schaeffer, N. C., and J. Dykema. Forthcoming. Advances in the science of asking questions. *Annual Review of Sociology*.

van der Zouwen, J., and J. H. Smit. 2004. Evaluating survey questions by analyzing patterns of behavior codes and question-answer sequences: A diagnostic approach. In: *Methods for Testing and Evaluating Survey Questionnaires*, ed. S. Presser, J. M. Rothgeb, M. P. Couper, J. T. Lessler, E. Martin, J. Martin, and E. Singer, 109–130. New York: Wiley.

Willis, G. B. 2005. *Cognitive Interviewing: A Tool for Improving Questionnaire Design*. Thousand Oaks, CA: Sage.

Willis, G. B., and J. T. Lessler. 1999. *Question Appraisal System: QAS-99*. Rockville, MD: National Cancer Institute.

Yan, T., and K. Olson. 2013. Analyzing paradata to investigate measurement error. In: *Improving Surveys with Paradata: Analytic Uses of Process Information*, ed. F. Kreuter, 73–96. Hoboken, NJ: John Wiley & Sons.

Yan, T., and R. Tourangeau. 2008. Fast times and easy questions: The effects of age, experience and question complexity on web survey RTs. *Applied Cognitive Psychology* 22(1):51–68.

19

Accuracy and Utility of Using Paradata to Detect Question-Reading Deviations

Jennifer Kelley

CONTENTS

19.1 Introduction

Words matter, especially in survey research. Interviewer deviations from question word-ing can change question meaning, thus undermining validity (Groves, et al. 2009; Krosnick, Malhotra, and Mittal 2014; Schuman and Presser 1996). For this reason, interviewers are instructed to read questions exactly as worded in standardized interviews. However, question-reading deviation estimates in telephone interviews range from a low of 4.6% (Mathiowetz and Cannell 1980) to a high of 36% (Cannell, Lawson, and Hausser 1975), and in face-to-face interviews, they can be as high as 84% (Ackermann-Piek and Massing 2014).

Given the importance of reading questions verbatim, and variability of interviewers' question-reading behavior, monitoring interviewers' behavior is arguably one of the most important procedures for quality control. In telephone interviews, live monitoring is often used, but in face-to-face interviews, monitoring is typically done by listening to interview recordings. Listening to interviews is resource-intensive and not always feasible (e.g., some interviews cannot be recorded because of technical limitations), and thus some organiza-tions are looking for alternative tools to make existing quality control methods more effi-cient. One such tool is paradata.

Most modern survey software can easily and cheaply capture survey *process* data, or paradata, throughout the survey life cycle. Organizations are looking for ways to lever-age these data to reduce costs, increase efficiency, and improve data quality. For example, time stamps have long been used to calculate interview length and detect respondent comprehension issues with individual questions, but they are also increasingly being used to monitor interviewers' question-reading behavior (Mneimneh, et al. 2014; Mneimneh, et al. 2018; Sun and Meng 2014).

To monitor interviewers' question-reading behavior using timing data, organizations estimate the *expected* question administration time to establish minimum or maximum (or both) question administration time thresholds (QATTs). The question duration is compared to the QATTs to identify and flag for further review questions that violate the thresholds or interviewers with high rates of such questions. Violations of minimum QATTs may indicate that interviewers omitted words from the question text, while violations of maximum QATTs may indicate that they added words.* This method gives organizations a tool for identifying *likely* problematic interviewers who should be the primary focus of monitoring efforts, thus saving time and money.

There are two known published methods of creating minimum QATTs: (1) the words per second (WPS) method, which consists of dividing the number of words in the question by a specified reading pace (e.g., ten words in the question divided by two words per second yields a cut point of 5 seconds – Sun and Meng 2014); and (2) using an *a priori* cutoff, such as one second (Mneimneh, et al. 2014). Little is known about how well these methods detect actual question misreadings. These studies also use only minimum QATTs, ignoring question readings that might be too long, such as if interviewers add words to the question text. There may be more accurate methods, such as using a WPS range or deriving QATTs from standard deviations of the mean question-reading times, so that both a minimum and maximum QATT can be used to identify questions read "too fast" or "too slow."

This study will take advantage of a unique data set from Wave 3 of the Understanding Society Innovation Panel that includes question timing paradata and behavior codes identifying misread questions to test how well QATT methods detect actual interviewer question-reading deviations. Three methods of developing QATTs are tested: (1) WPS point estimates (minimum QATT only), (2) WPS ranges (minimum and maximum QATTs), and (3) standard deviations of mean question-reading times (minimum and maximum QATTs). Because there is no compelling evidence on which cutoff value is most accurate for a said method, various threshold values will be tested for each method: (1) Words per Second: Point Estimate Method: 2 WPS; 3 WPS; 4 WPS; (2) Words per Second: Range Method: 2–3 WPS; 1–3 WPS; 2–4 WPS; 1–4 WPS; (3) Standard Deviation Method: 0.5; 1.0; 1.5; 2.0.

19.2 Background

19.2.1 Interviewers' Behavior and Measurement Error

Interviewer behavior is known to affect measurement error (Axinn 1991; Groves, et al. 2009). To reduce interviewer effects, organizations train standardized interviewers to read questions verbatim as past research shows that deviations in question wording can change the meaning of the question (Groves, et al. 2009; Krosnick, Malhotra, and Mittal, 2014; Schuman and Presser 1996). Yet interviewers still deviate from the scripted question wording in many ways. They may make simple errors – such as substituting "the" for "a" – or

---

* Interviewers also substitute words, but substitutions are not likely to affect the question administration timing.

they may intentionally change wording because they think they are helping respondents comprehend questions (Schober and Conrad 2002). Some interviewers may tailor the question to the respondent to signal they are listening to the respondent's previous answers (Dijkstra and Ongena 2006). In more extreme cases, interviewers may shorten questions intentionally by omitting words or skip questions to possibly shorten the interview.

To try to discourage intentional question misreadings, organizations traditionally monitor interviewers' question-reading behavior in face-to-face interviews by listening to interview recordings. Many organizations listen to the first few recordings for each interviewer in their entirety and then randomly select later interviews to be listened to, in entirety or in part, mostly due to resource limitations (Thissen and Myers 2016; Viterna and Maynard 2002). Using this method may not be as efficient as it could be because the method is applied equally to all interviewers rather than using existing information to focus on the interviewers who are most likely problematic. Additionally, this method prioritizes early cases over later cases despite research showing that as interviewers gain experience with the survey, interview duration decreases (Couper and Kreuter 2013; Olson and Peytchev 2007). The shorter durations may be due to interviewers becoming more efficient with experience or due to increased use of nonstandardized interview behaviors (Bradburn, et al. 1979; Fowler and Mangione 1990; Olson and Smyth, Chapter 20, this volume). Further, some surveys do not record interviews or only record certain questions or sections because of software or hardware limitations or because the interview contains sensitive questions. Finding an alternative, less resource-intensive quality control tool to use instead of monitoring when recordings are unavailable or to help identify the most likely problematic interviewers to make monitoring efforts more efficient would benefit survey organizations.

Paradata, specifically question-level time stamps, may be used to develop such a tool. Question durations, computed from time stamps, are used as proxies to indicate problems with question administration (Mneimneh, et al. 2014; Sun and Meng 2014). The theory is that too short of a duration may indicate omitting words, paraphrasing or skipping questions entirely, and too long of a duration may indicate adding words. However, question durations generally encompass all question-asking and question-answering activity, including both interviewer and respondents' contributions. Hence, organizations use irregular question durations to flag cases as possible question-reading deviations, but also acknowledge some of the irregularity could be attributed to respondents' behavior and thus conduct a further investigation. The investigations may correctly identify interviewer falsification, but they could also detect interviewers who need more training in standardized interviewing techniques.

While the above-cited studies are using QATTs in an attempt to increase efficiency in quality control, their use is not widespread and little is known about the accuracy and utility of said methods. Furthermore, for studies that do not audio-record interviews, having an automated flagging system is arguably even more crucial, as irregular question durations may be the only way to detect when interviewers are not following protocol or who need more training.

There are several possible methods for determining QATTs. The first method is by setting WPS point estimates. For example, Mneimneh, Pennell, Lin, and Kelley (2014) flagged questions read under one second in an attempt to detect skipped questions, but did not report the accuracy of this method. Sun and Meng (2014) created minimum QATTs using the number of words in the question divided by a reading pace of 110 milliseconds per Chinese character to flag questions as possibly containing deviations. The authors report the rate of true positive detections (i.e., the method correctly identifies deviations), but not false negatives or the overall accuracy of their method.

However, there is no known literature on which WPS rate to use for creating QATTs. Should the rate be set at the speech rate the interviewers are instructed to use? Some organizations instruct interviewers to read questions at 2–3 WPS, while others instruct interviewers to read at a normal conversation pace (Viterna and Maynard 2002). Or should the rate be determined by comprehension rates? Normal conversation rates can go as high as 250 words per minute (WPM) or 4.1 WPS (Foulke 1968), but listeners' comprehension starts to decrease at 212 WPM or 3.5 WPS (Omoigui, et al. 1999). Given the variability of interviewers' question-reading pace and the natural variability of speech rates in normal conversation, it is prudent to test different WPS point estimate values for developing QATTs.

Using a minimum single point estimate as in the WPS point estimate method may not capture reading deviations where interviewers add words to a question. Thus, a second QATTs method is to establish a WPS range that designates minimum and maximum QATTs to flag both questions read "too fast" and questions read "too slow." A WPS maximum QATT may incorrectly flag questions as question-reading deviations when in fact the longer question duration is due to the respondent's behavior (e.g., asked a question, thinking about answer). However, the risk of increasing false positives may be acceptable if the WPS range method detects more deviations than the WPS point estimate method (or other methods). Further, if the ultimate goal is to identify problematic interviews, then a false negative on a particular question may be acceptable if other questions in the same interview are correctly identified as containing reading deviations. Thus, the WPS range method is worth investigating, even though other detection methods that factor in respondents' behavior may be more accurate.

In light of limits to the WPS methods, using measures of dispersion to determine QATTs may be more effective. For example, QATTs could be set for each question at the mean question duration across all administrations plus or minus one standard deviation. This method would account for variability in interviewers' speech rate and respondents' behavior. However, it may have its own weaknesses. For one, it requires that sufficient data be available to reliably estimate a mean duration and standard deviation for each question. Thus, this method may not be feasible early in data collection unless timing data can be used from a previous wave in longitudinal surveys. Second, the behavior we want to detect (i.e., extreme question durations) influences means and standard deviations, thus influencing the QATTs. Nevertheless, to evaluate this method, several standard deviations (0.5; 1.0; 1.5; 2.0) will be tested in this study.

In summary, three methods of developing QATTs with varying thresholds are tested:

- Words per Second: Point Estimate Method: 2 WPS; 3 WPS; 4 WPS
- Words per Second: Range Method: 2–3 WPS; 1–3 WPS; 2–4 WPS; 1–4 WPS
- Standard Deviation Method: 0.5; 1.0; 1.5; 2.0

19.3 Data and Methods

19.3.1 Data

This study uses data from Wave 3 of the Understanding Society Innovation Panel (IP), which is a panel used for methodological research to inform the design of the

Understanding Society household panel study in the UK (Jäckle, et al. 2017). The IP uses a multi-stage probability sample (for more information on sample design, see Jäckle, et al. 2017). For Wave 3, 1,526 eligible households were identified and 1,027 household interviews were completed with a response rate of 67% (AAPOR RR1). All eligible adults (age 16+) in the household were selected to complete an individual, face-to-face, computer-assisted personal interview (CAPI) in which interviewers were instructed to read all questions verbatim. Conditional on the household response rate, the individual response rate was 82%, for a total of 1,621 completed interviews. Average interview length was 37.5 minutes. Selected sections* of the interview were audio-recorded with the permission of the respondent (72% consent rate). However, due to procedural and technical difficulties, only 820 interview recordings were available for analysis. The instrument was programmed in Blaise and the timing file contained timestamps for all interviews.

Behavior coding was used to determine which questions were read verbatim and which were misread. These data are used as the "gold standard" that the QATT deviation detection methods will be tested against for accuracy. To select a subset of the recorded files for behavior coding, two interviews were randomly selected from each of the 80 interviewers. In a few cases, the selected interviews were missing recordings at the section level, resulting in only a few recorded questions in the interview. When this happened, an additional interview was randomly selected from the same interviewer to ensure that each interviewer had at least 50 questions that could be coded.† This procedure yielded 168 interviews selected for behavior coding. Within the selected interviews, 402 questions were selected for analysis based on the following criteria:

- Question was intended to be read out loud
- Did not contain "fills"
- Were administered to both males and females
- Had one-to-one matching with timing file questions (i.e., did not loop)
- Had the same response options for all regions

Due to question routing, not all questions were administered to all respondents. The total sample size for coding and analysis is 10,386 question administrations.

The behavior coding was done directly from the audio files (no transcription) by a single coder. The coding builds on Cannell, Lawson, and Hausser's (1975) behavior coding scheme. The interviewer's first reading of each question was coded as (a) question read verbatim, (b) contains only minor deviations, or (c) contains at least one major deviation. Building on Cannell, Lawson, and Hausser (1975), explicit rules were created to evaluate if the deviation was minor or major (see Appendix 19A) with the primary distinction being the assumption that minor deviations most likely do not change the meaning of the question, but major deviations are likely to change the meaning of the question. Coding results show 34.5% of questions had minor deviations and 13.0% of questions had major deviations.

---

* A series of experiments (question wording, branching, and show card) were carried out in IP Wave 3 and the corresponding questionnaire sections were recorded (Jäckle, et al. 2017).

† This data set is used in multiple studies, including examinations of question characteristics and interviewer effects. To increase analytic power, a minimum of 50 questions per interviewer was established.

19.3.2 Dependent Variable and Variables for QATT Detection Methods

Because minor deviations most likely do not change the meaning of the question, the focus in this chapter is on how to best detect major deviations. Thus, the dependent variable, derived from the behavior codes, is coded as 0 = Verbatim/Minor Deviation and 1 = Major Deviation.

Next, variables for each of the QATT detection methods were created. First the WPS Point estimate thresholds were calculated at 2 WPS, 3 WPS, and 4 WPS by dividing the total number of words in the question text (not including optional text) by 2, 3, and 4, respectively. Any question duration that was faster than (i.e., below) the point estimate was flagged as a possible deviation in which words may have been omitted. Next, the WPS range thresholds (2–3 WPS; 2–4 WPS; 1–3 WPS; 1–4 WPS) were calculated using the same procedure. For example, the lower and upper thresholds for a ten-word question at 2–3 WPS would be 3.33 seconds (10 divided by 3) and 5 seconds (10 divided by 2), respectively. Question durations lower than the lower threshold or higher than the higher threshold were flagged as possible major deviations. Lastly, the QATT thresholds based on standard deviations were calculated. The mean and standard deviation of the duration for each question were calculated across all interviewers. Four different sets of thresholds were then set at ±0.5, ±1.0, ±1.5, and ±2.0 standard deviations from the mean question duration, and durations outside these thresholds were flagged as possible deviations.

19.3.3 Analysis

A crosstab was performed between the indicator of possible deviations produced by each QATT method for each question reading and the behavior coding indicator for major deviations to establish the rates of false negatives (i.e., question deviations incorrectly identified by the QATT method as verbatim), false positives (i.e., verbatim questions incorrectly identified as deviations), true negatives (i.e., verbatim question correctly identified), and true positives (i.e., question deviations correctly identified).

Additional analyses assessed the utility of each QATT method. First, the overall accuracy rate, which captures the percent of cases correctly classified as having or not having misreadings, was calculated for each QATT method. Next, the overall detection rate, which captures the percent of cases correctly classified as having a major deviation was calculated for each QATT method (1,353 questions were read with major deviations). The detection rate gives a clearer picture of how well each method detects major reading deviations than an overall accuracy rate because an overall accuracy rate can be very high for methods that correctly categorize true negatives (i.e., questions read verbatim or with only minor deviations) but fail to detect major deviations.

Second, the data were aggregated to the interview level to assess if false positive and false negatives can be mitigated at this level. When questions are flagged as suspicious by quality control procedures, it is illogical to think that only flagged questions should be investigated. In most cases, the activity leading up to the suspicious question(s) and the subsequent behavior is assessed, and, in some cases, the entire interview is reviewed. If an interview has any questions flagged as possibly misread, listening to the interview should catch the deviations that the QATT method missed (i.e., false negatives) and rule out flagged questions that were actually read verbatim (i.e., false positives).

The aggregated data sets (one for each QATT detection method) contain the interview number (rows; n = 168) and four variables (columns): the count of false negatives, the count of false positives, the count of true of negatives, and the count of true positives. From these, two new variables were created: (1) the interview has a true deviation (i.e., at least one question in the interview was behavior-coded as having a major deviation) and (2) the QATT method detected a deviation (at least one question in the interview violated the minimum or maximum QATTs). Crosstabs were used to determine interview-level false negatives, false positives, true negatives, and true positives. Overall classification accuracy rates were calculated as were rates of detecting at least one major deviation. Finally, the percent of interviews the method flagged for further investigation was calculated.

Per the behavior coding, all interviews contained at least one minor deviation and all but 29 interviews contained at least one major deviation. That almost 83% of the interviews contain at least one major deviation and thus would require further investigation makes the following discussion somewhat moot. However, only 168 interviews were behavior-coded. It could be that a larger data set or a different subsample of the interview recordings would have produced fewer interviews with at least one major deviation. Also, one could argue that at the start of field operations, the first interviews completed by each interview may have a high rate of interviews with major deviations. Thus, ruling out 17% of the incoming interviews needing review would reduce quality control efforts. We can learn from this analysis if we focus on the accuracy rate and the detection rate for correctly identifying the 139 interviews with major deviations.

19.4 Results

Table 19.1 shows the frequency (count and percentages) of question readings classified as violating QATTs by each of the QATT detection methods. As a reminder, the behavior coding found that questions were read with major deviations at a rate of 13%. The various QATT methods estimated as few as 8.2% and as many as 68.5% of question readings had

TABLE 19.1

Count and Percentages of Question Readings Classified as Violating QATTs by QATT Method (n = 10386)

Detection Method	Question Readings Classified as Violating QATTs	
	Count	%
2 WPS	5,304	51.1
3 WPS	2,347	22.6
4 WPS	1,255	12.1
2–3 WPS	7,112	68.5
1–3 WPS	3,713	35.8
2–4 WPS	6,020	58.0
1–4 WPS	2,621	25.2
SD 0.5	5,421	52.2
SD 1.0	2,408	23.2
SD 1.5	1,379	13.3
SD 2.0	982	9.5

deviations. The QATT methods with the closest rates to the behavior-coded data are SD 1.5 (13.3%) and 4 WPS (12.1%). These detection rates may be close to the behavior-coded rate, but the question of the methods' accuracy remains.

Table 19.2 shows how well each QATT method classified the question readings. The overall accuracy rate was highest for 4 WPS (87.1%) and lowest for 2–3 WPS (39.6%), with the remainder of the methods falling in between (46.1–80.6%). The results suggest that using point estimates are better than using a range.

However, is overall classification accuracy the best metric? A particular method may have a high overall classification accuracy rate because it accurately identifies true negatives (i.e., verbatim) but detects few or no true positives. In fact, this is the case for the 4 WPS method, which had the highest overall classification accuracy rate (87.1%). However, this method only correctly identified 46.9% of major question-reading deviations. Seven other methods correctly identified more major deviations, with the 2–3 WPS method identifying the most major deviations (81.0%). Yet, increasing the rate of detecting deviations comes at a price; the number of false positives can increase dramatically. The 2–3 WPS detects 81% of the major deviations, but the false positive rate soars to 57.9%, largely because this method cannot differentiate between minor and major deviations and thus flags minor deviations as problematic. This may be acceptable if the goal is to target major deviations only; however, quality control staff would spend a significant amount of time chasing down and ruling out the false positives.

Deciding on which QATT to use comes down to finding a balance between the method that maximizes the detection rate and minimizes the false positive rates. Looking at the results from this perspective, moving from 4 WPS to 3 WPS, accuracy drops 6.5 percentage points (from 87.1% to 80.6%) due to an increase in false positives (from 6.0% to 14.5%). However, we gain 15.6 percentage points in the detection rate (from 46.9% to 62.5%). If we moved from 3 WPS to 2 WPS, the detection rate would increase further (increases 17.8 percentage points), but the false positives soar to 40.6%, thus decreasing accuracy by 23.8 percentage points. Thus, 3 WPS has the best balance between the detection rate, false positive rate, and overall accuracy.

Table 19.3 shows the results from the interview-level analysis that compares the QATT indicator of any question misreadings in the interview to the behavior code indicator of

TABLE 19.2

Comparison of QATT Question Misread Indicator to Behavior Coding Indicator of Major Question Misreadings (n = 10386)

Detection Method	% False -	% False +	% True -	% True +	Overall Accuracy (%)	Detection Rate (%)
2 WPS	2.6	40.6	46.4	10.4	56.8	80.3
3 WPS	4.9	14.5	72.5	8.1	80.6	62.5
4 WPS	6.9	6.0	81.0	6.1	87.1	46.9
2–3 WPS	2.5	57.9	29.0	10.6	39.6	81.0
1–3 WPS	4.5	49.4	37.6	8.5	46.1	65.4
2–4 WPS	4.2	26.9	60.0	8.9	68.9	67.9
1–4 WPS	6.2	18.4	68.6	6.8	75.4	52.3
SD 0.5	3.9	43.1	43.9	9.1	53.0	69.9
SD 1.0	8.2	18.4	68.6	4.8	73.4	37.0
SD 1.5	10.9	11.0	75.9	2.2	78.1	16.6
SD 2.0	11.8	8.3	78.7	1.2	79.9	9.2

TABLE 19.3

Interview-Level Comparison of QATT Indicator of Any Question Misreadings to Behavior Coding Indicator of Any Major Question Misreadings

Detection Method	% False −	% False +	% True −	% True +	Overall Accuracy (%)	% of Interviews w/Misreadings Detected (n = 139)	% of Interviews Method Flagged for Review (n = 168)
2 WPS	0.0	17.3	0.0	82.7	82.7	100.0	100.0
3 WPS	1.2	13.7	3.6	81.5	85.1	98.6	95.2
4 WPS	7.1	4.2	10.1	78.6	88.7	95.0	82.7
2–3 WPS	0.0	17.3	0.0	82.7	82.7	100.0	100.0
1–3 WPS	0.0	17.3	0.0	82.7	82.7	100.0	100.0
2–4 WPS	0.0	17.3	0.0	82.7	82.7	100.0	100.0
1–4 WPS	0.6	14.9	2.4	82.1	84.5	99.3	97.0
SD 0.5	0.0	17.3	0.0	82.7	82.7	100.0	100.0
SD 1.0	0.0	15.5	1.8	82.7	84.5	100.0	98.2
SD 1.5	3.0	11.3	6.0	79.8	85.7	96.4	91.1
SD 2.0	8.9	9.5	7.7	73.8	81.5	89.2	83.3

any major question misreadings. The overall accuracy rate ranges from 81.5% (SD 2.0) to 88.7% (4 WPS). Similar to the question-level analysis, having a high accuracy rate does not mean the method is best at identifying interviews with any major deviations; the methods with the lower overall accuracy rates detect higher rates of major deviations, but also flag almost all, if not all, interviews as needing further review. If an organization's goal is to detect all interviews with any major deviation, no matter the false positive rate, there are several methods that accomplish this: 2 WPS, 2–3 WPS, 1–3 WPS, 2–4 WPS, SD 0.5, and SD 1.0. If the goal is to reduce quality control efforts, while acknowledging that some interviews that contain major deviations may not be detected, then the 4 WPS may be the best method; 17 (10.1%) interviews can be ruled out of needing further review and the method only incorrectly identifies 7 (4.2%) interviews as containing deviations (i.e., false positives) and 12 (7.1%) interviews as having no deviations (i.e., false negatives). If the goal is to identify interviewers for retraining (or to detect falsification), the data could be further aggregated to the interviewer level. Indeed, these data were aggregated (not shown) and 77 of the 80 interviewers had a least one major deviation across their interviews. This suggests that for this pool of interviewers, more training is needed that emphasizes reading questions verbatim.

19.5 Summary

This study finds interviewers engaged in question-reading deviations, either minor or major, at a rate of 48% of questions read. Deviations were mostly minor, but almost 13% of the question readings were classified through behavior coding as major deviations. Three methods for developing QATTs for classifying major deviations were tested for accuracy. Results show that the most accurate QATT method for classifying question readings is 4 WPS (87.1%), but the 2 WPS method is best at detecting questions read with major deviations (80.3%). However, along with failing to detect actual deviations (i.e., false negatives), the 2 WPS QATT produces the highest rate of false positives, raising questions about its utility. Finding the best method is then about identifying the method that maximizes the detection rate and minimizes false positives. The 3 WPS method does this best.

Aggregating the data up to the interview level reduces the rate of false positives and false negatives. The method that arguably shows the most utility at the interview level is 4 WPS. The 4 WPS method has the highest rate of correctly identifying interviews with no major deviations (10.1%), while only incorrectly identifying 0.04% of interviews as containing deviations and 0.07% of those with any deviations as having no deviations.

While more research is needed before providing a definitive recommendation for which QATT detection method to use for detecting major question-reading deviations, the 3 WPS method does show promise, finding the optimal balance between overall accuracy and detection rate of the methods tested. Aggregating the data to the interview (or interviewer) level would allow quality control teams to rank interviews (or interviewers) by the proportion of interviews that have been classified as having major deviations. We find that the 4 WPS method shows the most promise at this level. This targeted, automated approach should save time and money by reducing the need to listen to all interviews and concentrating quality control efforts on those interviews (or interviewers) with high rates of questions (or interviews) flagged as having major deviations.

Whereas these methods show considerable promise in this study, there is still a significant amount of research that can be done in this area, including developing and testing additional QATT methods. These might include more precise increments of WPS (e.g., 4.1 WPS, 4.2 WPS) or other methods, such as model-based approaches that utilize question, respondent or interviewer characteristics (or a combination), to generate QATTs. Developing QATTs for surveys conducted in different languages is another area of research that has not been explored. Would 4 WPS still show the most promise for other languages as it does for English?

This is the first known study to show that survey paradata, which is relatively inexpensive to collect, can be used to develop QATTs that can identify major question misreadings with reasonable success. This method has considerable potential to improve the efficiency of field monitoring. However, the study does have limitations. First, while the behavior coding was a unique feature of the data that allowed the study to be conducted, it was only performed on a subset of the interview recordings due to technical, administrative, and resource limits. While random sampling should ensure that the coded interviews are a representative subsample of all recorded interviews, there is a risk that the interviews that were not recorded differ from those that were recorded. Interviewers who engage in more question-reading deviations may not want to be recorded and thus may take steps to manufacture a "technical" issue (e.g., unplug or turn off the microphone) or falsely indicate that the respondent refused to be recorded. Thus, having a more complete sample or a different sample may change the results. Second, due to resource limits, the behavior codes that were used as the gold standard measure of reading deviations were produced by only one coder, thus precluding any test of inter-coder reliability. Also, even with a carefully developed coding scheme and coding criteria, behavior coding as a method does involve some subjectivity. Even with these limitations, this study suggests that establishing and using QATTs is a promising method to improve quality control processes in interviewer-administered surveys and is deserving of additional research.

References

Ackermann-Piek, D., and N. Massing. 2014. Interviewer behavior and interviewer characteristics in PIAAC Germany. *Methods, Data, Analyses* 8(2):199–222.

Axinn, W. G. 1991. The influence of interviewer sex on responses to sensitive questions in Nepal. *Social Science Research* 20(3):303–318.

Bradburn, N. M., S. Sudman, E. Blair, W. Locander, C. Miles, E. Singer, and C. Stocking. 1979. *Improving Interview Method and Questionnaire Design: Response Effects to Threatening Questions in Survey Research*. San Francisco: Jossey-Bass.

Cannell, C. F., S. Lawson, and D. Hausser. 1975. *A Technique for Evaluating Interviewer Performance*. Ann Arbor: University of Michigan Press.

Couper, M. P., and F. Kreuter. 2013. Using paradata to explore item level response times in surveys. *Journal of the Royal Statistical Society: Series A, (Statistics in Society)* 176(1):271–286.

Dijkstra, W., and Y. Ongena. 2006. Question-answer sequences in survey-interviews. *Quality and Quantity* 40(6):983–1011.

Foulke, E. 1968. Listening comprehension as a function of word rate. *The Journal of Communication* 18(3):198–206.

Fowler Jr, F. J., and T. W. Mangione. 1990. *Standardized Survey Interviewing: Minimizing Interviewer-Related Error*. Newbury Park: Sage Publications.

Groves, R. M., F. J. Fowler Jr, M. P. Couper, J. M. Lepkowski, E. Singer, and R. Tourangeau. 2009. *Survey Methodology: Second Edition*. Hoboken: John Wiley & Sons.

Jäckle, A., A. Gaia, T. Al Baghal, J. Burton, and P. Lynn. 2017. *Understanding Society–the UK Household Longitudinal Study, Innovation Panel, Waves 1–9, User Manual*. Colchester: University of Essex.

Krosnick, J. A., N. Malhotra, and U. Mittal. 2014. Public misunderstanding of political facts: How question wording affected estimates of partisan differences in birtherism. *Public Opinion Quarterly* 78(1):147–165.

Mathiowetz, N. A., and C. Cannell. 1980. Coding interviewer behavior as a method of evaluating performance. In: *Proceedings of the Section on Survey Research Methods*, 525–528. American Statistical Association.

Mneimneh, Z. N., L. Lyberg, S. Sharma, M. Vyas, D. B. Sathe, F. Malter, and Y. Altwaijri. 2018. Case studies on monitoring interviewer behavior in international and multinational surveys. In: *Advances in Comparative Survey Methods: Multinational, Multiregional, and Multicultural Contexts (3MC)*,ed. T. P. Johnson, B. E. Pennell, I. A. Stoop, and B. Dorer, 731–770. Hoboken: John Wiley & Sons, Inc.

Mneimneh, Z. N., B. Pennell, Y. Lin, and J. Kelley. 2014. Using paradata to monitor interviewers' behavior: A case study from a national survey in the Kingdom of Saudi Arabia. *Paper Presented at Comparative Survey Design and Implementation (CSDI) Workshop*.

Olson, K., and A. Peytchev. 2007. Effect of interviewer experience on interview pace and interviewer attitudes. *Public Opinion Quarterly* 71(2):273–286.

Omoigui, N., L. He, A. Gupta, J. Grudin, and E. Sanocki. 1999. Time-compression: Systems concerns, usage, and benefits. In: *Proceedings of the SIGCHI Conference on Human Factors in Computing Systems*, 136–143. New York: ACM.

Schober, M. F., and F. G. Conrad. 2002. A collaborative view of standardized survey interviews. In: *Standardization and Tacit Knowledge: Interaction and Practice in the Survey Interview*, ed. D. Maynard, N. C. Schaeffer, H. Houtkoop-Steenstra, and J. van der Zouwen, 67–94. New York: Wiley.

Schuman, H., and S. Presser. 1996. *Questions and Answers in Attitude Surveys: Experiments on Question Form, Wording, and Context*. Thousand Oaks: Sage Publications, Inc.

Sun, Y., and X. Meng. 2014. Using response time for each question in quality control on China Mental Health Survey (CMHS). In: *Proceedings in Comparative Survey Design and Implementation (CSDI) Workshop*.

Thissen, M. R., and S. K. Myers. 2016. Systems and processes for detecting interviewer falsification and assuring data collection quality. *Statistical Journal of the IAOS* 32(3):339–347.

Viterna, J., and D. W. Maynard. 2002. How uniform is standardization? Variation within and across survey research centers regarding protocols for interviewing. In: *Standardization and Tacit Knowledge: Interaction and Practice in the Survey Interview*, ed. D. Maynard, N. C. Schaeffer, H. Houtkoop-Steenstra, and J. van der Zouwen, 365–400. New York: Wiley.

20

What Do Interviewers Learn?: Changes in Interview Length and Interviewer Behaviors over the Field Period

Kristen Olson and Jolene D. Smyth

CONTENTS

20.1 Introduction

Interviewers are important actors in telephone surveys. By setting the pace for an interview, interviewers communicate the amount of time and cognitive effort respondents should put into their task. It is well-established that interviewers vary widely in the time they spend administering a survey, and that this time changes over the course of the data collection period as interviewers gain experience (Böhme and Stöhr 2014; Kirchner and Olson 2017; Loosveldt and Beullens 2013a, 2013b; Olson and Bilgen 2011; Olson and Peytchev 2007). In particular, interviewers get faster as they gain experience over the field period of a survey.

The within-survey effect of experience on interview length is generally attributed to interviewer learning effects. In particular, a learning effect occurs when interviewers learn how to change their behaviors to more quickly administer questions. This can include positive changes in behaviors over the field period such as error-free administration of questions or negative changes such as shortening questions (i.e., non-standardization) or avoiding positive, time-consuming behaviors like probing or verifying answers (e.g., Bohme and Stohr 2014; Kirchner and Olson 2017; Loosveldt and Beullens 2013a, 2013b; Olson and Peytchev 2007). Other hypotheses about why the length of interview changes over the course of the data collection period, including characteristics of the respondents or interviewers or differential respondent motivation correlated with their response propensity, have not explained away the learning effect (e.g., Kirchner and Olson 2017). However, Kirchner and Olson (2017) found that a measure of the interaction between interviewers and respondents – the number of words spoken by the interviewer and by the respondent – partially mediated the interviewer learning effect.

Despite the well-replicated finding that interviewers speed up over the field period, what behaviors change and whether they explain the decrease in interview length over the course of data collection has not been previously examined in published articles. This chapter examines two research questions:

RQ1: What standardized, nonstandardized, and inefficient interviewer behaviors change over the course of the data collection period?

RQ2: Do these behaviors account for changes in interview length over the course of the data collection period?

To answer these questions, we draw on two nationally representative US telephone surveys of adults. Both surveys were audio-recorded and transcribed. Interviewer and respondent behaviors were coded at the conversational turn level, allowing a detailed examination of the changes in interviewer behaviors over the course of the field period. We focus on interviewer behaviors, as the learning hypothesis focuses primarily on changes by the interviewer, although interviewer behaviors inevitably affect respondent behaviors as well.

20.2 Hypotheses for Behaviors Affected by Interviewer Learning

There are three main hypotheses about what interviewers may "learn" as they conduct interviews over the course of the field period. First, *interviewers may learn to omit or shorten certain standardized interviewer behaviors (i.e., "good" behaviors)*. Standardized behaviors include reading questions exactly as worded, using nondirective probes, repeating the respondents' answers to verify what they said, clarifying the question wording, and providing appropriate feedback to the respondent (Fowler and Mangione 1990). The standardized "good" behaviors may be eliminated as interviewers learn what may be shortcut, become bored or frustrated with certain questions, think that certain questions are emotionally draining for follow-up, or think that they remember the question wording, and thus do not read the item on the questionnaire directly (Kaplan and Yu Chapter 5; Ongena and Dijkstra 2007). As the field period progresses and interviewers learn from previous respondents' answers, they also may be more likely to enter a response that is not directly codable rather than probe nondirectively for a codable response (Ongena and Dijkstra

2007). Finally, interviewers may reduce their use of trained techniques that are used less frequently during interviews (e.g., probing), especially experienced interviewers for whom training is more distant (Olson and Bilgen 2011; Tarnai and Moore 2008; van der Zouwen, Dijkstra, and Smit 1991).

Second, *interviewers may learn to become more efficient at administering questions by reducing or eliminating seemingly extraneous behaviors*, including stuttering or disfluencies while reading questions (Olson and Peytchev 2007). Interviewers may also reduce or eliminate extraneous laughter in an effort to shorten their interactions with respondents. They may do so because they place a greater premium on efficiency than rapport, or because their own enthusiasm conducting the survey wears thin over time (Cleary, Mechanic, and Weiss 1981; Houtkoop-Steenstra 1997). For the same reasons, interviewers may reduce their use of verbal pleasantries, personal disclosures, flattery, and digression. Interviewers may also reduce or eliminate task-related feedback (e.g., "let me just get this down") as early bugs in the interview hardware or software are corrected or as they become more efficient in navigating the interview system or entering responses. Task-related feedback may also be reduced if interviewers think it is not helpful for maintaining rapport or guiding respondents through the interview. Notably, these inefficiency-related behaviors may not be part of interviewer training, but happen as part of normal conversation.

Finally, *interviewers may learn to increase the use of nonstandardized, time-saving behaviors* such as changing the question wording (including making major changes or skipping questions), directively probing inadequate answers, changing answers when verifying them, and interrupting respondents. Although interviewers are specifically trained to avoid these behaviors, nonstandardized behaviors are ubiquitous in standardized interviews (Edwards, Sun, and Hubbard Chapter 6; Ongena and Dijkstra 2006). For instance, interviewers may be more likely to adopt practices, such as directively probing an uncodable answer, in order to advance through the interview more quickly (van der Zouwen, et al. 1991).

It is also possible that these behaviors will differ for landline versus cell phone interviews, as previous research has illuminated differences in interviewer and respondent conversational behaviors across these devices (Timbrook, Smyth, and Olson 2018).

20.3 Data and Methods

This chapter builds on Kirchner and Olson (2017), using the same two telephone surveys. First, the Work and Leisure Today 1 (WLT1) Survey was a landline random digit dial (RDD) telephone survey conducted by AbtSRBI between July 31 and August 28, 2013 (n = 450, AAPOR RR3 = 6.3%). WLT1 contained questions about the respondents' employment, leisure activities, technology use, and demographics. It was deliberately designed to have some highly problematic questions, including difficult and unknown terms, sensitive items, and complex questions. To facilitate model estimation (van Breukelen and Moerbeek 2013), we restricted analyses to the 19 interviewers who conducted at least 10 interviews (n = 435 respondents).

Second, the Work and Leisure Today 2 (WLT2) Survey was a dual frame RDD telephone survey conducted by AbtSRBI during September 2015 (n = 902, landline = 451, AAPOR RR3 = 9.4%; cell phone = 451, AAPOR RR3 = 7.1%). This survey also contained questions about work, leisure, technology use, and demographics, but it did not include many of the highly problematic questions found in WLT1. Although these surveys are called WLT1 and WLT2, the samples are fully independent; that is, there is no longitudinal component. The WLT2 questionnaire contained two versions with alternative experimental question-naire designs and question wording on many questions. As with WLT1, we restricted the analysis to the 26 interviewers with at least 10 completed interviews (n = 896 respondents).

Each of the surveys was audio-recorded, transcribed, and behavior-coded at the con-versational turn level using Sequence Viewer (Dijkstra 2016). Eight fields were coded by trained undergraduate coders, with a 10% subsample of interviews in each study coded by two master coders; we use seven codes in this chapter. For each conversational turn, coders identified the actor (e.g., interviewer), the initial action (e.g., question asking), an assessment of that initial action (e.g., question read with changes), details of that action (e.g., changes were major), whether a particular actor laughed, either on its own or as part of a conversa-tional turn, whether there were any disfluencies (including uhs, ums, and stuttering), and whether one actor interrupted the other actor. Kappa values exceeded 0.8 for most codes; assessments of the initial action exceeded 0.5 (see Online Appendix for details).

20.3.1 Creating Behavior Measures

Because we are interested in explaining total interview length, we aggregate behaviors to the interview level. There are two approaches to examining summary measures of behav-iors at the interview level. First, we can examine the *total number of conversational turns* on which each type of behavior occurred. The number of conversational turns with a given behavior is a measure of how much conversation occurred due to this behavior within a single interview (i.e., an interview-level count). This measure accounts for all behaviors that occurred during the interview (e.g., multiple probing turns on the same question will be counted), but obscures whether the behaviors occurred on only a few questions or on many questions during the interview. The second approach is to use a count of the *total number of questions* on which an individual behavior occurred within a single interview. This question-level count cannot account for multiple occurrences of a behavior within a question, but it provides a measure of whether the behavior occurred on only a few ques-tions or on many questions throughout the interview. The two measures are highly cor-related. We use the question-level count in the current analysis. Results are similar for the count of the number of conversational turns across the interview (available on request).

20.3.2 Dependent Variables

We examine two sets of dependent variables. The first, corresponding to RQ1, are the *interviewer behaviors*. We counted the total number of questions on which each behav-ior occurred at least once across the entire questionnaire (an average of 50 questions per respondent in WLT1 and 51 questions per respondent in WLT2). Our five measures of standardized "good" behaviors include exact question reading, nondirective probes, exact verification, appropriate clarification, and appropriate feedback. Our five measures of inef-ficiency behaviors include stuttering and repairs during question reading, disfluencies, "pleasant talk," task-related feedback, and laughter. Finally, we have five measures of non-standardized behaviors, including (major) changes in question wording, directive probes, inadequate verification (paraphrasing), and interruptions. The operationalization and dis-tribution for each of these behaviors are shown in Table 20.1.

The second dependent variable is *interview length in minutes*. To address outliers (Yan and Olson 2013), interview length was trimmed at the 1st and 99th percentiles. The mean interview length was 12.65 minutes for WLT1 and 13.36 minutes for WLT2.

TABLE 20.1

Definition and Mean Number of Questions with Each Interviewer Behavior

	Definition		WLT1		WLT2	
	Action	Assessment	Mean	%	Mean	%
Standardized behaviors						
Exact question reading	Question asked	Question asked exactly as worded	23.36	46.23	36.99	72.22
Nondirective probes	Probe	Repeat entire question, Repeat part of question, Repeat response options, "Take your best guess," or Ask for explicit response	8.37	16.94	7.73	15.20
Exact verification	Verification	Verifies by repeating respondent's answer exactly	7.92	15.84	7.17	14.01
Appropriate clarification	Clarify	"Whatever it means to you," Provide definition exactly as worded, or Clarifying unit	1.94	3.84	0.45	0.86
Appropriate feedback	Actor = Interviewer + Feedback	Affirmation, Short acknowledgment, or Long motivational feedback	19.52	39.16	22.57	44.14
Inefficiency behaviors						
Stuttering during question reading	Question asked	Read question with stutters, or Read response options with stutters	2.74	5.51	2.42	4.74
Disfluencies	Any disfluencies		13.31	26.77	11.72	22.96
Pleasant talk	Feedback	Personal disclosure, Flattery, or Digression	0.49	0.99	0.65	1.27
Task-related feedback	Feedback	Task-related feedback, Telephone quality, or Time-related feedback	0.88	2.07	1.37	2.77
Laughter	Interviewer laughed + Both laughed		2.32	4.63	3.12	6.05
Nonstandardized behaviors						
Minor changes in question wording	Question asked	Read question with changes – Slight changes	15.40	30.79	4.88	9.60
Major changes in question wording	Question asked	Read question with changes – Major changes	5.44	11.01	6.27	12.15
Directive probes	Probe	Question repeated with changes, or Other directive probes	2.63	5.36	1.14	2.28
Inadequate verification	Verification	Repeated respondent's answer with changes	3.23	6.53	1.83	3.58
Interruptions	Interviewer interrupts respondent		5.63	11.35	3.30	6.49

Note: All behaviors evaluated only on conversational turns during which the interviewer was identified as the actor

20.3.3 Primary Independent Variable: Within-Survey Experience

Within-survey experience is the primary measure of whether interviewers learn over the course of the field period. Because we expect that the effect of learning will be larger at the beginning of the field period than at the end of the field period (Olson and Peytchev 2007), we include a log-transformed ordinal counter for interview order (i.e., 1 for the first interview for an interviewer, 2 for the second, etc.). This counter ranges from 1 to 27 in WLT1 and from 1 to 79 in WLT2.

20.3.4 Control Variables

Because respondent characteristics and response propensity also differ over the field period and across the two studies, we include the following control variables: an overall measure of interviewer experience (i.e., less than one year versus one year or more), the interviewer-level cooperation rate, other interviewer characteristics (race, gender, worked primarily weekday evening shifts), respondent characteristics (sex, age, education, employment status, income, household size, parental status, volunteer status, computer usage), and measures of response propensity (item nonresponse rate, whether the household ever refused, whether the interview was completed at first contact, number of call attempts, time of day the interview was completed). Finally, the number of answers that were changed by the interviewer as recorded in the paradata are included as control variables for both studies. In WLT2, we also included indicators for which experimental questionnaire was used and whether the interview was conducted on a landline or a cell phone.

20.3.5 Analytic Strategy

We estimate hierarchical two-level random intercept models accounting for the clustering of respondents within interviewers (e.g., Raudenbush and Bryk 2002). For the interviewer behaviors, we estimate two-level Poisson models with a log link and a random intercept due to interviewers, with the number of questions asked to each respondent as the exposure variable (see the online supplementary materials). These models are estimated using the mepoisson procedure in Stata 15.1. For interview length, we estimate a two-level linear model using the mixed procedure in Stata 15.1 with a random intercept due to interviewers (see the online supplementary materials). In these models, the interview behaviors are grand-mean centered. They are initially included as separate groups (standardized, inefficiency, nonstandardized) and then combined into a single model.

20.4 Results

20.4.1 RQ1: What Interviewer Behaviors Change over the Course of the Data Collection Period?

We start by addressing RQ1. We focus only on the interview order (within-survey experience) coefficient in our discussion below. The full models are in the online supplementary materials. Table 20.2 contains the coefficients from the log(interview order) term in both WLT1 and WLT2.

We start with standardized interviewing behaviors. We see notable differences across the two surveys. In WLT1, there is no change in the number of questions on which standardized interviewer behaviors occur across the data collection period at traditional $p<.05$ levels. In WLT2, on the other hand, there are statistically significant decreases in the number of questions on which nondirective probes, exact verification, and appropriate feedback occur as interviewers gain within-study experience. The difference in coefficients between WLT1 and WLT2 is statistically significant for nondirective probes ($z=2.23$, $p=0.026$) and exact verification ($z=2.55$, $p=0.011$). To understand the magnitude of these changes, we examine predicted marginal effects. The average workload among interviewers in WLT2 who conducted at least ten interviews was 34.5 interviews. As such, we examine changes from the 1st to the 30th interview. On average, the predicted number of questions in which interviewers use nondirective probes decreases 11% from 8.2 in the 1st interview to 7.3 in the 30th interview. The use of exact verification decreases 16% from a predicted average of 9.2 questions in the 1st interview to 7.7 in the 30th interview. Appropriate feedback decreases about 11% from being used on an average of 24.4 questions in the 1st interview to 21.6 in the 30th interview.

We now turn to inefficiency behaviors, shown in the middle of Table 20.2. In both WLT1 and WLT2, there are fewer questions with inefficiency behaviors as interviewers gain within-study experience. The coefficients do not statistically significantly differ across the two studies. In both studies, just over 4 questions are read with stutters on the 1st interview, compared to about 2.1 questions with stutters by the 30th interview. Although the number of questions on which a disfluency occurs differs for WLT1 and WLT2, the rate of

TABLE 20.2

Unstandardized Coefficients from Log(Interview Order) Predicting Count of Questions with Interviewing Behaviors

	WLT1		WLT2		z-Value for Test Across Surveys
	Coef.	SE	Coef.	SE	
Standardized interviewing behaviors					
Exact question reading	0.017	0.014	0.001	0.007	1.08
Nondirective probes	0.020	0.023	−0.033*	0.015	2.23*
Exact verification	0.020	0.026	−0.051**	0.016	2.55*
Appropriate clarification	0.091	0.050	−0.034	0.061	1.77
Appropriate feedback	−0.010	0.015	−0.035****	0.009	1.33
Inefficiency behaviors					
Stuttering during question reading	−0.201****	0.039	−0.201****	0.025	0.10
Disfluencies	−0.062**	0.018	−0.058****	0.012	0.01
Pleasant talk	−0.086	0.096	−0.098	0.051	−0.11
Task-related feedback	−0.151*	0.061	−0.052	0.035	−1.30
Laughter	−0.162****	0.043	−0.084****	0.024	−1.70
Nonstandardized behaviors					
Minor changes in question reading	0.032	0.018	0.050*	0.020	−0.98
Major changes in question reading	−0.050	0.028	0.036*	0.017	−2.49*
Directive probes	0.100*	0.042	−0.087*	0.039	4.04****
Inadequate verification	−0.176****	0.036	−0.106****	0.030	−1.28
Interruptions	−0.046	0.028	−0.069**	0.023	0.99

*p<.05, **p<.01, ***p<.001, ****p<.0001

decline is similar, 18–19%, across the field period in both studies – falling from a predicted average of 15.8 questions with some sort of disfluency on the 1st interview to 12.8 questions with disfluencies by the 30th interview in WLT1 (13.9 to 11.4 in WLT2). The number of questions with laughter also declines across the field period, from 3.9 (4.4, WLT2) questions on the 1st interview to 2.2 (3.3, WLT2) questions on the 30th interview in WLT1. The rate of task-related feedback declines from 1.5 questions on the 1st interview to 0.88 questions on the 30th interview in WLT1 but not WLT2. There is no statistical change in pleasant talk across the field period in either study.

Finally, we look at nonstandardized interviewing behaviors, shown at the bottom of Table 20.2. Across the two surveys, there are mixed changes in nonstandardized behaviors. In both studies, the rate of inadequate verification behaviors declines across the field period, from an average of 4.8 questions on the 1st interview in WLT1 (2.4, WLT2) to an average of 2.6 questions by the 30th interview (1.6, WLT2). The rate of interruptions also declines by about one question over the field period in both studies (WLT1: 6.4 questions to 5.4 questions; WLT2: 4.0 questions to 3.1 questions). In both studies, the number of questions with minor changes in question wording increases by about 1.5 questions over the field period (from 14.5 to 16.2 in WLT1 and from 4.3 to 5.1 in WLT2). None of the interview order coefficients differ between the two studies for these outcome variables.

There appears to be a trade-off between major changes in question reading and in directive probes in the two studies, with statistically significant differences in the interview order coefficients between the two studies. In WLT1, major changes in question reading decline by about 1 question over the field period (from 6.3 questions at the 1st interview to 5.3 questions at the 30th interview), whereas the use of directive probes increases by about 1 question (from 2.1 questions at the 1st interview to 3.0 questions at the 30th interview). In WLT2, the pattern is the opposite – major question reading changes increase (from 5.6 questions to 6.3 questions) and directive probes decrease (from 1.4 to 1.1 questions) (z-test for the difference between interview order coefficients in WLT1 and WLT2: directive probes: $z = 4.04$, $p < .0001$; major changes: $z = -2.49$, $p = 0.013$).

In sum, interviewers do change behaviors as they gain experience over the field period. They become more efficient in administering questions, having fewer questions with stutters, disfluencies, and laughter. Interviewer experience over the field period also changes both standardized and nonstandardized behaviors. In both studies, we see increases in minor changes in question wording and decreases in interruptions and use of inadequate verification behaviors. There is not a consistent increase in the use of adequate verification behaviors – rather, these behaviors go away. Other changes in nonstandardized behaviors are less consistent across the two studies, with a trade-off between major changes in question wording and directive probes.

20.4.2 RQ2: Do Interviewer Behaviors Account for Changes in Survey Length over the Course of the Data Collection Period?

We now turn to the question of whether the observed changes in interviewer behaviors explain the changes in survey length over the course of the data collection period. To answer this question, we examine whether the interview order coefficient predicting survey length changes in magnitude as groups of behaviors are included in the model (Aneshensel 2013, p. 184; mediation models for each behavior individually are included in Online Appendix 20C). As seen in Table 20.3, interviewer behaviors only partially explain the change in interview length over the course of the field period. Each group of behaviors

TABLE 20.3

Log(Interview Order) Coefficients Predicting Length of Interview with Interviewing Behaviors and Percent Change from Model with No Behaviors

	WLT1		WLT2	
	Log(Interview Order) Coefficient	% Reduction from No Behaviors	Log(Interview Order) Coefficient	% Reduction from No Behaviors
No behaviors, no controls	−0.189		−0.537****	
No behaviors, with controls	−0.443**		−0.855****	
Including only standardized behaviors	−0.627****	−41.5	−0.688****	19.5
Including only inefficiency behaviors	−0.211	52.4	−0.706****	17.4
Including only nonstandardized behaviors	−0.382*	13.8	−0.733****	14.3
Including all behaviors	−0.441**	0.5	−0.607****	29.0

Note: +p<.10, *p<.05, **p<.01, ***p<.001, ****p<.0001; negative percent reductions indicate an increase in the coefficient.

reduces the interview order coefficient by about 14–50%. The largest reduction in the coefficient for interview order comes with the inclusion of the inefficiency behaviors in WLT1, reducing the interview order coefficient by 52.4%. However, when all of the interviewer behaviors are included in the same model, this same magnitude reduction in the interview order coefficient is not observed, especially in WLT1. In WLT1, inclusion of the standardized behaviors *increases* the learning effect on length of interview, whereas the other behaviors *explain* the learning effect. Thus, the combined effects "cancel out" in the overall model. In sum, these 15 interviewer behaviors partially mediate the learning effect, but do not completely account for changes in the length of interview over the course of the field period.

20.4.3 Variance Components

There is significant variation across interviewers and respondents in the length of the interview. As shown in Table 20.4, the interviewer behaviors examined here explain between 21% and 32% of the variance in interview length at the interviewer level and between 42% and 54% of the variance in interview length at the respondent level. The inclusion of standardized behaviors alone actually increases the variance at the interviewer level in both studies, as does the inclusion of only inefficiency behaviors in WLT1. Nonstandardized behaviors explain the most variation in length across interviewers in both studies and across respondents in WLT2.

20.5 Conclusion

This chapter aimed to answer two questions – do interviewer behaviors change over the field period, and do the changes in these interviewer behaviors account for the learning

TABLE 20.4

Variance Components for Interview Length Models

	WLT1				WLT2			
	Interviewer		Respondent		Interviewer		Respondent	
	Var. comp.	% change	Var. comp.	% change	Var. comp.	% change	Var. comp.	% change
No behaviors	1.828		8.195		2.009		6.166	
Standardized behaviors only	2.847	56	4.463	−46	2.733	36	4.429	−28
Inefficiency behaviors only	2.980	63	6.480	−21	1.841	−8	5.074	−18
Nonstandardized behaviors only	0.899	−51	5.186	−37	1.597	−21	4.383	−29
All behaviors	1.237	−32	3.768	−54	1.594	−21	3.555	−42

effect observed in the shortening of the interview length over the field period? We found clear confirmation for the first question – interviewer behaviors do change over the course of the data collection period. We also found, reassuringly, that interviewer behaviors are related to interview length. However, interviewer behaviors do not fully explain the learning effect.

First, interviewers do not consistently lose standardized behaviors over the field period across these studies. This is good news. Where there are losses in standardized behaviors in WLT2, it appears to be in feedback behaviors (e.g., ok; thank you), as well as some minor decreases in nondirective probing and verification behaviors. These changes in standardized behaviors explain between none and about 20% of the change in interview length. Second, interviewers do become more efficient in administering surveys over the field period. These changes in inefficiency behaviors explain 17–44% of the change in interview length over the field period. Notably, the inefficiency behaviors alone render the interview order coefficient non-significant at traditional $p < .05$ levels. Finally, interviewers do change in their use of nonstandardized behaviors. There is evidence of an increase in minor changes of question wording over the field period in both studies, perhaps because interviewers are further away from training. Alternatively, interviewers may be learning that respondents have problems with certain questions and preemptively changing the question wording to anticipate where those problems occur. Other nonstandardized behaviors such as inadequate verification and interruptions decrease over the field period. We also see potential trade-offs between major changes in question wording and directive probes in these surveys. Collectively, nonstandardized behaviors explain about 14% of the change in interview length over the field period.

How interviewer behaviors are related to interview length is more complicated than simply the number of questions on which these behaviors occur. Other factors that we have not yet examined are whether interviewers become faster at selecting among the various behaviors they use after question asking or whether they are completing the behaviors themselves more quickly (e.g., do they probe or clarify with fewer words? with faster paced speech?). We also did not examine how question characteristics themselves affect the occurrence of these behaviors. There are clear differences in the prevalence of certain interviewer behaviors across WLT1 and WLT2. Although the content of the questionnaires was similar over these two studies, the questionnaires varied in difficult terms, sensitive questions, and other question characteristics. It may be that inefficiency behaviors are largely properties of individuals' conversational norms and basic linguistic practices, whereas standardized and nonstandardized behaviors are more sensitive to properties of the questions themselves. There remains much future research to do in this area.

We note several limitations with this study. First, we looked only at changes in interviewer behaviors, but many interviewer behaviors occur in reaction to respondent behaviors, either because of the requirements of standardized interviewing (e.g., probing to obtain an answer) or to maintain rapport (Garbarski, Schaeffer, and Dykema 2016). That is, inferential problems may arise because of how the behaviors themselves unfold during an interview, where one behavior may be a trigger for another behavior. As such, future research should examine changes in respondent behaviors as well. Second, although the results largely replicate when we aggregate behaviors to the level of number of conversational turns rather than the number of questions on which the behavior occurs, the number of conversational turns may be a better reflection of the length of the interview. Third, there is sensitivity in our conclusions depending on the collection of behaviors that are included in these models. Some of this is due to potential overlap of constructs represented by the various behaviors (e.g., different types of question-asking behaviors), creating issues of multicollinearity if the number of questions asked in the survey is also included in the duration models. Despite these limitations, a significant strength of this study is that it pursued the goal of replication by examining two surveys conducted two years apart with different interviewing teams. Also, this study used behavior codes to directly evaluate what is happening in the survey interview itself, an incredibly time-consuming and expensive method to produce, for both surveys. However, both surveys were conducted by the same organization. Future research will examine surveys conducted by a different organization.

Even with these limitations, our findings do yield practical implications. Most notably, interviewing efficiency may be gained by changing interviewer training practices. Currently, survey-specific interviewer trainings often involve round robins where interviewers read aloud a single question, but may not read through the entire questionnaire more than once or twice before the start of the field period (Tarnai and Moore 2008). These round robins may not give interviewers enough question-specific practice to reduce stuttering and disfluencies prior to live interviewing. Requiring interviewers to complete entire practice interviews multiple times could help eliminate some inefficiencies prior to live interviewing, thus ensuring more efficient delivery even on early interviews and mitigating the change in inefficient behaviors over the field period. Some organizations also retrain interviewers during the field period. This study suggests retraining on nonstandardized and inefficient behaviors could further reduce the length of the interview.

This study is the first to evaluate how a wide range of interviewer behaviors change over the course of a field period, both individually and as related to interview length. We find that interviewers do change behaviors and that these behaviors partially explain changes in interview length over the field period. We see notable decreases in a wide variety of interviewer behaviors over the course of the data collection period. These decreases suggest less interaction overall between an interviewer and a respondent later in the field period. However, that these 15 theoretically derived interviewer behaviors do not fully explain changes in the length of the interview suggests that there is more about the interaction that is important for future research.

Acknowledgments

This work was supported by the National Science Foundation Grant No. SES-1132015. Any opinions, findings, and conclusions or recommendations expressed in this material are those of the authors and do not necessarily reflect the views of the National Science Foundation. Thanks to Antje Kirchner, Beth Cochran, Jinyoung Lee, Jerry Timbrook, Amanda Ganshert, and Alexis Swendener for research assistance. Thanks to all of our transcriptionists and behavior coders for their amazing work. Previous versions of this chapter were presented at the 2015 AAPOR conference and the 2018 Joint Statistical Meetings.

References

Aneshensel, C. S. 2013. *Theory-Based Data Analysis for the Social Sciences*. Thousand Oaks, CA: Sage Publications.

Böhme, M., and T. Stöhr. 2014. Household interview duration analysis in capi survey management. *Field Methods* 26(4):390–405.

Cleary, P. D., D. Mechanic, and N. Weiss. 1981. The effect of interviewer characteristics on responses to a mental health interview. *Journal of Health and Social Behavior* 22(2):183–193.

Dijkstra, W. 2016. Sequence Viewer.

Fowler, F. J., and T. W. Mangione. 1990. *Standardized Survey Interviewing: Minimizing Interviewer-Related Error*. Newbury Park, CA: Sage Publications.

Garbarski, D., N. C. Schaeffer, and J. Dykema. 2016. Interviewing practices, conversational practices, and rapport: Responsiveness and engagement in the standardized survey interview. *Sociological Methodology* 46(1):1–38.

Houtkoop-Steenstra, H. 1997. Being friendly in survey interviews. *Journal of Pragmatics* 28(5):591–623.

Kirchner, A., and K. Olson. 2017. Examining changes of interview length over the course of the field period. *Journal of Survey Statistics and Methodology* 5:84–108.

Loosveldt, G., and K. Beullens. 2013a. 'How long will it take?' An analysis of interview length in the fifth round of the European Social Survey. *Survey Research Methods* 7(2):69–78.

Loosveldt, G., and K. Beullens. 2013b. The impact of respondents and interviewers on interview speed in face-to-face interviews. *Social Science Research* 42(6):1422–1430.

Olson, K., and I. Bilgen. 2011. The role of interviewer experience on acquiescence. *Public Opinion Quarterly* 75(1):99–114.

Olson, K., and A. Peytchev. 2007. Effect of interviewer experience on interview pace and interviewer attitudes. *Public Opinion Quarterly* 71(2):273–286.

Ongena, Y. P., and W. Dijkstra. 2006. Methods of behavior coding of survey interviews. *Journal of Official Statistics* 22:419–451.

Ongena, Y. P., and W. Dijkstra. 2007. A model of cognitive processes and conversational principles in survey interview interaction. *Applied Cognitive Psychology* 21(2):145–163.

Raudenbush, S. W., and A. S. Bryk. 2002. *Hierarchical Linear Models: Applications and Data Analysis Methods*. Newbury Park, CA: Sage.

Tarnai, J., and D. L. Moore. 2008. Measuring and improving telephone interviewer performance and productivity. In: *Advances in Telephone Survey Methodology*, ed. J. M. Lepkowski, C. Tucker, J. M. Brick, E. D. de Leeuw, L. Japec, P. J. Lavrakas, M. W. Link, and R. L. Sangster, 359–384. Hoboken, NJ: John Wiley & Sons, Inc.

Timbrook, J., J. Smyth, and K. Olson. 2018. Why do mobile interviews take longer? A behavior coding perspective. *Public Opinion Quarterly* 82(3):553–582.

van Breukelen, G., and M. Moerbeek. 2013. Design considerations in multilevel studies. In: *The SAGE Handbook of Multilevel Modeling*, ed. M. A. Scott, J. S. Simonoff, and B. D. Marx, Chapter 11. London: SAGE Publications Ltd.

van der Zouwen, J., W. Dijkstra, and J. H. Smit. 1991. Studying respondent-interviewer interaction: The relationship between interviewing style, interviewer behavior, and response behavior. In: *Measurement Errors in Surveys*, ed. P. Biemer, R. M. Groves, L. Lyberg, N. A. Mathiowetz, and S. Sudman, 419–438. New York: John Wiley & Sons, Inc.

Yan, T., and K. Olson. 2013. Analyzing paradata to investigate measurement error. In: *Improving Surveys with Paradata: Analytic Uses of Process Information*, ed. F. Kreuter, 73–95. Hoboken, NJ: John Wiley & Sons, Inc.

Section VII

Estimating Interviewer Effects

21

Modeling Interviewer Effects in the National Health Interview Survey

James Dahlhamer, Benjamin Zablotsky, Carla Zelaya, and Aaron Maitland

CONTENTS

21.1 Introduction

Through the administration of survey questions, interviewers have the potential to influence the amount of measurement error in the resulting estimates from these questions. Interviewers can have either a positive or a negative effect on the data. On the one hand, interviewers can clarify concepts and probe responses to collect accurate data (Conrad and Schober 2000; Schober and Conrad 1997). On the other hand, differences in how interviewers administer survey questions and probe inadequate responses can lead to systematic deviations or biases in the level of a statistic, such as a mean, across interviewers and increase the overall amount of variance associated with that statistic. For example, some interviewers may elicit more affirmative responses to a question (say, doctor-diagnosed hypertension) relative to other interviewers. While we assume that the systematic biases in doctor-diagnosed hypertension across interviewers cancel out so that they have a mean of zero, the variability caused by the systematic biases remains. This increase in variance due to interviewers is often called the interviewer effect and is measured by a statistic called the interviewer intraclass correlation coefficient, or IIC (Kish 1962). As the IIC increases, so does the variance of the sample statistic. Hence, the increase in variance due to interviewers reduces the ability to detect true differences between groups of respondents.

Interviewer effects have many potential causes, two of which include the behavior of the interviewer and the characteristics of the questions administered by the interviewer. Standardized interviewing relies on all interviewers administering questions in essentially the same manner to all survey respondents. In reality, interviewer behavior may vary within and across interviewers in ways that lead to interviewer effects. For example, interviews become shorter as the field period progresses, suggesting that interviewers may read questions faster as they conduct more interviews (Olson and Peytchev 2007; Holbrook, et al., Chapter 17, this volume; Garbarski, et al., Chapter 18, this volume). Successful interviewers are also required to master at least two often contradictory sets of skills: one being the ability to improvise in doorstep communication to recruit households that are reluctant to participate, and a second being the ability to read a script exactly as worded once the interview has started (Groves and Couper 1998). It is possible that interviewers who are better at tailoring communications at the doorstep are less likely to read questions as worded in the interview given the diverse nature of these skills. Hence, interviewers with higher cooperation rates may be prone to larger interviewer effects (Brunton-Smith, Sturgis, and Williams 2012).

Questions that are more complex, lengthy, or have higher reading levels can be more difficult to administer and harder for respondents to answer, leading to interviewer effects (Mangione, Fowler, and Louis 1992; Pickery and Loosveldt 2001). For example, longer questions include more text for interviewers to read and require respondents to hold more information in working memory, which may lead to more requests for clarification and probing. Sensitive questions that may be embarrassing for interviewers are also prone to interviewer effects (Mangione, Fowler, and Louis 1992; Schnell and Kreuter 2005). Several studies have also found that attitudinal or subjective questions appear to be more prone to interviewer effects relative to factual questions (see West and Blom 2017 for a review). In addition, questions with optional text may be administered inconsistently, with different interviewers applying different rules for when the text should be read.

An ideal design for measuring interviewer effects is an interpenetrated design, where sample units are randomly assigned to interviewers (Mahalanobis 1946). An interpenetrated design ensures that the interviewers' workload assignments are similar in terms of the expected responses to the survey questions. Interpenetrated designs are rare in most large and ongoing in-person surveys that utilize geographic clustering to control sampling costs, because it is often infeasible and cost-prohibitive to randomly assign sample units to interviewers across a large geographic area. It is more realistic to achieve partial interpenetration within a smaller geographic area. In recent years, it has become more common to study interviewer effects in the absence of interpenetrated assignment with multilevel models, where sample units are clustered within interviewers and the models adjust for the characteristics of the respondents and geographic areas being worked by interviewers (Davis, et al. 2010; Hox, de Leeuw, and Kreft 1991; Schaeffer, Dykema, and Maynard 2010; West and Blom 2017).

In this chapter, we describe the use of multilevel models to estimate interviewer effects in the U.S. National Health Interview Survey (NHIS). We first use multilevel models, adjusting for characteristics of the respondents and areas that make up the interviewers' assignments, to estimate the amount of clustering in survey responses by interviewer. We then conduct a second stage of analysis to examine how these interviewer effects vary by question characteristics (e.g., sensitive versus non-sensitive questions), and across groups of interviewers defined by interviewer characteristics (e.g., interviewer cooperation rates).

This is the first study of interviewer variance in the NHIS since the early 1960s (Koons 1973).

21.2 Methods

21.2.1 Data

The data used in these analyses come from 26,742 sample adults aged 18 and over who participated in the 2017 NHIS. A multipurpose, nationally representative health survey of the civilian, noninstitutionalized U.S. population, the NHIS is conducted annually by the National Center for Health Statistics. Interviewers with the U.S. Census Bureau administer the questionnaire using computer-assisted personal interviewing (CAPI). Telephone interviewing is permitted to complete missing portions of the interview.

The sample adult is randomly selected from all adults aged ≥18 years in the family and answers for himself/herself (unless physically or mentally unable to do so, in which case a knowledgeable adult serves as a proxy respondent). The Sample Adult Core module collects information on adult sociodemographics, health conditions, health status and limitations, health behaviors, and health care access and use. The final sample adult response rate for 2017 was 53.0% (National Center for Health Statistics 2018).

21.2.2 Measures

Outcome variables. A total of 125 outcomes, based on 102 questions, were included in the analysis (see supplemental Table A21.1). Thirteen questions with a nominal scale produced multiple outcomes, using the approach described by Mangione, Fowler, and Louis (1992). For these 13 questions, response categories with a prevalence of 5% or greater were made into separate outcomes.

Twenty-five of the 102 questions came from the Family Core module that collects data on topics such as disability status, health insurance coverage, and income. For this module, a respondent aged 18 or older answers for himself/herself and all other members of the family. In roughly 30% of interviews, the sample adult was not the family respondent; hence, outcomes generated from these 25 questions include a mix of self-report and proxy report. Preliminary analyses did show proxy responses to be associated with slightly larger interviewer effects than self-reports. However, to avoid a substantial loss of data, we include a control variable for self (versus proxy) report in all models of outcomes based on questions from the Family Core module.

Given that a primary focus of our research is how interviewer effects vary by question characteristics, the 102 questions used in this analysis were purposively selected based on question characteristics shown to be associated with interviewer effects in past research*: difficult/complex topics (e.g., details of private health insurance plans) versus easier/less complex topics (e.g., demographics such as age, sex, and marital status); long versus short questions; definitions/clarifying text versus no definitions/clarifying text; optional text

---

* It is important to note that this is not an exhaustive nor representative set of NHIS questions. How later results would be impacted if all NHIS questions were included is unknown.

versus no optional text; sensitive versus not sensitive; and demographic/factual versus attitudinal/subjective.

For this analysis, we defined difficult questions or questions on complex topics to be those that required respondents to recall behaviors or events that may be difficult to remember (e.g., ever received the hepatitis A vaccine) or dealt with a complicated topic to which respondents may have given little or no thought (e.g., details of public and private forms of health insurance plans). As proxies for difficult or complex topic questions, we hypothesized that these questions would generate a greater number of inadequate responses and more requests for clarification by the respondent. In turn, this may provide more opportunities for interviewers to improvise and deviate from the interview script, potentially leading to larger interviewer effects.

We also hypothesized that questions including definitions of key terms or clarifying information would present greater comprehension problems for respondents. In turn, these items would likely elicit more requests for clarification or to have the question repeated, again providing more opportunities for interviewers to improvise and stray from the interview protocol (e.g., shortening or simplifying the question). Similarly, we hypothesized that questions with lower Flesch reading ease scores (Flesch 1948), i.e. questions that are more difficult to read, would lead to variations across interviewers in how the questions were administered. Additionally, these questions may lead to more comprehension problems among respondents. Flesch reading ease scores were grouped into the following categories for analysis: very easy/easy/fairly easy (70.0 ≤ scores ≤ 100.0), standard (60 ≤ scores < 70.0), and fairly difficult/difficult/very confusing (0.0 ≤ scores < 60.0). Finally, we hypothesized that questions with optional text would lead to variability across interviewers as to when this text was read, in turn leading to larger interviewer effects.

Sensitive questions may lead to larger interviewer effects compared to non-sensitive items (Mangione, Fowler, and Louis 1992; Schnell and Kreuter 2005). To identify sensitive questions for this study, we turned to earlier research using NHIS data to explore the effects of respondent, question, and interviewer characteristics on item response times and item nonresponse (Dahlhamer, et al. 2019). In that research, each author (5 total) rated the sensitivity of 270 questions using two rating items: "This question is very personal (1 = completely disagree to 5 = completely agree)" and "I would be uncomfortable asking this question (1 = completely disagree to 5 = completely agree)". For each question under analysis, the five raters' scores on the first rating item were summed and then the five raters' scores on the second rating item were summed. Since the summed scores based on the two rating items were highly correlated (0.94), the two rating scores for each question under analysis were further summed to create an index of sensitivity.* The index was then recoded into four discrete categories using quartiles as cut points. Questions falling in quartile four had the highest sensitivity scores (index scores ranging from 18 to 37). A subset of these questions was selected for inclusion in this study.

Finally, we selected a mix of attitudinal/subjective questions and factual/demographic questions, again using Dahlhamer, et al. (2019) as a guide. (Please see online supplemental Table A21.1 for all questions included in the analysis and how they were coded with respect to question characteristics.)

---

* For example, for a given question under analysis, the following scores were assigned by each of the five raters using rating item #1: 1, 3, 3, 2, and 1. The total score assigned to that question using rating item #1 would be 10. Each of the five raters then assigned the following scores to the question under analysis using rating item #2: 1, 3, 4, 1, and 2. The total score assigned to that question using rating item #2 would be 11. The two rating scores were then summed to produce an index score of 21 for that survey question.

Covariates included in multilevel models. For each outcome variable, we fitted a multilevel model that included fixed effects of respondent characteristics, interviewer characteristics, and county characteristics. Respondent characteristics included age (18–24, 25–44, 45–64, 65+), sex, race/ethnicity and language (Hispanic–English, Hispanic–Spanish, non-Hispanic white–any language, non-Hispanic black–any language, non-Hispanic other–any language), education (less than high school diploma/General Educational Development (GED) credential, high school diploma/GED, some college/Associate of Arts (AA) degree, bachelor's degree or higher), marital status (married/living with partner, never married, divorced/separated, widowed), reported health status (poor/fair, good, very good/excellent), cognitive difficulties (whether or not sample adult is limited in any way because of difficulty remembering or because he/she experiences periods of confusion), and mode of interview (telephone versus in-person). As noted earlier, for outcomes based on questions in the Family Core module, an additional control was included for whether the sample adult response was self-reported or reported by a proxy. Finally, we included the sample adult sampling weight (log-transformed) as an additional control variable in the full models. (Please see supplemental Table A21.2 for descriptive statistics for respondent characteristics included in the multilevel models.)

The fixed effects of county-level control variables were also included in the final models. These measures largely mimicked the respondent-level demographic measures and represent five-year rolling estimates taken from the American Community Survey (2012–2016). Measures included the proportion of the county population that was: aged 65 and over; under the age of 30; female; aged 25 and older that has less than a high school diploma or GED; aged 25 and older that has at least a bachelor's degree; married; never married; without health insurance coverage; with a physical or mental disability; Hispanic; and non-Hispanic black. The last two measures on this list are categorical (quartiles), while all other measures are retained as continuous measures and grand mean centered. (Please see online supplemental Table A21.3 for descriptive statistics for county-level measures included in the multilevel models.)

Finally, the fixed effects of four interviewer characteristics were included in the final models, including two measures of interviewer experience: whether the interviewer worked on the NHIS in 2016 and interviewing experience in 2017. The latter is a chronological count of the sample adult interviews conducted by the interviewer for 2017. For both measures, we hypothesize that greater interviewing experience will lead to greater mastery of interviewing skills and the interview protocol which, in turn, may be associated with smaller interviewer effects on estimates. A third measure captured the interviewer's cooperation rate for 2017. We included this measure to explore the link between interviewers' abilities to secure respondent participation and subsequent data quality. Finally, we included a measure of within-interview interviewer behavior: pace of sample adult interviews for 2017. Pace was defined as the mean number of seconds per question. We hypothesize that a faster interview pace may lead to greater deviations from the interview protocol (e.g., reading questions exactly as worded) and, therefore, may be associated with larger interviewer effects on estimates. (Please see supplemental Table A21.4 for descriptive statistics for interviewer characteristics included in the multilevel models.)

21.2.3 Data Structure

The analytic data set has a complex nested structure. Each of the 125 outcomes was measured at the respondent level. Respondents (level 1) are uniquely nested within interviewers (level 2) as well as counties (level 2). Interviewers, however, are not uniquely nested within counties, resulting in a cross-classified data structure. Since interviewers and counties are treated as level 2 units in this analysis, interviewers and counties are said

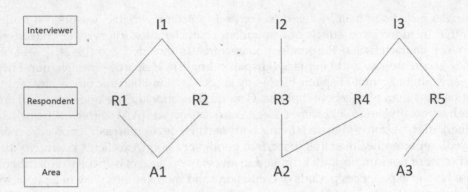

FIGURE 21.1
Network graph depicting cross-classification of respondents (125 outcomes) by interviewer and county.

to be cross-classified for each of the 125 outcomes. Figure 21.1 depicts the cross-classified data structure. In total, 26,742 sample adults were nested within 991 interviewers and 815 counties. Therefore, on average, there are roughly 27 respondents per interviewer and 33 respondents per county.* We did not constrain the number of respondents (level 1 units) per interviewer (level 2 units) or county (level 2 units) to some minimum (e.g., 10 per group). Research has shown that the number of groups or level 2 units (of which we have large sample sizes) is more important in determining statistical power for multilevel modeling than the number of level 1 units within groups (Snijders 2005; West, Chapter 23, this volume), and that small numbers of level 1 units within groups are not detrimental to point and interval estimates of parameters (Bell, et al. 2014; Maas and Hox 2005).

21.2.4 Statistical Analyses

For each of the 125 outcomes, we started with an unconditional model that included only the random effects for interviewers. In step two, we added the random effects for counties. In step three, we added the fixed effects of respondent characteristics to the model. In steps four and five, we added the fixed effects of the county-level characteristics and then the interviewer characteristics, respectively. Hence, step five was the final step in the modeling process in which all random and fixed effects were included in the model.[†] The IIC based on the final model for each outcome is the focus of all analyses reported hereafter.

Of the 125 outcomes, 119 are dichotomous and were therefore modeled using logistic regression. Following Beretvas (2010), the full logistic regression model predicts a logit transformation of the probability that a dichotomous outcome for respondent i is equal to

1, $\ln\left(\dfrac{p_{i(s,t)}}{1-p_{i(s,t)}}\right)$, as a function of an overall mean (γ), a set of p respondent characteristics

$\left(\displaystyle\sum_{a=1}^{p}\beta_a \text{Respondent\_char}_{ai}\right)$, a set of q county characteristics $\left(\displaystyle\sum_{b=1}^{q}\beta_b \text{County\_char}_b\right)$, a set of r

---

* Note that the number of sample adults, interviewers, and counties vary across outcomes given item nonresponse and question universe.

† Due to minimal or no variance at the county level, final models did not converge for some outcomes. To achieve convergence in these cases, random effects for county were dropped and the model was re-estimated (county-level fixed effects were retained).

interviewer characteristics $\left(\sum_{c=1}^{r}\beta_c\text{Interviewer\_char}_c\right)$, a random effect due to county s (u_{0s}), and a random effect due to interviewer t (u_{0t}), where u_{0s} and u_{0t} are normally distributed with mean zero and variances σ_{u0s} and σ_{u0t} respectively:

$$\ln\left(\frac{p_{i(s,t)}}{1-p_{i(s,t)}}\right) = \gamma + \sum_{a=1}^{p}\beta_a\text{Respondent\_char}_{ai} + \sum_{b=1}^{q}\beta_b\text{County\_char}_b$$

$$+ \sum_{c=1}^{r}\beta_c\text{Interviewer\_char}_c + u_{0s} + u_{0t} \qquad (21.1)$$

The following formula was used to approximate the value of the IIC for dichotomous outcomes:

$$\rho_{\text{interviewer}} = \frac{\sigma_{u_t}^2}{\sigma_{u_s}^2 + \sigma_{u_t}^2 + 3.29} \qquad (21.2)$$

The level 1 (respondent) variance is set to 3.29, which is the variance of the underlying standard logistic distribution (Snijders and Bosker 1999).

The remaining six outcomes were treated as continuous and modeled using linear regression. Again, following Beretvas (2010), the full linear regression model predicts a continuous outcome $(Y_{i(s,t)})$ as a function of an overall mean (γ), a set of p respondent characteristics $\left(\sum_{u=1}^{p}\beta_a\text{Respondent\_char}_{ai}\right)$, a set of q county characteristics $\left(\sum_{b=1}^{q}\beta_b\text{County\_char}_b\right)$, a set of r interviewer characteristics $\left(\sum_{c=1}^{r}\beta_c\text{Interviewer\_char}_c\right)$, a random effect due to county s (u_{0s}), a random effect due to interviewer t (u_{0t}), and a residual term $(e_{i(s,t)})$, where u_{0s} and u_{0t} are normally distributed with mean zero and variances $\sigma_{u_s}^2$ and $\sigma_{u_t}^2$, respectively, and $e_{i(s,t)}$ is normally distributed with mean zero and variance $\sigma_e{}^2$:

$$Y_{i(s,t)} = \gamma + \sum_{a=1}^{p}\beta_a\text{Respondent\_char}_{ai} + \sum_{b=1}^{q}\beta_b\text{County\_char}_b$$

$$+ \sum_{c=1}^{r}\beta_c\text{Interviewer\_char}_c + u_{0s} + u_{0t} + e_{i(s,t)} \qquad (21.3)$$

The IIC equation for continuous outcomes is as follows:

$$\rho_{\text{interviewer}} = \frac{\sigma_{u_t}^2}{\sigma_{u_s}^2 + \sigma_{u_t}^2 + \sigma_e^2} \qquad (21.4)$$

The two-level, cross-classified logistic regression models were estimated with SAS PROC GLIMMIX (v9.4) using maximum likelihood estimation (method = laplace), while the two-level, cross-classified linear regression models were estimated with SAS PROC MIXED (v9.4) using restricted maximum likelihood estimation.

IICs by question characteristics. As noted previously, 13 of the 102 questions under analysis produced two or more outcomes. To analyze results at the question level, IICs for multiple

TABLE 21.1

Descriptive Statistics for Measures Used in Analyses of IICs by Question Characteristics

Measure	Number of Questions	Percent
Question on a complex topic		
Yes	22	21.6
No	80	78.4
Question length		
Quartile 1 (<68 characters)	24	23.5
Quartile 2 (68–94 characters)	26	25.5
Quartile 3 (95–155 characters)	26	25.5
Quartile 4 (≥156 characters)	26	25.5
Flesch reading ease score		
Very easy/easy/fairly easy (scores of 70.0–100.0)	55	53.9
Standard (scores of 60.0 to <70.0)	25	24.5
Fairly difficult/difficult/very confusing (scores of 0.0 to <60.0)	22	21.6
Question includes definitions/clarifying statements		
Yes	25	24.5
No	77	75.5
Question includes optional text		
Yes	52	51.0
No	50	49.0
Type of question		
Factual/demographic	85	83.3
Attitudinal/subjective	17	16.7
Question is deemed to be sensitive		
Yes	17	16.7
No	85	83.3

Source: National Health Interview Survey, 2017.

outcomes from a single question were averaged; the average IIC was then assigned to that question. We took this approach to ensure that a characteristic associated with a single question would not be over-represented in the data set. This was especially important for our analysis of IICs by question characteristics. To explore differences in IICs by question characteristics, we computed median IICs across the 102 questions for each category of a question characteristic (e.g., the median IIC for questions that include optional text versus the median IIC for questions that do not include optional text; see Table 21.1 for descriptive statistics for the question characteristics). As discussed in the Results section, we focus on medians as opposed to means due to the right-skewed distribution of the estimated IICs. Therefore, to test for significant differences in median IICs by categories of a question characteristic, we used the following non-parametric tests: the Mann–Whitney–Wilcoxon two-sample test for characteristics with two categories (e.g., sensitive questions versus non-sensitive questions) and the Kruskal–Wallis test for measures with three or more categories (e.g., length of question, broken into quartiles). Given the small sample size ($n = 102$) for these analyses, we used an alpha level of .10 to determine if differences in median IICs were statistically significant.

IICs by interviewer characteristics. We are also interested in the associations between characteristics of interviewers or their behaviors and interviewer effects. For interviewer pace, we were particularly interested in the tails of the distribution, that is, interviewers

TABLE 21.2

Descriptive Statistics for Measures Used in Analyses of IICs by Interviewer Characteristics

Measure	Number of Interviewers	Percent of Interviewers	Number of Interviews	Percent of Interviews
Pace of interview (mean seconds per question)				
Group 1: <6.81	165	16.6	4,041	15.1
Group 2: ≥6.81 to ≤10.57	632	63.8	18,709	70.0
Group 3: >10.57	194	19.6	3,992	14.9
Cooperation rate				
Group 1: <64.02	296	29.9	3,980	14.9
Group 2: ≥64.02 to ≤87.85	530	53.5	18,793	70.3
Group 3: >87.85	165	16.6	3,969	14.8
Number of sample adult interviews				
Group 1: 1–20	515	52.0	4,157	15.5
Group 2: 21–40	236	23.8	6,977	26.1
Group 3: 41 or more	240	24.2	15,608	58.4
Worked on the NHIS in 2016?				
Yes	855	86.3	1,768	6.6
No	136	13.7	24,974	93.4

Source: National Health Interview Survey, 2017.

who went the fastest and interviewers who went the slowest. We created a trichotomous measure of pace, whereby the fastest interviewers (who worked roughly 15% of interviews) comprised one group, the slowest interviewers (who also worked roughly 15% of interviews) comprised the second group, and the remainder of interviewers comprised the third group. A similar measure was created for interviewer cooperation rates. Interviewers with the lowest cooperation rates (who worked roughly 15% of interviews) were assigned to one group, interviewers with the highest cooperation rates (who also worked roughly 15% of interviews) were assigned to a second group, and the remainder of interviewers comprised the third group. Finally, we focused on two measures of interviewer experience: whether the interviewer worked on the NHIS in 2016 and the total number of sample adult interviews the interviewer worked in 2017. The latter measure defined three groups of interviewers: 1–20 sample adult interviews, 21–40 sample adult interviews, and 41 or more sample adult interviews. (See Table 21.2 for descriptive statistics for these measures.)

Taking interviewer pace as an example, the full model described in Equation 21.1 (bivariate outcomes) or Equation 21.3 (continuous outcomes) was then estimated and the IIC was computed (Equation 21.2 for bivariate outcomes, Equation 21.4 for continuous outcomes) for each of the 39 questions analyzed (see details regarding selection of questions below) using data collected by the fastest interviewers.* This process was repeated for the slowest interviewers, and then for the interviewers with medium pace. A data set was then constructed that included a total of 117 IICs: 39 for interviewers with the slowest pace, 39 for interviewers with the fastest pace, and 39 for interviewers with a medium pace. The same

---

* Convergence errors and/or non-positive definite random effect covariance matrices emerged for some outcomes due to insufficient variance at the county level. Dropping the random county effects from the models generally solved this problem.

steps were undertaken for the three interviewer cooperation rate groups, the three groups of interviewers defined by the number of sample adult interviews completed in 2017, and the two groups of interviewers defined by whether they worked on the NHIS in 2016. (See online supplemental Table A21.5 for a description of the data set used in these analyses.)

We then took the computed IICs and tested for differences in median IIC for the groups defined by each interviewer characteristic using either a Kruskal–Wallis test (for interviewer measures broken into three groups) or a Mann–Whitney–Wilcoxon two-sample test (for the "worked on the NHIS in 2016" measure). Again, the resulting distributions of IICs within groups of interviewers tended to be right-skewed; hence, the focus on medians as opposed to means. In addition, where suggested by the initial results, we collapsed two groups (for the trichotomous measures) and re-tested for a significant difference using Mann–Whitney–Wilcoxon two-sample tests. Again, given the small sample sizes for these analyses, we used an alpha level of .10 to determine if differences in median IICs were statistically significant.

As a check on the robustness of the findings of these analyses, we also estimated two-level models in which random intercepts were included for question and fixed effects were estimated for a specific interviewer characteristic (e.g., pace of interview defined as fastest, slowest, and medium). The results of these models corroborate the findings of the analyses described in this section and reported in Section 21.3.2.

Selection of questions for analysis of interviewer characteristics. We used the following criteria to select questions for this analysis. First, to ensure adequate sample sizes, we focused on questions where all sample adults were in universe. Second, for dichotomous outcomes, we limited the analysis to those with at least a 90%/10% split to avoid unstable models resulting from empty cells. And third, for questions that produced multiple outcomes, we selected the most prevalent response category to represent the question. In all, 45 of the 102 questions met these criteria. Due to model convergence errors, however, six items were eliminated, leaving 39 questions for analysis.

21.3 Results

Across the 102 questions, the IICs ranged from a low of .0011 to a high of .2273. The mean IIC was .0507 and the median was .0372. As these results indicate, the distribution of IICs was right skewed. As a result, we present medians in all subsequent analyses of IICs by question and interviewer characteristics.

21.3.1 IICs by Question Characteristics

Table 21.3 presents median IICs by seven question characteristics, two of which were significantly associated with interviewer effects: complex or difficult topic and length of question. Questions on complex or difficult topics (e.g., details of private health insurance plans) produced a median IIC of .081 (n = 22) compared to a median IIC of .031 (n = 80) for questions on less complex/difficult items (e.g., demographics such as age and sex; Mann–Whitney–Wilcoxon test, p-value <.001).

While the median IICs across the first three quartiles of question length were similar (quartile 1 median IIC = .029; quartile 2 median IIC = .023; quartile 3 median IIC = .036), the longest questions (quartile 4, or >155 characters) had the largest median IIC at .052

TABLE 21.3

Median IICs by Question Characteristics (n = 102 Questions)

Question Characteristic	n	Median IIC	p-Value[a]
Question on a complex topic			<0.001[b]
Yes	22	0.081	
No	80	0.031	
Question length			0.056[c]
Quartile 1 (<68 characters)	24	0.029	
Quartile 2 (68–94 characters)	26	0.023	
Quartile 3 (95–155 characters)	26	0.036	
Quartile 4 (≥156 characters)	26	0.052	
Flesch reading ease score			0.134[c]
Very easy/easy/fairly easy (scores of 70.0–100.0)	55	0.031	
Standard (scores of 60.0 to <70.0)	25	0.034	
Fairly difficult/difficult/very confusing (scores of 0.0 to <60.0)	22	0.052	
Question includes definitions/clarifying statements			0.243[b]
Yes	25	0.041	
No	77	0.034	
Question includes optional text			0.565[b]
Yes	52	0.034	
No	50	0.040	
Type of question			0.177[b]
Factual/demographic	85	0.031	
Attitudinal/subjective	17	0.041	
Question is deemed to be sensitive			0.584[b]
Yes	17	0.034	
No	85	0.038	

Source: National Health Interview Survey, 2017.

[a] An alpha level of .10 was used to determine if differences in median IICs were statistically significant.

[b] p-Value based on a Mann–Whitney–Wilcoxon test.

[c] p-Value based on a Kruskal–Wallis test.

(Kruskal–Wallis test, p-value = .056). Collapsing quartiles one through three results in a median IIC of .032 (n = 76), which is significantly lower than the median IIC for quartile 4 (Mann–Whitney–Wilcoxon test, p-value = .011). We observe a similar finding for median IIC by the Flesch–Kincaid reading ease score. Collapsing the last four categories – "standard," "fairly difficult," "difficult," or "very confusing" items – results in a median IIC of .048, which is significantly higher than the median IIC (.031) for the "very easy," "easy," or "fairly easy" items at the .10 alpha level (Mann–Whitney–Wilcoxon test, p-value = .085).

21.3.2 Interviewer IICs by Interviewer Characteristics

As described earlier, 39 outcomes or questions were selected for the analysis of median IICs by interviewer characteristics. The overall IICs for these 39 items ranged from .0071 to .2113. Table 21.4 presents median IICs for each group of interviewers defined by each of four interviewer characteristics. For pace of interview, the fastest interviewers (group 1)

TABLE 21.4

Median IICs for 39 Items by Interviewer Characteristic

Interviewer Characteristic	Median IIC	p-Value[a]
Pace of interview (mean seconds per question)		0.073[2]
Group 1: <6.81	0.049	
Group 2: ≥6.81 to ≤10.57	0.034	
Group 3: >10.57	0.032	
Cooperation rate		0.104[b]
Group 1: <64.02	0.047	
Group 2: ≥64.02 to ≤87.85	0.039	
Group 3: >87.85	0.020	
Number of sample adult interviews		0.021[c]
Group 1: 1–20	0.047	
Group 2: 21–40	0.038	
Group 3: 41+	0.030	
Worked on the NHIS in 2016?		0.208[c]
No	0.035	
Yes	0.042	

Source: National Health Interview Survey, 2017.
[a] An alpha level of .10 was used to determine if differences in median IICs
    were statistically significant.
[b] p-Value based on a Kruskal–Wallis test.
[c] p-Value based on a Mann–Whitney–Wilcoxon test.

had the highest median IIC (.049) among the three groups, with groups 2 and 3 (slowest) having similar median IICs (.034 and .032 respectively). A Kruskal–Wallis test reveals a difference in medians significant at the .10 level (p = .073). For 22 of the 39 items, the fastest interviewers have the largest IIC. Collapsing groups 2 and 3, given their similar median IICs, and then comparing to group 1 results in a significant difference in medians at the .05 level (Mann–Whitney–Wilcoxon test, p-value = .023). Hence, interviewers performing at the fastest pace appear to be associated with a significantly higher median IIC compared with slower paced interviewers.

Regarding interviewer cooperation rates, we observe a consistent but nonsignificant decline in the median IIC across the three groups, whereby interviewers with the lowest cooperation rates have the highest median IIC (.047) and the interviewers with the highest cooperation rates have the lowest median IIC (.020). We collapsed groups 1 and 2 into a single category. This collapsed group of interviewers had cooperation rates of approximately 88% or less and a median IIC of .040 across the 39 items. When compared to the median IIC (.020) of the interviewers with the highest cooperation rates, the difference is significant at the .05 level (Mann–Whitney–Wilcoxon test, p-value = .035).

For the number of completed sample adult interviews measure, we observe a significant decline in the median IIC as interviewers conduct more sample adult interviews (Kruskal–Wallis test, p-value = .021). In pairwise comparisons, no significant difference in median IIC is observed between group 1 and group 2, but significant differences are observed between groups 1 and 3 (Mann–Whitney–Wilcoxon test, p-value = .022), and groups 2 and 3 (Mann–Whitney–Wilcoxon test, p-value = .013). To further underscore these findings, the smallest IIC was observed for group 3 interviewers (completed 41 or more sample adult interviews) for 28 of the 39 questions.

21.4 Discussion

Interviewers are integral to the success of face-to-face surveys through the administration of questions during the interview. During the question-and-answer process, interviewers may positively reduce bias by clarifying concepts and probing responses effectively (Conrad and Schober 2000; Schober and Conrad 1997). Conversely, interviewers can negatively impact the variability of estimates through differences in how they ask questions and probe inadequate responses (Kish 1962). The interviewer effect, or the increase in variance of a sample statistic that is attributable to interviewers, is one way to measure this influence on survey responses. This increase in variance in the resulting survey estimate reduces the ability to detect true differences between groups of respondents. In this chapter, we used multilevel models to estimate interviewer effects on several NHIS outcomes. We also investigated the extent to which interviewer effects vary by question and interviewer characteristics.

Question characteristics were shown to be associated with interviewer effects. Questions on complex topics, long questions, and relatively more difficult questions to read (as measured by Flesch–Kincaid reading ease score) were associated with larger interviewer effects than questions on less complex topics, shorter questions, and relatively easy questions to read. Questions with these characteristics are likely to be more difficult for interviewers to administer, leading to more requests for clarification by survey respondents as well as more interviewer intervention to obtain adequate answers. In other words, difficult questions consistently introduce interviewer effects (West and Blom 2017).

Interviewer characteristics are also associated with interviewer effects. Interviewers who conduct the interview at a faster pace tend to have larger interviewer effects compared to those who conduct the interview at a slower pace. This finding suggests that having interviewers slow down when reading questions may be beneficial for data quality. We also found that interviewers with the highest cooperation rates tend to have the lowest interviewer effects. In one of the few prior studies to address the interaction between interviewer cooperation rates and interviewer variance, Brunton-Smith, Sturgis, and Williams (2012) identified a curvilinear relationship, whereby interviewers with the lowest and highest cooperation rates produced the largest interviewer effects. While more research in this area is necessary, our results suggest that high-performing interviewers can master the tailoring and improvisational skills needed to counter respondent reluctance and gain cooperation at the doorstep and adhere to a standardized interviewing protocol once the interview starts.

This research is not without limitations. In order to study interviewer effects in an ongoing survey like the NHIS, we used multilevel models to isolate the effect of the interviewer from the area in which the interviews are conducted. Although this is a commonly used approach, given the cost of implementing interpenetrated samples, it is possible that we have not adequately adjusted for respondent and area effects, leading to a misallocation of some of the variance in our models. In addition, we reported simple bivariate results for the relationships of question and interviewer characteristics with interviewer effects. The reported findings may differ if the simultaneous effects of question and interviewer characteristics on interviewer effects are explored. (For a more thorough description of statistical models that would enable such analyses, see Loosveldt and Wuyts, Chapter 22 in this volume). Finally, we were limited to interviewer measures that could be constructed from the available survey data. While the findings of past research are somewhat mixed (see West and Blom 2017), observable (e.g., age, sex, and race) and unobservable (e.g., attitudes about interviewing) characteristics of interviewers may also play an important role in explaining the effects of interviewers on NHIS estimates.

In conclusion, we found features of questions and characteristics and behaviors of interviewers to be significantly associated with interviewer effects on NHIS estimates. In addition to understanding the relative influence of these characteristics in a multivariable context, future research could leverage other methods to help isolate the underlying causes of the interviewer effects we observed. For example, behavior coding of computer-assisted recorded interviews (CARI) could clarify the interviewer or respondent behaviors, as well as question features, which lead to interviewer effects. In addition, future work could focus on how best to use results from the multilevel models described herein to identify interviewers with the largest impact on IICs (hence, as a tool for monitoring interviewer performance). Following a method described by Kreuter (2002), for example, IICs would be estimated for a subset of questions for the entire set of interviewers. Next, the IICs would be re-estimated dropping all the cases from one interviewer from the data set. If the IIC drops significantly for a question when an interviewer's cases are removed, that interviewer has a substantial impact on the IIC and the interviewer effect for that item. The interviewer in question would be flagged for follow-up and possible re-training. Another approach would be to identify interviewers with extreme predicted values of random effects (EBLUPs) during a given period of data collection and follow up with those interviewers to understand if they are struggling with particular questions.

References

Bell, B. A., G. B. Morgan, J. A. Schoenberger, J. D. Kromrey, and J. M. Ferron. 2014. How low can you go? An investigation of the influence of sample size and model complexity on point and interval estimates in two-level linear models. *Methodology* 10(1):1–11.

Beretvas, S. N. 2010. Cross-classified and multiple-membership models. In: *Handbook of Advanced Multilevel Analysis*, ed. J. J. Hox, and J. K. Roberts, 313–334. New York: Routledge.

Brunton-Smith, I., P. Sturgis, and J. Williams. 2012. Is success in obtaining contact and cooperation correlated with the magnitude of interviewer variance? *Public Opinion Quarterly* 76(2):265–286.

Conrad, F. G., and M. F. Schober. 2000. Clarifying question meaning in a household telephone survey. *Public Opinion Quarterly* 64(1):1–28.

Dahlhamer, J. M., A. Maitland, H. Ridolfo, A. Allen, and D. Brooks. 2019. Exploring the associations between question characteristics, respondent characteristics, interviewer characteristics and survey data quality. Pp. 153-192 in: *Advances in Questionnaire Design, Development, Evaluation and Testing*, ed. P. Beatty, D. Collins, L. Kaye, J. L. Padilla, G. Willis, and A. Wilmot. New York: Wiley.

Davis, R. E., M. P. Couper, N. K. Janz, C. H. Caldwell, and K. Resnicow. 2010. Interviewer effects in public health surveys. *Health Education Research* 25(1):14–26.

Flesch, R. 1948. A new readability yardstick. *The Journal of Applied Psychology* 32(3):221–233.

Groves, R. M., and M. P. Couper. 1998. *Nonresponse in Household Interview Surveys*. New York: Wiley.

Hox, J. J., E. D. de Leeuw, and I. G. G. Kreft. 1991. The effect of interviewer and respondent characteristics on the quality of survey data: A multilevel model. In: *Measurement Errors in Surveys*, ed. P. Biemer, R. M. Groves, L. E. Lyberg, N. A. Mathiowetz, and S. Sudman, 439–461. New York: Wiley.

Kish, L. 1962. Studies of interviewer variance for attitudinal variables. *Journal of the American Statistical Association* 57(297):92–115.

Koons, D. A. 1973. *An Experimental Comparison of Telephone and Personal Health Interview Surveys Vital and Health Statistics Series 2, No. 54.* Hyattsville: National Center for Health Statistics.

Kreuter, F. 2002. *Kriminalitatsfurcht: Messung und methodische probleme.* Berlin: Leske and Budrich.

Maas, C. J. M., and J. J. Hox. 2005. Sufficient sample size for multilevel modeling. *Methodology* 1(3):86–92.

Mahalanobis, P. C. 1946. Recent experiments in statistical sampling in the Indian statistical institute. *Journal of the Royal Statistical Society* 109:325–378.

Mangione, T. W., F. J. Fowler, Jr., and T. A. Louis. 1992. Question characteristics and interviewer effects. *Journal of Official Statistics* 8(3):293–307.

National Center for Health Statistics. 2018. *2017 National Health Interview Survey (NHIS) Public Use Data Release: Survey Description.* Hyattsville: National Center for Health Statistics.

Olson, K., and A. Peytchev. 2007. Effect of interviewer experience on interviewer pace and interviewer attitudes. *Public Opinion Quarterly* 71(2):273–286.

Pickery, J., and G. Loosveldt. 2001. An exploration of question characteristics that mediate interviewer effects on item nonresponse. *Journal of Official Statistics* 17(3):337–350.

Schaeffer, N. C., J. Dykema, and D. W. Maynard. 2010. Interviewers and interviewing. In: *Handbook of Survey Research*, 2nd Edition, ed. P. V. Marsden, and J. D. Wright, 437–470. Bingley: Emerald Group Publishing Limited.

Schnell, R., and F. Kreuter. 2005. Separating interviewer and sampling-point effects. *Journal of Official Statistics* 21(3):389–410.

Schober, M. F., and F. G. Conrad. 1997. Does conversational interviewing reduce survey does conversational interviewing reduce survey measurement error? *Public Opinion Quarterly* 61(4):576–602.

Snijders, T. 2005. Power and sample size in multilevel linear models. In: *Encyclopedia of Statistics in Behavioral Science*, volume 3, ed. B. S. Everitt and D. C. Howell, 1570–1573. Hoboken: John Wiley & Sons.

Snijders, T., and R. Bosker. 1999. *Multilevel Analysis.* London: Sage.

West, B. T., and A. Blom. 2017. Explaining interviewer effects: A research synthesis. *Journal of Survey Statistics and Methodology* 5:175–211.

22

A Comparison of Different Approaches to Examining Whether Interviewer Effects Tend to Vary Across Different Subgroups of Respondents

Geert Loosveldt and Celine Wuyts

CONTENTS

22.1 Introduction

It is common knowledge in the field of survey methodology that interviewers in face-to-face or telephone interviews can have undesirable effects on the obtained answers. Interviewers may introduce these effects in an active way, for example by asking suggestive questions. The effects may also arise in a passive way, if certain interviewer characteristics ultimately elicit socially desirable answers.

These active and passive effects may be systematic across all interviewers, resulting in "pure" interviewer bias, or they may differ from interviewer to interviewer, resulting in additional variance in the data that is attributable to systematic differences between interviewers rather than differences between respondents. In this chapter, we focus on the latter type of interviewer effects. One can analyze interviewer effects of this type by decomposing the variance in a (substantive) variable into variance that can be explained by the differences between interviewers, termed "between-interviewer variance," and residual or "within-interviewer" variance. Between-interviewer variance reduces the statistical precision of survey estimates, and in some cases, this variance can seriously affect data quality.

There is a long tradition of analyzing interviewer effects by means of interviewer variance analysis (for two reviews, see Schaeffer, Dykema, and Maynard 2010 and West and Blom 2017). However, very few papers have examined whether interviewer effects are stronger or weaker in specific subgroups of respondents. One might argue, for example, that cognitive difficulties, presumably experienced to a greater extent by lower-educated respondents (Knäuper, et al. 1997; Krosnick and Alwin 1987; Narayan and Krosnick 1996), obstruct the paradigmatic question–response process and provoke additional interventions of the interviewers. In this situation, the interviewers' task to make the question–response process go smoothly becomes more complex and the risk of interviewers influencing respondents' answers in a systematic way increases. Following this reasoning, the interviewer–respondent interaction is considered as a stepping stone to studying the link between interviewer effects and respondents' education level, and one can expect that interviewer effects are likely to be stronger among lower-educated respondents. Identifying groups of respondents who are more prone to interviewer effects allows for a more focused analysis of interviewer behavior, and can provide useful information to develop interviewer training programs with special attention to respondent groups that tend to engender higher interviewer effects.

We discuss the basic statistical model used in the research tradition of analyzing interviewer effects in the next section. We will argue that this basic model is not suitable for comparing interviewer effects across different respondent groups. In this chapter, we show how one can extend the basic model to allow for such investigations. In particular, we will demonstrate that lower-educated respondents elicit larger interviewer effects.

22.2 The Basic Model

A frequently used metric for evaluating interviewer variance is the intra-class correlation coefficient (ICC), also referred to as the intra-interviewer correlation (IIC) in this context because the classes are defined by interviewers (Hox 1994). This coefficient expresses the homogeneity of the obtained answers within interviewers, or equivalently the ratio of the between-interviewer variance in a variable of interest to the total variance in that variable. To calculate the within-interviewer and the between-interviewer variance components that are critical for computing the IIC, it is necessary to take into account the two-level hierarchical data structure in which respondents are nested within interviewers. A two-level random intercept model with Y as the dependent variable and no independent variables (i.e., a null model) is generally the starting point to estimate the within- and between-interviewer variance components.

The standard expression of this model is

$$Y_{ij} = \gamma_{00} + u_{0j} + e_{ij}. \tag{22.1}$$

In this model, Y_{ij} is the value of the dependent variable Y for respondent i ($i = 1, ..., N$) interviewed by interviewer j ($j = 1, ..., J$), γ_{00} is the fixed (overall) intercept, u_{0j} is the interviewer-specific part of the random intercept for interviewer j (interviewer level), and e_{ij} is the residual error term for respondent i interviewed by interviewer j (respondent level), with $e_{ij} \sim N\left(0, \sigma_e^2\right)$ and $u_{0j} \sim N\left(0, \sigma_u^2\right)$.

There are significant differences between interviewers when σ_u^2 differs significantly from zero. The IIC for each variable Y is estimated as the proportion of the total variance in the dependent variable Y that is explained by the differences between interviewers:

$$IIC = \frac{\sigma_u^2}{\sigma_u^2 + \sigma_e^2}.$$

This quantitative expression of the interviewer effect and interpretation of the IIC is correct under the "comparable respondent groups" assumption. This means that the differences between interviewers are not arising because interviewers did not interview comparable groups of respondents. To control for possible differences between the interviewers in terms of the groups of respondents interviewed, one can expand the basic model in Equation 22.1 by including relevant background characteristics of respondents as predictors of the variable Y. Accordingly, the interviewer effects express the variability between interviewers *after adjusting for these respondent characteristics*. Notice that the interpretation of the relationships of the independent variables with the dependent variable in the expanded basic model concerns explaining the variability in the dependent variable, and not explaining the differences between interviewers. In fact, the relationships between respondent characteristics and interviewer effects are not specified in the model.

In addition to the elaboration of the basic model with respondent characteristics, one could also expand the model by including interviewer characteristics (e.g., experience, workload, gender, etc.). This further extension of the basic model allows us to explain the observed differences between interviewers. If these characteristics partly explain the interviewer variance, they provide some insights into the mechanisms underlying interviewer effects. This elaboration of the basic model to include respondent and interviewer characteristics results in the following model:

$$Y_{ij} = \gamma_{00} + \sum_{r=1}^{R} \gamma_{rj} X_{rij} + \sum_{t=1}^{T} \gamma_{0t} W_{tj} + u_{0j} + e_{ij} \tag{22.2}$$

with X_{rij} indicating respondent characteristics X_r ($r = 1, \ldots, R$) measured for respondent i by interviewer j, and W_{tj} indicating interviewer characteristics W_t ($t = 1, \ldots, T$) measured for interviewer j. Notice that the inclusion of respondent characteristics (the "control by modeling" approach) is only a partial solution to the fact that in most cases there is no interpenetrated sample assignment to the interviewers (the "control by design" approach).

The extended basic model with respondent and interviewer characteristics (Equation 22.2) allows analysts to assess the relationships between the interviewer-specific means and interviewer characteristics, but not the relationships between interviewer effects and respondent characteristics. However, it is reasonable to assume that some groups of respondents have more difficulties in understanding and answering the questions, which intensifies the interaction between interviewer and respondent, increasing the risk of interviewer effects. This can result in higher IICs in these respondent groups. The specification of the basic multilevel model does not allow us to investigate whether the occurrence of interviewer effects varies across different respondent subgroups.

In this chapter, we focus on two procedures based on the basic model that allow for a direct investigation of the relationships between respondent characteristics and the occurrence of interviewer effects (as measured by the IICs). We do not discuss the ability of interviewer characteristics to explain interviewer effects. We use the respondent's education level to illustrate the two procedures, examining whether interviewer effects differ according to the respondent's education level. In the next section, we present the data used in the analysis and the results of a preliminary analysis.

22.3 Data and Preliminary Analysis

22.3.1 Data

We used data from 21 countries* that participated in Round 8 of the European Social Survey, or ESS (Austria, Belgium, Czech Republic, Germany, Estonia, Finland, France, Hungary, Iceland, Ireland, Italy, Lithuania, the Netherlands, Norway, Poland, Portugal, Slovenia, Spain, Sweden, Switzerland, and the United Kingdom).

We computed intra-interviewer correlations for 14 substantive questions from the two rotating modules of the Round 8 ESS questionnaire. Specifically, we selected eight questions from the climate change and energy module (Module D) and six questions from the welfare attitudes module (Module E) (see Online Appendix 22A). Each of the questions used an 11-point response scale, ranging, for example, from 0 = Not at all likely to 11 = Extremely likely, or from 0 = Extremely bad to 11 = Extremely good.

We used the variable *EISCED* (highest level of education, ES-ISCED) to create the key independent variable for the analysis. This seven-category, harmonized variable is constructed from detailed, country-specific variables and a common ISCED-based coding frame, along with bridging specifications in the ESS. The *EISCED* variable was recoded into three categories: up to lower secondary education (Level 1: *EISCED* = 1–2), upper secondary education or advanced vocational education (Level 2: *EISCED* = 3–5), and tertiary education (Level 3: *EISCED* = 6–7). This recoding into three categories ensured that a sufficient number of units were available to reliably estimate the IICs in each category, while maintaining a relevant ordering of educational attainment.

The respondent-level control variables included in the models used to calculate the IICs for each education category were as follows: respondent's age (centered around 30) and gender (0 = Female, 1 = Male), whether the language of the interview was the respondent's home language (0 = No, 1 = Yes), and self-reported degree of urbanization of the respondent's domicile, included with dummy variables (1 = Big city [as reference category], 2 = Suburbs or outskirts of big city, 3 = Town or small city, 4 = Country village, 5 = Farm or home in countryside).

The interviewer report, completed by the interviewers at the end of each interview, includes four items describing the respondent's behaviors during the interview. Two of these four assessments are related to the respondent's cognitive abilities: "Did the respondent ask for clarification of any questions?" (RESCLQ) and "Overall, did you feel that the respondent understood the questions?" (RESUNDQ). Both questions were answered on a 5-point scale (1 = Never; 2 = Almost never; 3 = Now and then; 4 = Often; 5 = Very often). These variables can give some indications about whether the interaction proceeded smoothly, and they were used in a preliminary analysis of respondents' understanding of survey questions, described in the following section.

22.3.2 A Preliminary Analysis

A first preliminary step in the analysis consists of the calculation and assessment of the overall IICs. How large are the IICs and are there differences between questions? The basic two-level model with a random intercept and relevant respondent-level control variables

---

* Due to the specific circumstances of the data collection, Russia and Israel are not taken into account.

that we used to obtain an overall IIC for each of the 14 selected substantive questions in each country was specified as follows:

$$y_{ij} = b_0 + a_2 \left(\text{Education} = 2\right) + a_3 \left(\text{Education} = 3\right)$$

$$+ b_1 \text{ Age} + b_2 \text{ Male} + b_3 \text{ Same language} \tag{22.3}$$

$$+ \sum_{k=2}^{5} d_k \text{ Domicile type}_k + u_{0j} + e_{ij}$$

with $e_{ij} \sim N\left(0, \sigma_e^2\right), u_{0j} \sim N\left(0, \sigma_u^2\right)$ and $\text{IIC} = \dfrac{\sigma_u^2}{\sigma_u^2 + \sigma_e^2}$.

Figure 22.1 shows the boxplots of the IICs for the 14 questions in each country. The high IICs in some of the countries and the variability across countries are striking. Nine countries (Estonia, Spain, Poland, Ireland, Austria, Czech Republic, Italy, Hungary, and Lithuania) have mean IICs higher than 0.10; there are four countries with a mean IIC in the interval [0.05–0.1] and eight countries have a mean below 0.05. The variability of the IICs within a country tends to be larger for countries with a higher mean IIC. This means that in countries with large interviewer effects, there are also large differences in the interviewer effects for different questions. Overall, the pattern of variability in the IICs suggests that a further evaluation of whether interviewer effects tend to vary across different subgroups of respondents is relevant. As was already argued in the Introduction section, education level seems to be an obvious respondent characteristic to create the subgroups.

Indirect support for the expected link between interviewer effects and respondents' education level is derived from the interviewers' reports on respondents' understanding of survey questions. A higher frequency of asking for clarification and explaining the questions is expected for lower-educated respondents than for higher-educated respondents. Inadequate understanding and requests for clarification intensify the interaction between the respondent and the interviewer, resulting in a higher risk of interviewer effects.

A descriptive analysis of the mean frequency of understanding questions and asking for clarification (Table 22.1) shows that the patterns of the means are consistent with

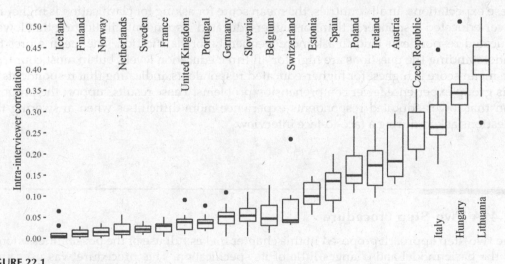

FIGURE 22.1
Boxplots of the IICs for the 14 questions by country.

TABLE 22.1

Mean Frequency of Respondents Understanding the Questions and Asking for Clarification by Education Level

Country	Understanding the Questions			Asking for Clarification		
	Up to Lower Secondary	Upper Secondary	Tertiary	Up to Lower Secondary	Upper Secondary	Tertiary
Finland	4.6	4.8	4.9	2.0	1.7	1.8
Iceland	4.5	4.7	4.8	2.2	2.1	2.1
Norway	4.5	4.7	4.8	2.0	1.8	1.8
Netherlands	4.5	4.7	4.8	2.1	1.7	1.7
Sweden	4.5	4.7	4.7	2.2	1.9	2.0
France	4.4	4.8	4.9	2.2	1.7	1.6
Portugal	4.5	4.9	5.0	2.3	1.8	1.8
United Kingdom	4.4	4.7	4.8	2.4	2.0	2.0
Germany	4.5	4.7	4.8	2.1	1.9	1.8
Slovenia	4.2	4.6	4.8	2.0	1.6	1.4
Belgium	4.3	4.6	4.8	2.1	1.8	1.7
Switzerland	4.4	4.8	4.8	2.3	1.6	1.6
Estonia	4.2	4.5	4.7	2.5	2.2	1.9
Spain	4.3	4.6	4.7	2.5	2.1	2.0
Poland	4.4	4.7	4.9	2.1	1.7	1.4
Ireland	4.3	4.6	4.8	2.5	2.1	1.9
Austria	4.6	4.7	4.9	2.0	1.7	1.6
Czech Republic	4.3	4.6	4.8	2.4	2.0	1.8
Italy	4.1	4.5	4.7	2.7	2.1	1.9
Hungary	4.1	4.5	4.7	2.4	1.8	1.4
Lithuania	3.9	4.2	4.5	2.8	2.3	2.0

Note: 1 = Never; 2 = Almost never; 3 = Now and then; 4 = Often; 5 = Very often, i.e., higher values indicate a higher frequency.

these expectations. In all countries, the mean score for asking for clarification is higher for lower-educated respondents than for higher-educated respondents, indicating that lower-educated respondents tend to ask more frequently for clarification. The mean scores for understanding the questions are high for all three education levels, but in most countries the mean score is highest for higher-educated respondents, indicating that respondents in this group experience fewer comprehension problems. These results support the proposition that lower-educated respondents experience more difficulties when answering the questions asked during a face-to-face interview.

22.4 A Two-Step Procedure

The two-step approach proposed in this chapter makes full use of the possibilities offered by the basic model and changes little of its specification. This procedure was used in a paper about increased interviewer effects among older respondents (Beullens, Loosveldt, and Vandenplas 2018). The reported results supported the expectation that interviewer effects tend to be larger for older respondents.

In the first step of the procedure, the IICs are calculated within the categories of a particular respondent characteristic. In the current application, the respondent education level was used. In general, we define X as the respondent characteristic that is used to create the respondent groups with G categories ($g = 1, ..., G$). The IICs are calculated within the G different groups for different questions V, using the extended basic model presented in the previous section that includes the respondent characteristics as control variables (Equation 22.2, without respondents' education level). The questions V are identified by the index q ($q = 1, ..., Q$). IIC_{qg} is the intra-interviewer correlation calculated for question q in category g of X. The output of the first step is a data set with Q*G IICs. Each record of this meta–data set contains the value of the IIC, the identification of the question q, and the group g for which the IIC was calculated. This basic information can be expanded with some question characteristics (e.g., the number of scale points, topic of the question, rank order of the question in the questionnaire, etc.). Because G IICs are calculated for each question, the data set has a hierarchical structure. The IICs calculated for each category of the respondent characteristic are nested within the questions.

In the second step of the procedure, the IICs are used as the dependent variable in the following multilevel model, with the IICs nested within the questions:

$$IIC_{qg} = \gamma_{00} + \sum_{g=2}^{G} \beta_g X_g + u_{0q} + e_{qg} \tag{22.4}$$

with $e_{qg} \sim N\left(0, \sigma_e^2\right)$ and $u_{0q} \sim N\left(0, \sigma_u^2\right)$

In this linear model with a random intercept, the categories of the group variable are treated as dummy variables, with the first category as the reference category. Based on the estimated parameters β_g, the direct effect of the respondent characteristic (group variable) on the IICs can be evaluated. This is in fact the ultimate objective of the procedure. A significant parameter β_g means that there are significant differences between the mean IICs in the reference category and the specific category g. The general intercept (γ_{00}) expresses the expected IIC for the reference group, for a randomly selected survey question. The random intercepts u_{0q} (the question-specific part of the intercept) allow the different survey questions to deviate from this general intercept, and the variance σ_u^2 expresses the magnitude of these deviations from the general intercept for survey questions.

In the current application, we calculated an IIC within the three categories of the education variable for each of the 14 selected questions. Only interviewers who have interviewed at least three respondents in a given respondent group were included in the first step. The group of interviewers in each category was not completely the same, but since interviewers were not systematically assigned to certain respondent groups, we can assume that the group of interviewers in each education level represented a random selection from the general population of interviewers.

The IICs were calculated separately for each country. The implementation of the first step resulted in 882 IICs (14 questions × 3 respondent groups × 21 countries). Each IIC is identified by the question, the education level, and the country.

In the second step of the procedure, the relationship between IICs and education level was formally evaluated by the specification of a cross-classified multilevel model with the IICs (expressed as percentage of explained variance) as the dependent variable. The cross-classified multilevel model is an extension of the model presented in Equation 22.4 and is appropriate because the same questions were asked in each of the countries. This extended model contains a random intercept for the questions (u_{0q}) and a random intercept for the countries (u_{0c}). Education level is included as a fixed categorical variable (two dummies) with the lowest education level as the reference category. This results in the following specification of the model (where the variance of the random country intercepts is an additional parameter):

$$\text{IIC}_{qgc} = b_0 + b_1\left(g = 2\right) + b_2\left(g = 3\right) + u_{0c} + u_{0q} + e_{qgc} \tag{22.5}$$

The estimates of the parameters in this model are presented in Table 22.2. The mean IIC for the low-educated respondent group is quite high (14%). As expected, there is a significant decrease for higher education levels. For the moderately and highly educated respondent groups, the mean IICs are about 2 and 3 percentage points smaller, respectively. The significant variance component for the random effect of country supports the large cross-country variability observed in Figure 22.1. The variance of the random effects of questions is much smaller.

The large differences between countries concerning the IICs give rise to the question of whether the effect of education level differs depending on the general level of interviewer effects in a country. Are differences between the education levels larger in countries with higher IICs? To answer that question, the countries were grouped into two categories: countries with a mean IIC below 0.1 (*HighIIC country* = 0) and countries with a larger mean IIC (*HighIIC country* = 1), and interaction effects with the two education level dummies were specified. This expanded model was specified as follows:

$$\text{IIC}_{qgc} = b_0 + b_1\left(g = 2\right) + b_2\left(g = 3\right) + b_3 HighIIC\ country$$

$$+ b_4\left(g = 2\right) * HighIIC\ country$$

$$+ b_5\left(g = 3\right) * HighIIC\ country + u_{0c} + u_{0q} + e_{qgc} \tag{22.6}$$

TABLE 22.2

Estimated Fixed Effects and Variance Components in the Cross-Classified Model for the IICs with Respondent Education Group as a Predictor, Comparing the Two-Step Procedure and the Conditional Random Interviewer Effect Model

	Two-step Procedure (Procedure 1)			Conditional Random Interviewer Effect Model (Procedure 2)		
	Estimate	SE	*p*-Value	Estimate	SE	*p*-Value
Fixed effects						
(Intercept)	14.069	2.694	< 0.001	14.096	2.677	< 0.001
Education level 2	−1.862	0.486	< 0.001	−1.976	0.480	< 0.001
Education level 3	−3.289	0.486	< 0.001	−3.416	0.480	< 0.001
Variance components						
Country	143.803		< 0.001	141.949		< 0.001
Question	4.085		< 0.001	4.060		< 0.001
Residual	34.720			33.825		

Note: Statistical significance of the variance components is derived from likelihood ratio tests comparing the model fit with and without the given variance component.

The interaction between the high education level indicator and the indicator of countries with a mean IIC above 0.1 is negative and statistically significant ($b_5 = -4.114$, $p < 0.001$). This means that in countries with high IICs, there is an additional reduction in the interviewer effects for the higher-educated respondents when compared with lower-educated respondents. Consequently, the differences in IICs between lower- and higher-educated respondents increase. It is possible that in these countries higher-educated respondents are less influenced by the interviewers or that lower-educated respondents are more influenced. The latter would mean that interviewers in those countries experience more difficulty interviewing lower-educated respondents.

It is clear that to apply the two-step procedure, a sufficient number of interviewers and a sufficient number of respondents for each interviewer are needed in order to be able to calculate the IICs. The number of IICs should also be sufficiently large to reliably estimate the coefficients in the second step of the procedure. The number of IICs increases with the number of substantive questions and the number of respondent groups included in the analysis. Increasing the number of substantive questions included in the analysis is generally preferred. Increasing the number of categories in the respondent group variable is substantively not always meaningful and may well result in too few respondents and interviewers in particular categories to calculate the IICs in the first step.

22.5 The Conditional Random Interviewer Effect Model

As shown in the previous section, conditional IICs are essential in the evaluation of interviewer effects across different respondent groups. In the two-step procedure, conditional IICs are obtained by applying the same basic model within the categories of a respondent characteristic. In this second possible approach, the basic model with a substantive variable Y as dependent variable is adapted to allow the variance of the random intercept and the residual error to differ across subgroups. This model has been used to study whether conversational interviewing yields higher interviewer effects than standardized interviewing (West, et al. 2018a; West, et al. 2018b), and is also discussed in more detail in Chapter 23 of this volume. The authors concluded that significant increases in interviewer effects due to the conversational interviewing style are rare.

For simplicity, assume a respondent characteristic (X) with two categories ($g = 1, 2$). The following model can then be specified:

$$y_{ij} = \gamma_{00} + \beta_1 (g = 1) + u_{1j} (g = 1) + u_{2j} (g = 2) + e_{ij} \tag{22.7}$$

$$u_{1j} \sim N\left(0, \sigma_{u1}^2\right); u_{2j} \sim N\left(0, \sigma_{u2}^2\right); e_{ij} \sim N\left(0, \sigma_{e1}^2\right) \text{ if } g = 1; e_{ij} \sim N\left(0, \sigma_{e2}^2\right) \text{ if } g = 2$$

In this model, β_1 is the fixed effect of the group variable: the differences in the expected value of Y between the two groups of respondents created by X. The random variables u_{1j} and u_{2j} are the random effects for the interviewers in groups 1 and 2 defined by the respondents, with variance components σ_{u1}^2 and σ_{u2}^2. The random intercepts are calculated separately for the two categories of the group variable; hence the use of the term "conditional random interviewer effect model." In addition, the variance of the residual errors is calculated within the groups, respectively: σ_{e1}^2 and σ_{e2}^2. The assumption of the basic model concerning the homoscedasticity of the residual errors at the first level is thereby explicitly relaxed. One can perform a likelihood ratio test of whether the variance components in the two groups of respondents are significantly different when fitting this model.

With these conditional residual error variances and the conditional variances of the random intercept, one can calculate conditional IICs (e.g., $IIC_1 = \dfrac{\sigma_{u1}^2}{\sigma_{u1}^2 + \sigma_{e1}^2}$). As with the two-step procedure, these conditional IICs make clear whether or not the interviewer effects differ for different categories of respondents, which is exactly the aim when evaluating the direct link between respondent characteristics and interviewer effects.

The first and second approaches have in common that between-interviewer and within-interviewer variances are calculated for the categories of the respondent characteristics. The main difference is the way the variance components are estimated. In the first procedure, the estimates are obtained by a separate analysis for each of the education levels. In the second procedure, the variance components are estimated simultaneously for the three education levels in the same analysis. This allows for a homogeneity test (via likelihood ratio testing) of the covariance matrices for each question. The null hypothesis of this test specifies that the two variance components, the between-interviewer variance and the residual variances, are equal across the three educational groups. The rejection of this null hypothesis indicates that the interviewer effects vary according to the respondent groups. Compared with the two-step procedure, an additional step to model the IICs is not necessary to evaluate the effect of a respondent characteristic.

To evaluate the relationship between education level of the respondent and interviewer effects, we specified the following conditional random effect model for each of the 14 selected questions and separately for each country:

$$y_{ij} = b_0 + a_2 \left(\text{Education} = 2 \right) + a_3 \left(\text{Education} = 3 \right)$$

$$+ b_1 \, \text{Age} + b_2 \, \text{Male} + b_3 \, \text{Same language} + b_4 \, \text{Domicile type}$$

$$+ u_{1j} \left(\text{Education} = 1 \right) + u_{2j} \left(\text{Education} = 2 \right) + u_{3j} \left(\text{Education} = 3 \right) + e_{ij} \qquad (22.8)$$

with $e_{ij} \sim N\left(0, \sigma_{e1}^2\right)$ if Education $= 1$
$e_{ij} \sim N\left(0, \sigma_{e2}^2\right)$ if Education $= 2$
$e_{ij} \sim N\left(0, \sigma_{e3}^2\right)$ if Education $= 3$ and
$u_{1j} \sim N\left(0, \sigma_{u1}^2\right), u_{2j} \sim N\left(0, \sigma_{u2}^2\right), u_{3j} \sim N\left(0, \sigma_{u3}^2\right)$

This model yields an estimate of the between-interviewer variance and the residual variance in the three education groups, for each variable and each country. Both conditional variance components can be used to calculate conditional IICs for each variable in each country. This means that for each question in each country, three conditional IICs were obtained, equivalent to the IICs obtained by the two-step procedure.

In the current application with the three education levels, the homogeneity hypothesis was tested separately for each variable in each country. For 180 of the 294 tests (14 questions in 21 countries), the null hypothesis was rejected, indicating significant differences in the between-interviewer and/or residual variance between the three education level groups in more than half of the tests (61%). This means that for many questions, the variance components that one needs to calculate the IIC are significantly different for the three educational groups at the 0.05 significance level. This also supports the hypothesized relationship between respondents' education levels and interviewer effects.

In order to compare the results of the first and second approaches, we can model the estimated IICs obtained by applying the second procedure with the same cross-classified model used in the second step of the first procedure. The results are presented in Table 22.2. Given the almost perfect correlation observed between the IICs obtained from the first procedure and the IICs obtained from the second procedure ($r = 0.994$, $p < 0.001$), it is not surprising to find that the results of the second procedure are very similar to the results of the first procedure. The estimated differences between the three education levels are as expected and similar to the first procedure. The IICs are high in the lowest education level but significantly decrease when the education level of the respondent increases. Similar to the two-step procedure, the differences between countries are larger than the differences between questions. The analysis with interaction effects between country classification (low vs. high overall IICs) with the two education-level dummies also confirms the previously observed additional decrease of the IICs for the higher-educated respondents in countries with high IICs. In countries with large IICs, the differences between lower- and higher-educated respondents are more pronounced.

22.6 Conclusion

The calculation of intra-interviewer correlation coefficients is a common practice to evaluate the effects of interviewers on survey responses. In this context, the question of whether these interviewer effects are larger or smaller for particular groups of respondents is relevant. This question implies the evaluation of the relationship between a respondent characteristic and the IICs. We argue that the extended basic model with a random intercept used to calculate the IICs is not appropriate to evaluate this kind of relationship, because respondent characteristics in this model are used to explain the variability in a substantive variable and not to explain the variability in the IICs. A two-step procedure, based on this extended basic model, and a conditional random interviewer effect model are both suitable to evaluate the direct effect of a respondent characteristic on the IICs. In both procedures, conditional IICs are calculated to evaluate the impact of a respondent characteristic on the IICs.

To apply these procedures, there must be a sufficient number of interviewers per respondent group to apply the multilevel modeling (see Chapter 23 of this volume for guidance on this issue). As a result, the number of categories of the respondent characteristic will usually be limited. Both procedures test the effect of the respondent characteristic on the IICs in a different way. In the conditional random interviewer effect model, the homogeneity across the respondents of the covariance matrix for the observed responses, including the between-interviewer variance and the residual variance, may be tested at the question level. In the two-step procedure, the direct effect of the respondent characteristic is tested in the model of the second step in which the conditional IICs are the dependent variable and the respondent characteristic is an independent variable.

Both procedures were applied to evaluate the effect of the respondent's education level on interviewer effects in the ESS. Fourteen questions of the ESS round 8 questionnaire and data from 21 countries were used in the analysis. We found remarkably large differences in IICs between countries. However, the results of both procedures support the expectation that interviewer effects are significantly higher in the group defined by lower-educated respondents. The results of a focused interviewer-respondent interaction analysis of interviews with lower-educated respondents may provide insight into the typical problems during an interview with these respondents. This can result in additional points of attention that can be addressed during interviewer training.

References

Beullens, K., G. Loosveldt, and C. Vandenplas. 2018. Interviewer effects among older respondents in the European Social Survey. *International Journal of Public Opinion Research* 31(December):1–17.

Hox, J. J. 1994. Hierarchical regression models for interviewer and respondent effects. *Sociological Methods and Research* 22(3):300–318.

Knaüper,B., R. Belli, D. Hill, and A. Herzog. 1997. Question difficulty and respondents' cognitive ability: The effect on data quality. *Journal of Official Statistics* 13:181–199.

Krosnick, J., and D. Alwin. 1987. An evaluation of a cognitive theory of response – order effects in survey measurement. *Public Opinion Quarterly* 51(2):201–219.

Narayan, S., and J. Krosnick. 1996. Education moderates some response effects in attitude measurement. *Public Opinion Quarterly* 60(1):58–88.

Schaeffer, N. C., J. Dykema, and D. W. Maynard. 2010. Interviewers and interviewing. In: *Handbook of Survey Research*, 2nd edition, ed. J. D. Wright, and P. V. Marsden, 437–470. Bingley: Emerald Group Publishing Limited.

West, B., and A. Blom. 2017. Explaining interviewer effects: A research synthesis. *Journal of Survey Statistics and Methodology* 5:175–211.

West, B., F. Conrad, F. Kreuter, and F. Mittereder. 2018a. Nonresponse and measurement error variance among interviewers in standardized and conversational interviewing. *Journal of Survey Statistics and Methodology* 6(3):335–359.

West, B., F. Conrad, F. Kreuter, and F. Mittereder. 2018b. Can conversational interviewing improve survey response quality without increasing interviewer effects? *Journal of the Royal Statistical Society: Series A: Statistics in Society* 181(1):181–203.

23

Designing Studies for Comparing Interviewer Variance in Two Groups of Survey Interviewers

Brady T. West

CONTENTS

23.1 Introduction

Interviewer effects, defined generally as the varying effects that human interviewers can have on the responses ultimately recorded in an interviewer-administered survey, continue to be a difficult problem for survey researchers to address in all surveys that employ human interviewers to collect data (Friedel 2019; Kibuchi et al. 2019; Vandenplas et al. 2018; West and Blom 2017; West et al. 2018a). Whether they are interacting with survey respondents in a face-to-face setting, asking questions over the telephone, or conducting interviews via online video systems such as Skype, interviewers may use different techniques to recruit sampled units and conduct survey interviews. This in turn may lead to variance among interviewers in the survey responses that they collect. Interviewer effects become a problem when the estimate produced by a specific interviewer working on their assigned sample deviates systematically from the "true" estimate for their sample. These deviations can either vary among different interviewers, introducing the interviewer variance that is the focus of this chapter, or they can occur in the same direction for all interviewers, introducing a systematic bias in the survey responses recorded.

Myriad studies have noted how variance among interviewers for a survey estimate results in unwanted inflation of that estimate's variance, ultimately reducing statistical power and lowering effective sample sizes (see West and Blom 2017 for a recent review). Ideally, interviewers will not introduce variability in estimates beyond random variation due to the sample assignment process. If interviewer variance emerges as a problem for selected estimates in a given survey, whether they be descriptive (e.g., means, proportions) or analytic (e.g., regression coefficients; see Fischer, et al. 2019) in nature, survey researchers need to take steps to mitigate this problem. One possible solution is for survey researchers to try to understand observable interviewer-level features that correlate with the survey measures of interest and are therefore contributing to the between-interviewer variance.

Many of the empirical studies on interviewer effects have focused on this objective of identifying interviewer-level predictors that may *explain* differences among interviewers in some quantity of interest (e.g., Berk and Bernstein 1988; Blohm, Hox, and Koch 2007; Catania, et al. 1996; Jäckle, et al. 2013; Németh and Luksander 2018). For example, interviewer experience may explain a substantial portion of the *overall* variation among interviewers in response rates, where more senior interviewers and interviewers with more experience working on a given survey tend to have higher response rates (e.g., Jäckle, et al. 2013; see also Lipps and Pollien 2011). Put differently, years of experience may have a positive correlation with a binary indicator of responding to the survey request. Studies that aim to understand sources of interviewer effects on response rates (e.g., Durrant, et al. 2010) may view experience as an *explanatory factor* for the observed interviewer variance in the response rates, and attempt to quantify the fraction of this variance attributable to interviewer experience (e.g., Vassallo, et al. 2015).

Far fewer studies have considered a different approach to conducting empirical research on interviewer effects: allowing for the possibility that *different groups* of interviewers may be producing interviewer variance components of *different magnitude* (e.g., Brunton-Smith, Sturgis, and Williams 2012; Freeman and Butler 1976; Lipps and Pollien 2011; West, et al. 2018a). In other words, the interviewer group *moderates the magnitude* of the interviewer-related variance for a survey estimate. In this case, intervention from the survey researcher may only be necessary for a particular group of interviewers producing substantial interviewer variance, rather than the entire interviewing corps. For example, there may only be significant interviewer variance in response rates among *inexperienced* interviewers, and *experienced* interviewers may not vary in any meaningful way; one could then examine potential explanatory factors for the variance *within inexperienced interviewers only*.

This type of empirical result has the potential to conserve study resources, in that additional training may not be necessary for *all* interviewers. Survey methodologists may also have explicit research interest in understanding whether different groups of interviewers defined by observable characteristics (e.g., socio-demographics or experience; Freeman and Butler 1976; Lipps and Pollien 2011), observable ability to make contact or obtain cooperation (e.g., Brunton-Smith, Sturgis, and Williams 2012), or different interviewing techniques assigned experimentally (e.g., conversational vs. standardized interviewing; West, et al. 2018a) are producing different amounts of interviewer variance. The design of these latter types of methodological studies is the focus of the present chapter.

This chapter lays out critical study design considerations that survey researchers interested in comparing interviewer variance components between different groups of interviewers need to consider before undertaking such investigations. Careful design of these types of studies will lead to meaningful conclusions about the heterogeneity across different groups of survey interviewers in interviewer-related variance for different types of survey estimates, and ultimately to more meaningful and cost-efficient interventions that aim to improve the overall quality of the survey data collected by interviewers.

23.2 Critical Study Design Considerations

This section presents critical design considerations for these types of comparative studies. We use a case study from West, et al. (2018a) to provide concrete illustrations of each consideration.

23.2.1 Interpenetrated Sample Assignment

Arguably, the most difficult aspect of designing studies of interviewer effects is the random assignment of subsamples of the larger overall sample to different interviewers, or *interpenetrated sample assignment*. In this scenario, the survey estimate of interest will be equal in expectation among interviewers at the time of subsample assignment, and any variance among interviewers in the estimates is likely attributable to the work done by the interviewers, whether it be recruitment or measurement (e.g., Mangione, Fowler, and Louis 1992). Unfortunately, financial and practical constraints often prohibit this ideal form of random subsample assignment. In face-to-face studies, for example, interviewers often only work in the areas where they live, which clearly is more cost-efficient for the overall data collection. Interpenetrated sample assignment is much more feasible within calling shifts of a telephone survey data collection. However, even in this case, "final" assignments of nonrespondents to interviewers for the purpose of studying nonresponse error variance among interviewers can be a very challenging problem (West and Olson 2010).

For face-to-face data collections, one potential design solution for this problem is *partially interpenetrated assignment* (e.g., O'Muircheartaigh and Campanelli 1998; Dahlhamer, et al. Chapter 21, this volume), where assigned interviewer workloads cross different (and usually adjacent) geographic areas, and analysts can use cross-classified multilevel models to disentangle interviewer and area effects on the survey responses. Another potential solution, if purely random subsample assignment is not possible (say, in a survey of a small city where transportation is not costly), is to assign *multiple* interviewers to a given area, and then assign each interviewer random subsamples of the sampled cases within each small area (e.g., Schnell and Kreuter 2005). West et al. (2018a) recruited 60 interviewers for a study of 15 areas in Germany, and assigned four interviewers to work in each of the 15 areas. Within each area, data collection managers assigned each of the four interviewers a random subsample of 120 of the 480 addresses randomly sampled from that area.

When employing this second solution, interviewers are clearly not assigned random subsamples of the full sample, since multiple interviewers are all working in the same area (rather than in all areas under study). However, when analyzing the data, simple control for fixed effects of the study areas in multilevel models including random interviewer effects will *explain* the variance among interviewers in the measures collected on a given survey variable due to the areas, isolating the effects of the individual interviewers working on random subsamples in each area. West, et al. (2018a) included 14 fixed area effects in their multilevel models (treating one of the 15 areas as an arbitrary reference area), effectively accounting for any between-area variance that may have contributed to the between-interviewer variance.

While these two solutions are not as ideal as fully interpenetrated sample assignment, they do offer practical and cost-efficient design approaches to isolating the effects of interviewers on survey measures of interest. In the case where no form of interpenetrated assignment is feasible (e.g., face-to-face surveys with only one interviewer working in each geographic area), specialized analytic approaches are needed to isolate the interviewer effects in the groups being compared. West and Elliott (2015) presented one idea based on *anchoring variables*, which are variables that are unlikely to be subject to interviewer effects at the time of their measurement (e.g., age) and correlate with other survey variables that may be subject to interviewer effects when measured (e.g., self-reported health). When subsamples of cases are not randomly assigned to interviewers, this will result in variance among interviewers in values of the anchoring variable prior to any recruitment or measurement occurring. In a statistical model for the two measures of the anchoring variable and the survey variable of interest collected from each responding sample case, one could allow the model-based residuals for the two measures to be correlated within each case, and include random interviewer effects only for the survey variable of interest (see West and Elliott 2015 for details). This approach can help to isolate unique effects of the interviewers on the survey variable. Additional development of these types of analytic techniques for non-interpenetrated samples is clearly needed in the literature.

23.2.2 Geographic Balance

Ensuring that the groups of interviewers being compared are equally represented across geographic areas is also critical when designing studies focused on comparing interviewer variance components across the groups. For example, if a researcher aims to compare variance components between experienced and inexperienced interviewers, the researcher should take all possible steps to ensure that each geographic area under study is being worked by both experienced and inexperienced interviewers. This design approach will help to avoid possible confounds of the comparative factor of interest (e.g., interviewer experience) with geography. Ideally, there will be an even number of interviewers assigned to work in each geographic area, and half of the interviewers will either have a particular observed feature (e.g., low experience) or be assigned to one experimental treatment (e.g., conversational interviewing), while half of the interviewers will have the complementary observed feature or be assigned to a competing experimental treatment. This would prevent confounding of the factor of interest with the geographic areas under study.

For example, West, et al. (2018a) randomly assigned two of the interviewers in each geographic area to implement conversational interviewing, and the remaining two interviewers to implement standardized interviewing (see Groves and McGonagle 2001 for another example of this type of effort to achieve geographic balance). This procedure resulted in balance within each area in terms of the experimental factor, in addition to overall balance for the two experimental groups *across* the geographic areas (i.e., the two groups were equivalent overall in terms of socio-demographic distributions and prior survey experience). If all the interviewers working in a particular area were assigned to the *same* experimental treatment, it would have been impossible to disentangle area effects from treatment effects on the variables of interest. Furthermore, estimates of interviewer variance due to one of the experimental factors would only be based on selected areas, rather than all possible areas. This design consideration tends to be less of an issue for telephone surveys, where interviewers making phone calls from a centralized location will likely work sampled cases from multiple geographic areas.

23.2.3 Randomized Assignment to Groups

Ideally, for studies focused on the comparison of interviewer variance components across groups of interviewers, interviewers should be assigned at random to different levels of the experimental factor (e.g., Mangione, Fowler, and Louis 1992). This was the case in the West, et al. (2018a) study, where the objective of the study was to compare the interviewer variance introduced by conversational and standardized interviewing techniques (see also Schober and Conrad 1997). In these types of experimental studies, one should employ *stratified randomization*, and assign interviewers to the alternative treatment groups at random within strata defined by relevant characteristics for the study (e.g., areas, interviewer experience, aggregate respondent characteristics, etc.). The experimental design literature often refers to these characteristics as *blocking variables*. This will ensure balance between the groups of interviewers defined by the experimental factors in terms of these characteristics (e.g., Billiet and Loosveldt 1988). While clearly important for studies focused on comparing interviewer variance components across groups of interviewers defined by levels of an experimental factor, random assignment of *interviewers* (as opposed to *respondents*; e.g., Cannell, Miller, and Oksenberg 1981) to different levels of the experimental factor is also important for avoiding confounds of interviewer effects with the effects of the factor (Lavrakas, Kelly, and McClain 2019).

Clearly, this type of random assignment is not possible for observable interviewer demographics or other characteristics. One cannot assign interviewers at random to different race/ethnicity groups, experience groups, etc., since these traits are not under the control of the researcher and survey organizations cannot exclusively hire interviewers with specific socio-demographic characteristics. In studies focused on comparing two groups of interviewers defined by these types of observable characteristics (e.g., West and Elliott 2014), one needs to carefully analyze any differences between the two groups in terms of other observable characteristics that may confound the comparison of interest. For example, older interviewers may differ from younger interviewers in terms of other observable features, like years of interviewing experience. Any attempted inferences about differences in interviewer variance components between groups of interviewers defined by age need to carefully qualify other differences between these groups that may be responsible for the differences in variance components observed.

There may be other unobservable characteristics that vary between the groups of interviewers being compared that are very difficult to account for without additional analysis or data collection activities (e.g., analysis of audio-recordings). This may impact experimental comparisons of the interviewer variance components between the two groups. With large samples of interviewers, it may be possible to create groups of interviewers defined in part by these additional confounding factors (for purposes of comparing the variance components), but for smaller samples, these uncontrolled characteristics may need to be accounted for in the models used to estimate the variance components.

23.2.4 Power Analysis

Also fundamental to the design of these types of studies comparing variance components across independent groups of interviewers is the careful execution of a power analysis. The study needs to have enough statistical power to detect hypothesized differences in variance components between the two groups of interviewers. In general, large samples of interviewers with large subsample assignments for each interviewer will be necessary to detect moderate differences in variance components for both continuous and binary survey variables of interest. For example, West, et al. (2018a) found (using code at the link provided below) that having 30 interviewers in each experimental group, with 30 respondents per interviewer, would provide adequate statistical power for detecting differences in variance components that were consistent with the range of variance components reported in the literature. A study of this size – 900 respondents in each experimental group – is larger than many studies traditionally used for detecting differences in *means* across experimental conditions. Unfortunately, "canned" statistical software for performing these types of power analyses is not presently available, and empirical simulation of the power of a given study design is required.

The basic steps in this type of simulation study for estimating the power of a hypothetical comparative study design are as follows:

1. Define the variable of interest for which an estimate of an interviewer variance component is desired in each group, and the expected distribution of that variable (e.g., normal, Bernoulli, etc.).

2. Define the size of the hypothesized difference in the interviewer variance components between the two groups that the researcher would be interested in detecting as significant/important (including any hypothesized differences between the groups in means of the variable of interest or residual variance components, if applicable).

3. Repeatedly simulate data from a multilevel statistical model defined by the hypothesized differences in variance components (and any fixed effects capturing differences in means) between the two groups for the variable of interest. Create data sets with the desired/feasible sample sizes specified by the researcher, in terms of both the number of interviewers in each group and the number of respondents (or assigned sample units, in the case of studying interviewer variance in response rates) per interviewer.

4. In each simulated data set, fit the same multilevel model to the simulated data, and test the difference in the variance components for significance at a pre-specified significance level (West and Elliott 2014). Save an indicator of whether the test was found to be significant at the pre-specified level.

5. For many simulated data sets (say, 500), compute the proportion of simulated data sets where the difference in variance components was found to be significant at the pre-specified significance level. This is the estimated power of the study design specified by the researcher, in terms of the number of interviewers per group and the sample size per interviewer.

6. Repeat steps 2 through 5 for any additional variables for which a comparison of interviewer variance components is desired.

This general simulation approach is relatively straightforward to program in general-purpose statistical software. Readers can download a flexible macro implementing this simulation approach in the SAS software from the following GitHub repository: https://github.com/bradytwest/SimStudiesSAS. The SAS macro implements this approach for both continuous and binary variables, and one could easily extend it to accommodate other types of variables. An example of executing this macro follows:

```
%varcompsim(nreps=100, nint=30, numobs=30, intvar1=1, intvar2=7,
pintvar1=0.03, pintvar2=0.33, sigeps1=64, sigeps2=64);
            Analysis Variable
               : power_cont
                     Mean
              0.7900000
```

This result shows that for a planned study with 30 interviewers per group, 30 respondents (observations) per interviewer, a hypothesized difference in interviewer variance components for a continuous variable of 1.0 vs. 7.0, and a constant residual variance of 64.0, the study would have approximately 79% power (based on 100 simulations) to detect that difference in variance components as being significant at the 0.05 level. The pintvarX options in the macro are for the hypothesized variance components for a binary outcome (output not shown).

Any studies designed to *explore* potential differences in variance components rather than *test hypotheses* about differences in variance components do not necessarily require a power analysis. However, the exploratory nature of the study (and the possibility that the results could motivate larger studies with more statistical power) should be clearly described in any descriptions of the study results.

23.2.5 Analytic Approaches

West and Elliott (2014) describe alternative approaches to the analysis of these types of studies comparing interviewer variance components, all of which are easy to implement in modern statistical software. Essential to all the analytic approaches described is the extension of a standard multilevel model including random interviewer effects to allow for heterogeneity in the interviewer variance component across groups of interviewers. Many studies of interviewer variance, and especially those focused on *explaining* interviewer variance, assume that the between-interviewer variance component applies to all interviewers, rather than varying across interviewer groups. Studies comparing variance components across two independent groups of interviewers necessarily relax this assumption and allow the interviewer variance component to *differ* depending on the interviewer group of interest.

In the first *frequentist* approach to this type of analysis, one fits a (generalized) multilevel model appropriate for the dependent variable of interest (continuous, binary, count, etc.). This model includes:

1. A fixed effect for the interviewer group indicator variable (allowing expected values of the dependent variable to vary across the interviewer groups);

2. Any other essential fixed effects based on the study design (e.g., fixed area effects in the case of West, et al. (2018a), which accounted for between-area variance in the variable of interest and helped to isolate the interviewer effects);

3. *Two* random interviewer effects, one of which refers to interviewers in interviewer group 1, and one of which refers to interviewers in interviewer group 2, each arising from a normal distribution with a unique variance component; and

4. Potentially unique residual variance components for each group, if applicable based on the model being fitted.

One can fit this model using software procedures that enable maximum likelihood estimation of generalized multilevel models and permit variances of random effects to differ across groups of interviewers. One popular option, for example, is to fit the models using proc glimmix in SAS and include a group option in the random statement that includes the random effects of interviewers in the model. Note this option in the example code below, which requests estimates of variance components for two groups of interviewers defined by marital status (where int _ nvmarried is a binary indicator for never-married interviewers). West and Galecki (2011) provide a summary of other software enabling this type of modeling approach.

```
proc glimmix data = bayes.final_analysis asycov;
    class final_int_id int_nvmarried;
    model sexmain (event = "1") = int_nvmarried / dist = binary link =
        logit
      solution cl;
    random int / subject = final_int_id group = int_nvmarried;
    covtest homogeneity / cl (type = plr);
    nloptions tech=nrridg;
run;
```

Following the frequentist approach, one can then implement a classical likelihood ratio test to test the null hypothesis that the two groups of interviewers have equal variance components. This involves saving the –2 log-likelihood value of the fitted model allowing for heterogeneity in the variance components, refitting the model after *removing* the heterogeneity (i.e., only including one random interviewer effect with a single corresponding variance component, or constraining the interviewer variance components to be equal), saving the –2 log-likelihood value of this simplified model, and computing the positive difference in the two –2 log-likelihood values. Asymptotically, this positive difference is a test statistic that under the null hypothesis follows a chi-square distribution with one degree of freedom (corresponding to the difference in the number of parameters between the two models), and extreme values of the test statistic resulting in small p-values can be used as a basis for rejecting the null hypothesis.

This frequentist approach assumes large sample sizes for the number of interviewers in each group and the number of cases per interviewer (West and Elliott 2014). An attractive alternative for studies with smaller sample sizes (or studies where a variance component in a given group might be very close to zero, which are notorious for causing estimation problems in the aforementioned frequentist approach; see West and Elliott 2014) is Bayesian in nature. Following this Bayesian approach, one first specifies reasonable prior distributions for the parameters of interest (variance components and fixed effects) for each group, reflecting prior knowledge about expected distributions. In the absence of any prior information, one can specify non-informative priors for the parameters. Then, one specifies the likelihood for the observed data based on the model of interest and simulates many draws of the parameters of interest from their resulting posterior distribution. The simulation procedure could use several possible approaches, such as MCMC methods.

For each simulated draw of the two variance components, one would compute the difference in the two components from that draw, which would serve as a draw of this difference from the relevant posterior distribution. Finally, one can compute the 2.5% and 97.5% percentiles of the empirical distribution of these simulated differences in variance components and examine whether this 95% credible interval for the difference in variance components includes zero. A credible interval that excludes zero suggests that the difference in the variance components is statistically meaningful, above and beyond any sampling variance. West and Elliott (2014) provide example code implementing this Bayesian approach.

These two approaches will generally provide similar inferences regarding the differences in the two variance components of interest. However, the Bayesian approach is better-suited for smaller studies, and situations where an estimated variance component is "fixed" to be zero when following the frequentist approach. See West and Elliott (2014) for additional technical details.

23.3 Discussion

This chapter has provided an overview of decisions that survey researchers need to consider when designing studies that focus on comparing the variance components for a variable of interest across two groups of interviewers. Importantly, this chapter has focused on research questions related to interviewer characteristics that affect the *magnitude* of an interviewer variance component, rather than interviewer characteristics that *explain* an observed interviewer variance component. West, et al. (2018a) describe an experimental study of the type described in the chapter.

Many studies of interviewer effects in the literature have focused on interviewer variance in *respondent reports* on survey variables of interest (e.g., Dahlhamer, et al., Chapter 21, this volume; Loosveldt and Wuyts, Chapter 22, this volume). The types of studies described in this chapter could certainly examine differences in interviewer variance components based on respondent reports. However, the approaches described here could also be applied to studies that focus on differences in *nonresponse error variance* across groups of interviewers (e.g., West and Olson 2010; West, Kreuter, and Jaenichen 2013; West et al. 2018b). That is, given sampling frame information on sampled units and assuming interpenetrated sample assignment, is there evidence of differences in interviewer variance components between different groups of interviewers for variables that are available on the sampling frame for responding sampled units (prior to measurement)?

Suppose that interviewers in a study are assigned roughly equivalent subsamples of sampled units. Unfortunately, analyses conducted at a given point in the data collection indicate that the interviewer variance in a variable available on the sampling frame for the respondents (prior to measurement, e.g., age of the head-of-household from a linked commercial data source) is larger for one group of interviewers than another (e.g., inexperienced interviewers tend to vary in terms of the ages of recruited respondents). In this scenario, survey managers could intervene with the group of interviewers where there is more evidence of variability among the interviewers in terms of the features of recruited respondents. For example, West, et al. (2018b) showed that conversational interviewers varied substantially in terms of the ages of recruited respondents as measured from administrative records, and that this same variance remained when analyzing the *reported* ages of the respondents (as one might expect, if respondents do not struggle with reporting their ages). Any inference about conversational interviewers introducing measurement error variance in the *reported ages* of survey respondents would be erroneous in this context.

This chapter also provided guidance with respect to performing power analyses in these types of studies, and statistical tools for analyzing the data collected. Careful power analysis will ensure that adequate study resources are dedicated to selecting large enough samples of both interviewers and respondents, ensuring adequate statistical power for the comparisons of interest. Lack of power analyses in advance could lead to studies in which estimates of variance components and comparisons of these components across groups are unreliable.

Careful attention to all the design considerations outlined in this chapter will lead to rigorous and informative studies about differences in interviewer variance components between groups of interviewers for variables of substantive or methodological interest. These studies will enrich and extend the literature on interviewer effects, and open new doors for future studies that aim to improve the quality of the data collected in interviewer-administered surveys.

References

Berk, M. L., and A. B. Bernstein. 1988. Interviewer characteristics and performance on a complex health survey. *Social Science Research* 17(3):239–251.

Billiet, J., and G. Loosveldt. 1988. Improvement of the quality of responses to factual survey questions by interviewer training. *Public Opinion Quarterly* 52(2):190–211.

Blohm, M., J. Hox, and A. Koch. 2007. The influence of interviewers' contact behavior on the contact and cooperation rate in face-to-face household surveys. *International Journal of Public Opinion Research* 19(1):97–111.

Brunton-Smith, I., P. Sturgis, and J. Williams. 2012. Is success in obtaining contact and cooperation correlated with the magnitude of interviewer variance? *Public Opinion Quarterly* 76(2):265–286.

Cannell, C. F., P. V. Miller, and L. Oksenberg. 1981. Research on interviewing techniques. *Sociological Methodology* 12:389–437.

Catania, J. A., D. Binson, J. Canchola, L. M. Pollack, W. Huack, and T. J. Coates. 1996. Effects of interviewer gender, interviewer choice, and item wording on responses to questions concerning sexual behavior. *Public Opinion Quarterly* 60(3):345–375.

Durrant, G. B., R. M. Groves, L. Staetsky, and F. Steele. 2010. Effects of interviewer attitudes and behaviors on refusal in household surveys. *Public Opinion Quarterly* 74(1):1–36.

Fischer, M., B. T. West, M. R. Elliott, and F. Kreuter. 2019. The impact of interviewer effects on regression coefficients. *Journal of Survey Statistics and Methodology* 7(2):250–274.

Freeman, J., and E. W. Butler. 1976. Some sources of interviewer variance in surveys. *Public Opinion Quarterly* 40(1):79–91.

Friedel, S. 2019. What they expect is what you get: The role of interviewer expectations in nonresponse to income and asset questions. *Journal of Survey Statistics and Methodology* smz022:1–26. doi:10.1093/jssam/smz022.

Groves, R. M., and K. A. McGonagle. 2001. A Theory-guided interviewer training protocol regarding survey participation. *Journal of Official Statistics* 17:249–265.

Jäckle, A., P. Lynn, J. Sinibaldi, and S. Tipping. 2013. The effect of interviewer experience, attitudes, personality and skills on respondent co-operation with face-to-face surveys. *Survey Research Methods* 7:1–15.

Kibuchi, E., P. Sturgis, G. B. Durrant, and O. Maslovskaya. 2019. Do interviewers moderate the effect of monetary incentives on response rates in household interview surveys? *Journal of Survey Statistics and Methodology* smy026:1–21. doi:10.1093/jssam/smy026.

Lavrakas, P. J., J. Kelly, and C. McClain. 2019. Experimentation and possible confounding interviewer effects in survey research. In: *Experimental Methods in Survey Research: Techniques That Combine Random Sampling with Random Assignment*, ed. P. J. Lavrakas, M. W. Traugott, C. Kennedy, A. L. Holbrook, E. D. de Leeuw, and B. T. West. Hoboken: Wiley.

Lipps, O., and A. Pollien. 2011. Effects of interviewer experience on components of nonresponse in the European Social Survey. *Field Methods* 23(2):156–172.

Mangione, T. W., F. J. Fowler, and T. A. Louis. 1992. Question characteristics and interviewer effects. *Journal of Official Statistics* 8:293–307.

Nemeth, R., and A. Luksander. 2018. Strong impact of interviewers on respondents' political choice: Evidence from Hungary. *Field Methods* 30(2):155–170.

O'Muircheartaigh, C., and P. Campanelli. 1998. The relative impact of interviewer effects and sample design effects on survey precision. *Journal of the Royal Statistical Society: Series A* 161(1):63–77.

Schnell, R., and F. Kreuter. 2005. Separating interviewer and sampling-point effects. *Journal of Official Statistics* 21:389–410.

Schober, M. F., and F. G. Conrad. 1997. Does conversational interviewing reduce survey measurement error? *Public Opinion Quarterly* 61(4):576–602.

Vandenplas, C., G. Loosveldt, K. Beullens, and K. Denies. 2018. Are interviewer effects on interview speed related to interview effects on straight-lining tendency in the European Social Survey? An interviewer-related analysis. *Journal of Survey Statistics and Methodology* 6(4):516–538.

Vassallo, R., G. B. Durrant, P. W. F. Smith, and H. Goldstein. 2015. Interviewer effects on nonresponse propensity in longitudinal surveys: A multilevel modelling approach. *Journal of the Royal Statistical Society: Series A* 178(1):83–99.

West, B. T., and A. G. Blom. 2017. Explaining interviewer effects: A research synthesis. *Journal of Survey Statistics and Methodology* 5:175–211.

West, B. T., F. G. Conrad, F. Kreuter, and F. Mittereder. 2018a. Can conversational interviewing improve survey response quality without increasing interviewer effects? *Journal of the Royal Statistical Society: Series A* 181(1):181–203.

West, B. T., F. G. Conrad, F. Kreuter, and F. Mittereder. 2018b. Nonresponse and measurement error variance among interviewers in standardized and conversational interviewing. *Journal of Survey Statistics and Methodology* 6(3):335–359.

West, B. T., and M. R. Elliott. 2014. Frequentist and Bayesian approaches for comparing interviewer variance components in two groups of survey interviewers. *Survey Methodology* 40:163–188.

West, B. T., and M. R. Elliott. 2015. New methodologies for the study and decomposition of interviewer effects in surveys. *Invited Talk Presented at the Annual Meeting of the Statistical Society of Canada*, Halifax, Nova Scotia.

West, B. T., and A. T. Galecki. 2011. An overview of current software procedures for fitting linear mixed models. *The American Statistician* 65(4):274–282.

West, B. T., F. Kreuter, and U. Jaenichen. 2013. Interviewer effects in face-to-face surveys: A function of sampling, measurement error or nonresponse? *Journal of Official Statistics* 29(2):277–297.

West, B. T., and K. Olson. 2010. How much of interviewer variance is really nonresponse error variance? *Public Opinion Quarterly* 74(5):1004–1026.

Index

Printed in the United States
by Baker & Taylor Publisher Services